LEUKOCYTE RECRUITMENT, ENDOTHELIAL CELL ADHESION MOLECULES, AND TRANSCRIPTIONAL CONTROL
Insights for Drug Discovery

LEUKOCYTE RECRUITMENT, ENDOTHELIAL CELL ADHESION MOLECULES, AND TRANSCRIPTIONAL CONTROL
Insights for Drug Discovery

edited by

Tucker Collins
Harvard Medical School

KLUWER ACADEMIC PUBLISHERS
Boston / Dordrecht / London

Distributors for North, Central and South America:
Kluwer Academic Publishers
101 Philip Drive
Assinippi Park
Norwell, Massachusetts 02061 USA
Telephone (781) 871-6600
Fax (781) 681-9045
E-Mail <kluwer@wkap.com>

Distributors for all other countries:
Kluwer Academic Publishers Group
Distribution Centre
Post Office Box 322
3300 AH Dordrecht, THE NETHERLANDS
Telephone 31 78 6392 392
Fax 31 78 6546 474
E-Mail <services@wkap.nl>

 Electronic Services <http://www.wkap.nl>

Library of Congress Cataloging-in-Publication Data

Leukocyte recruitment, endothelial cell adhesion molecules, and transcriptional control:
Insights for drug discovery / edited by Tucker Collins.
     p. cm.
   Includes bibliographical references and index.
   ISBN 0-7923-7223-9 (alk. paper)
     1. Leukocytes. 2. Cell adhesion. 3. Cell adhesion molecules. 4.
Inflammation—Mediators. 5. NF-kappa B (DNA-binding protein) I. Collins, Tucker.

QP95.L647 2000
612.1'12—dc21
                                                                          00-048483

Printed on acid-free paper.
Printed in the United States of America

The Publisher offers discounts on this book for course use and bulk purchases.
For further information, send email to <joanne.tracy@wkap.com> .

TABLE OF CONTENTS

Leukocyte Recruitment, Endothelial Cell Adhesion Molecules, and Transcriptional Control: Insights for Drug Discovery

Preface: Transcriptional Control of Cell Adhesion Molecules
Tucker Collins and Mary Gerritsen

PRINCIPAL CONTRIBUTORS

Collins, Tucker M.D., Ph.D.
Department of Pathology
Brigham and Women's Hospital
221 Longwood Avenue
Boston, MA 02115
Tel: 617-732-5990
Fax: 617-278-6990
tcollins@rics.bwh.harvard.edu

Gerritsen, Mary Ph.D.
Genentech, Inc.
Cardiovascular Research
Department
460 Point San Bruno Blvd.
So. San Francisco, CA 94080
Tel: 650-225-8297
Fax: 650-225-6327
meg@gene.com

Lee, Frank M.D., Ph.D.
Department of Pathology and
Lab Medicine
University of Pennsylvania
School of Medicine
218 John Morgan Building
Philadelphia, PA 19104
Tel: 215-898-4701
Fax: 215-573-2272
franklee@mail.med.upenn.edu

Maniatis, Thomas Ph.D.
Harvard University
Department of Molecular and
Cellular Biology
7 Divinity Avenue
Cambridge, MA 02138
Tel: 617-495-1811
Fax: 617-495-3537

Manning, Anthony Ph.D.
Pharmacia Corporation
St. Louis, MO
Tel: 636-737-6717
Fax: 636-737-6772
anthony.m.manning@
stl.monsanto.com

McEver, Rodger M.D.
The University of Oklahoma
Health Sciences Center
825 N.E. 13th Street
Oklahoma City, OK 73104
Tel: 405-271-6480
Fax: 405-271-3137
rodger-mcever@ouhsc.edu

Neish, Andrew M.D.
Emory University
Woodruff Memorial Research
Building, Room 2337
1639 Pierce Street
Atlanta, GA 30322
Tel: 404-727-8545
Fax: 404-727-8538
aneish@emory.edu

Parks, Thomas Ph.D.
Cellegy Pharmaceuticals, Inc.
349 Oyster Point Boulevard
Suite 200
South San Francisco, CA 94080
Tel: 650-616-2276
Fax: 650-616-2230
tparks@cellegy.com

Parvin, Jeffrey M.D., Ph.D.
Department of Pathology
Brigham and Women's Hospital
75 Francis Street
Boston, MA 02115
Tel: 617-278-0818
Fax: 617-732-7449
jparvin@rics.bwh.harvard.edu

von Andrian, Ulrich M.D.
The Center for Blood Research
Harvard Medical School
200 Longwood Avenue
Boston, MA 02115
Tel: 617-278-3130
Fax: 617-278-3030
uva@cbr.med.harvard.edu
http://cbrweb.med.harvard.edu/~uva/

PREFACE: TRANSCRIPTIONAL CONTROL OF CELL ADHESION MOLECULES

More than a hundred fifty years ago, Julius Cohnheim made the first microscopic observations of small blood vessels in transparent tissues, such as the tongue and mesentery of the frog. He was the first to describe the initial changes in blood flow, the subsequent edema caused by increased vascular permeability, and the leukocyte adhesive interactions that characterize the acute inflammatory response.[1,2] The sequence of cellular events in the journey of leukocytes from the lumen of the vessel to the interstitial tissues, called extravasation, can be subdivided into a series of sequential events (Figure 1). In the lumen, marginated leukocytes initially roll, then become activated and adhere firmly to endothelium. This is followed by the transmigration of the leukocytes across the endothelial barrier and the underlying basement membrane. Once in the tissue, the leukocytes migrate toward a chemoattractant stimulus generated by local chemokines or chemoattractant peptides produced by the invading organism. Leukocytes then ingest offending agents, process antigens, kill microbes, and degrade necrotic tissue. In many instances, the inflammatory response succeeds in neutralizing the injurious stimulus and the site of the reaction is restored to its normal condition.

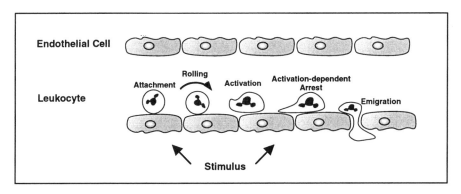

Figure 1. Schematic diagram of the cellular events in inflammation, shown here for neutrophils. (Modified from McEver, Chapter 1).

The localized attachment and sequestration of circulating leukocytes is now recognized as a central aspect of the inflammatory response. Localized attachment suggests that the luminal surface of the endothelium can become locally hyperadhesive, a phenomenon that has been a focus of intense investigation. It is now clear that leukocyte adhesion and transmigration are determined by the recognition and binding of adhesion molecules on the leukocyte to endothelial surfaces, and that chemical mediators such as cytokines and chemokines can modulate these processes.[3,4] In the past decade, our understanding of the structure and function of the vascular adhesion molecules that contribute to leukocyte-vessel wall interaction has increased dramatically. The key adhesion receptors belong to three molecular families: the selectins, the immunoglobulin supergene family, and the integrins.[5-10] The selectins, named for the N-terminal lectin domain, consist of E-selectin, which is largely restricted to endothelium; P-selectin, present in endothelium and platelets; and L-selectin, which is displayed by most leukocyte types. The immunoglobulin family is large, but includes several family members that are expressed by endothelial cells, notably intercellular adhesion molecule-1 (ICAM-1) and vascular cell adhesion molecule-1 (VCAM-1). Both of these molecules interact with integrin counter-receptors found on leukocytes. The importance of these molecules in the inflammatory response has been demonstrated by studies in patients with defects in leukocyte function, as well as in transgenic mouse models in which the genes for the adhesion molecules have been disrupted.

The induction of endothelial adhesion molecule expression is a key aspect in leukocyte recruitment during an inflammatory response. Some inflammatory mediators, notably cytokines such as tumor necrosis factor-α (TNF-α) and interleukin-1 (IL-1), induce the transcription and surface expression of endothelial adhesion molecules. For example, E-selectin, which is not present in normal endothelium, is dramatically induced at the transcriptional level by the inflammatory cytokines and mediates the adherence of neutrophils, monocytes, and certain types of lymphocytes by binding to its receptors. The same cytokines also induce the expression of VCAM-1 and upregulate levels of ICAM-1, which are present at low levels in normal endothelium. The central premise of this book is that by understanding the mechanisms regulating the expression of a select set of vascular cell adhesion molecules, insights might be obtained that will be useful in the treatment of inflammatory diseases.

The transcription factor nuclear factor-κB (NF-κB) has emerged as a dominant theme in the regulation of leukocyte adhesion genes. This transcription factor has been referred to as a "master switch" of the inflammatory response, and most likely plays a central role in the regulation of vascular adhesion molecules during acute inflammatory responses to injury and infection.[11-16] Recent studies have provided a correlation between the presence of activated NF-κB, the expression of NF-κB-dependent genes, and the presence of disease. Importantly, results from animal models have demonstrated that functional inhibition of

NF-κB activation can alter the course of disease. Collectively, these studies suggest that NF-κB inhibition may be a viable therapeutic strategy.

In this volume, active investigators have been asked to contribute their perspective on specific aspects of this field, as well as on related areas that may be important in understanding the regulation of expression of these molecules. Included are discussion of the structural and functional aspects of two categories of the adhesion molecules: the selectins and the members of the immunoglobulin supergene family. Following each overview is a more specific review on the regulation of relevant family members: E-selectin, VCAM-1, and ICAM-1, respectively. A chapter on NF-κB is included because of its importance in the cytokine-induced expression of each of these vascular adhesion molecules. Because transcriptional activators like NF-κB must interact with the basal transcription machinery to induce gene expression, an overview of the basal apparatus is also included. This book also provides an overview of anti-inflammatory agents and their effects on leukocyte recruitment, as well as a focused look at the NF-κB signaling system as a potential target for drug discovery. Taken together, the components of the book provide a unique description of a crucial step in the adhesion process – the regulation of endothelial cell adhesion molecule expression – and suggest that this step may be an important new area for manipulating the inflammatory response.

TC
MG

REFERENCES

1.   Cells, Tissues and Disease: Principles of General Pathology. Edited by Majno G, Joris I. Blackwell Science, Cambridge, MA, 1996.
2.   Robbins Pathologic Basis of Disease. Sixth Edition. Cotran RS, Kumar V, Collins T. Editors. W.B. Saunders Company, Philadelphia, PA, 1999.
3.   Butcher EC. Leukocyte-endothelial cell recognition: three or more steps to diversity and sensitivity. Cell 1991; 76:1033-1036.
4.   Springer TA. Traffic Signals for lymphocyte circulation and migration: the multistep paradigm. Cell 1994; 76:301-314
5.   Carlos TM, Harlan JM. Leukocyte-endothelial adhesion molecules. Blood 1994; 84:2068-2101.
6.   Cellular and Molecular Mechanisms of Inflammation - Vascular Adhesion Molecules. Edited by Cochrane CG, Gimbrone, MA. Academic Press Inc., London, 1991.
7.   Adhesion. Its role in inflammatory disease. Edited by Harlan JM, and Liu DY. W.H. Freeman and Company, New York, 1992.
8.   The Adhesion Molecule Facts Book. Edited by Pigott R, Power C. Academic Press, London, 1993.
9.   Frenette PS, Wagner DD. Adhesion molecules. N Engl J Med 1996; 334:1526-1529.
10.   Vestweber D, Blanks JE. Mechanisms that regulate the function of the selectins and their ligands. Physiol Rev 1999; 79:181-213.
11.   Ghosh S, May MJ, Kopp EB. NF-κB and Rel proteins: evolutionarily conserved mediators of immune responses. Ann Rev Immunol 1998; 16:225-260.
12.   Baldwin AS. The NF-κB and IκB proteins: new discoveries and insights. Ann Rev Immunol 1996; 14:649-683.
13.   Verma IM. Rel/NF-κB/IκB family: intimate tales of association and dissociation. Genes Dev 1995; 9:2723-2735.
14.   Baeuerle PA, Baltimore D. NF-κB: ten years after. Cell 1996; 87:13-20.
15.   Barnes PJ, Karin M. Nuclear factor-κB- a pivotal transcription factor in chronic inflammatory diseases. N. Engl. J. Med. 1997; 336:1066-1071.
16.   Mercurio F, Manning AM. Multiple signals converging on NF-κB. Curr Opin Cell Biol 1999; 11:226-232.

ACKNOWLEDGMENT

The authors gratefully acknowledge the efforts of John Boucher, the copy editor and page layout artist for this volume.

Chapter 1

THE SELECTINS IN LEUKOCYTE RECRUITMENT

Rodger P. McEver

W.K. Warren Medical Research Institute,
Departments of Medicine and Biochemistry and Molecular Biology,
University of Oklahoma Health Sciences Center
825 N.E. 13th Street, Oklahoma City, OK 73104
and the Cardiovascular Biology Research Program,
Oklahoma Medical Research Foundation,
Oklahoma City, OK 73104

INTRODUCTION

In response to infection or tissue injury, leukocytes roll along the endothelial lining of blood vessels, then stick more firmly, and finally migrate through the vessel wall into the underlying tissues. During the past fifteen years, remarkable progress has been made in elucidating the molecular mechanisms that underlie this multistep pathway of leukocyte recruitment. Specific combinations of adhesion and signaling molecules regulate the accumulation of distinct subsets of leukocytes into lymphatic tissues or inflammatory sites. Locally generated cytokines or other mediators induce the expression of adhesion molecules on the endothelial cell surface that promote the initial tethering and rolling of leukocytes on the vessel wall. The relatively slow velocity of rolling leukocytes allows them to become activated by locally generated chemokines or lipid autacoids. The activated leukocytes express other adhesion molecules that stabilize adhesion and promote emigration in response to chemotactic gradients. Adhesive interactions of leukocytes with other leukocytes or with platelets may enhance leukocyte accumulation.

In most situations, interactions of selectins with cell-surface carbohydrate ligands mediate the rolling of leukocytes on the vessel wall, whereas interactions of integrins with immunoglobulin-like ligands mediate the firm adhesion and emigration of activated leukocytes[1-3] (Fig. 1-1). Selectins are remarkable in their ability to form rapid, yet transient, interactions with their ligands, allowing a free-flowing leukocyte to tether to and then roll on the endothelial surface under the shear forces found in postcapillary venules. Physiologic leukocyte recruitment requires that the expression of selectins and their ligands be tightly regulated. This chapter focuses on the pivotal functions of selectins during the initial stages of leukocyte adhesion. Earlier reviews provide additional information and more complete citations of the primary literature.[4,5]

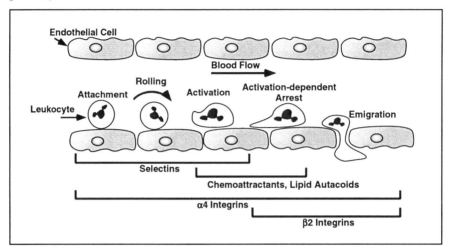

Figure 1-1. Multistep model of leukocyte recruitment. Free-flowing leukocytes attach to and then roll on postcapillary venules at inflammatory sites or on high endothelial venules of lymphoid tissues. Chemoattractants or lipid autacoids expressed on or near the endothelial cell surface activate the rolling leukocyte. The activated leukocyte then arrests, spreads, and finally emigrates between endothelial cells into the underlying tissues. Selectins mediate the initial attachment and rolling of leukocytes, whereas β2 leukocyte integrins mediate arrest, spreading, and emigration. Under some conditions, α4 integrins expressed on mononuclear cells and eosinophils, but not on neutrophils, mediate both rolling and arrest.

Selectin-Ligand Recognition

Figure 1-2 depicts a schematic representation of the three selectins. Table 1-1 illustrates the cells that express L-, E-, or P-selectin and the cells that display ligands for each selectin. Each selectin has an N-terminal carbohydrate-recognition domain characteristic of Ca^{2+}-dependent (C-type) lectins, followed by an epidermal growth factor (EGF)-like motif, a series of short consensus repeats (SCRs), a transmembrane domain, and a short cytoplasmic tail. L-selectin, expressed on most leukocytes, binds to constitutively expressed ligands on high endothelial venules (HEV) of lymph nodes, to inducible ligands on endothelium at sites of inflammation, and to ligands on other leukocytes. E-selectin, expressed on activated endothelial cells, and P-selectin, expressed on activated platelets and endothelial cells, bind to ligands on myeloid cells and subsets of lymphocytes. Like L-selectin, P-selectin may also bind to ligands on HEV or activated endothelial cells.

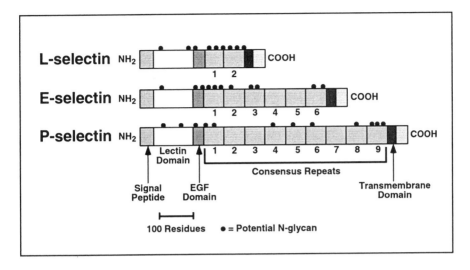

Figure 1-2. Domain organization of human selectins. Each molecule contains an N-terminal carbohydrate-recognition domain characteristic of Ca^{2+}-dependent lectins, followed by an EGF-like domain, a series of short consensus repeats, a transmembrane domain, and a short cytoplasmic tail. Each protein is modified with several N-linked oligosaccharides.

The amino acid sequences of the lectin and EGF domains of the selectins are highly conserved, suggesting that both domains contribute to ligand recognition. The lectin domain is directly involved in cell-cell contact through Ca^{2+}-dependent interactions with cell-surface glycoconjugates.[4,5] The three-dimensional structure of the lectin and EGF domains of E-selectin has been determined by X-ray crystallography;[6] molecular modeling suggests that P- and L-selectin have domains of similar structure.[7] The structural data, in conjunction with results from site-directed mutagenesis,[7-11] suggest that carbohydrate binds to a small, shallow region that overlaps a single Ca^{2+} coordination site opposite where the EGF domain is attached. Because no glycoconjugate has been co-crystallized with a selectin, the precise residues involved in carbohydrate recognition remain unclear. Deletion of the EGF domain impairs selectin binding to target cells, demonstrating its importance.[12,13] Chimeric selectins in which EGF domains have been exchanged were reported to alter ligand specificity in some studies,[14-16] but not in others.[17] Substitutions of individual residues in the EGF domain may also alter ligand recognition.[17] More recent data demonstrate that exchange of the EGF domain affects the kinetics but not the specificity of binding to selectin ligands.[18] Surprisingly, the EGF domain of E-selectin has only limited contacts with the lectin domain.[6] In the absence of further structural information, it is difficult to explain how alterations in the EGF domain affect binding of selectins to ligands.

Selectin	CD nomenclature	Previous names	Expressed by	Target cell
Table 1-1. The selectin family of adhesion molecules				
L-selectin	CD62L	LAM-1 LECAM-1	PMNs, monocytes, lymphocyte subsets	Activated ECs, HEVs of PLN and ML, other leukocytes, hematopoietic stem cells
E-selectin	CD62E	ELAM-1	Cytokine-activated ECs	PMNs, monocytes, eosinophils, lymphocyte subsets, some tumor cells
P-selectin	CD62P	GMP-140 PADGEM	Thrombin-activated platelets and ECs, cytokine-activated ECs	PMNs, monocytes, eosinophils, lymphocyte subsets, some tumor cells, HEVs, activated ECs?

EC, endothelial cell; ELAM, endothelial leukocyte adhesion molecule; GMP, granule membrane protein; HEV, high endothelial venule; LAM, lymphocyte adhesion molecule; LECAM, leukocyte-endothelial cell adhesion molecule; ML, mesenteric lymphoid tissue; PLN, peripheral lymph node; PADGEM, platelet activation dependent granule external membrane protein; PMN, polymorphonuclear cell or neutrophil.

Some studies suggest that the SCRs enhance binding of L- and E-selectin to their respective ligands.[16,19,20] However, these studies used selectins expressed on cell surfaces or as bivalent immunoglobulin-fusion constructs, making it difficult to distinguish the relative contributions of affinity and avidity to binding. The selectins are relatively rigid, extended proteins, which implies that the SCRs do not directly contact the lectin domain.[21-23] Therefore, it is not obvious how the SCRs could affect the structure or function of the lectin domain. Indeed, a soluble form of P-selectin containing only the lectin and EGF domains binds to its ligand with the same affinity and kinetics as a soluble, monomeric form of P-selectin that contains the lectin and EGF domains plus all nine SCRs.[24]

All three selectins bind sialylated and fucosylated oligosaccharides such as sialyl Lewis x (sLe^x; Neu5Acα2,3Galβ1,4[Fucα1,3]GlcNAc-R), a terminal component of glycans attached to glycoproteins and glycolipids on most leukocytes and some endothelial cells.[25] Consistent with this observation, target cells must be sialylated and fucosylated to interact with selectins.[25]

However, selectins bind with very low affinity to isolated sLe^x-related oligosaccharides.[25-27] Furthermore, L- and P-selectin, but not E-selectin, bind a variety of sulfated structures such as heparin and sulfatides that are not sialylated or fucosylated.[28,29] These observations suggest that selectins might bind with higher affinity to glycoproteins with a more specific arrangement of post-translational modifications. Indeed, only a small number of glycoproteins from leukocytes or endothelial cells bind with higher affinity to selectins (Fig. 1-3). These proteins must be sialylated and fucosylated to interact with selectins.[30-32] One glycoprotein ligand, E-selectin ligand-1 (ESL-1), binds specifically to E-selectin.[31] ESL-1 is N-glycosylated at a maximum of five sites, and has not been demonstrated to be O-glycosylated.[31,33] The structures of the N-glycans have not been determined; it is possible that they are linear, sialylated dimeric Le^x structures, which bind better to E-selectin than does sLe^x itself.[34]

The other glycoprotein ligands appear to interact with more than one selectin,[22,32,35] and there is some evidence that they bind better to L- or P-selectin than to E-selectin.[22,36] These molecules are sialomucins; they have many serines and threonines that are attachment sites for O-glycans,[32,37-39] and the O-glycans are required for selectin recognition.[40-42] Importantly, these glycoproteins must also be sulfated to bind L- and P-selectin.[43-47] The sulfate esters are attached to the C-6 position of Gal and GlcNAc residues in the O-glycans of the HEV L-selectin ligand, glycosylated cell adhesion molecule-1 (GlyCAM-1).[48-50] In contrast, the sulfate esters are attached to at least one of three clustered tyrosines near the N-terminus of P-selectin glycoprotein ligand-1 (PSGL-1), a selectin ligand on leukocytes.[42,45-47] Remarkably, only 14% of the O-glycans on PSGL-1 from human myeloid cells are fucosylated; most of these are novel core-2 structures with a β1,6 branch consisting of polylactosamine modified with three α1,3-linked fucoses and a terminal α2,3-linked sialic acid.[51] CD43, another sialomucin expressed on myeloid cells, has none of these structures.[51,52] The identified fucosylated glycans on GlyCAM-1 from HEV are short core-2 structures that are sulfated but lack polylactosamine.[48-50] Together, these data suggest that unique O-glycan structures may be created only at restricted sites on specific proteins. High affinity binding of L- and P-selectin to sialomucins may require a composite recognition site created by several post-translational modifications in the optimal spatial array. An N-terminal, high affinity binding site on human PSGL-1 consists of three tyrosine sulfates in proximity to a core-2 O-glycan capped with sLe^x.[53] High affinity binding sites on GlyCAM-1 may require appropriate spacing of two or more sulfated, sialylated, and fucosylated O-glycans. In support of the model for composite recognition sites, L-selectin binds only a subset of proteolytic fragments from sialomucins derived from colon carcinoma cells, although the O-glycans are apparently similar on both the binding and non-binding fragments.[54]

6

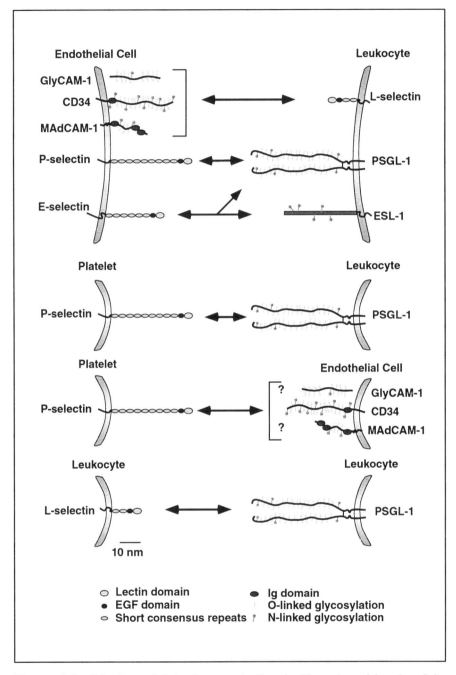

Figure 1-3. Selectins and their glycoprotein ligands. The estimated lengths of the selectins[21-23] and of PSGL-1[84] are based on hydrodynamic data and electron microscopy. The lengths of GlyCAM-1, CD34, and MAdCAM-1 are modeled from the dimensions of another sialomucin, CD43.[179] PSGL-1 has been studied primarily in humans, whereas the other glycoprotein ligands have been studied primarily in mice.

Regulation of post-translational modifications of specific proteins may specify when certain cells interact with selectins. This may be particularly important for T lymphocytes, where conversion to the memory phenotype enhances binding to selectins,[55-57] and for endothelial cells, where inflammatory mediators may induce sulfation of ligands for L-selectin.[44]

Selectin-Mediated Tethering and Rolling of Leukocytes under Hydrodynamic Flow

Many *in vitro* and *in vivo* studies have documented that leukocytes use selectins to attach to and roll on the vessel wall under the shear forces characteristic of postcapillary venules, where most leukocyte recruitment takes place (reviewed in refs. 2,4,5) (Fig. 1-4). Leukocytes roll on endothelial cells through interactions of L-selectin on the leukocyte with constitutively expressed ligands on lymphoid HEV or inducible ligands on inflamed endothelium. Leukocytes also roll on E-selectin or P-selectin expressed on activated endothelial cells. Leukocytes roll on P-selectin expressed on adherent activated platelets, and activated platelets use P-selectin to form platelet-leukocyte conjugates. P-selectin on activated platelets also binds to L-selectin ligands on HEV and perhaps on activated endothelium; thus, leukocytes adherent to activated platelets can attach indirectly to the endothelium.[58] Unstimulated platelets also roll on P- and E-selectin on activated endothelial cells.[59] Leukocytes use L-selectin to roll on adherent leukocytes[60] and to form leukocyte aggregates.[61] Under shear forces, therefore, selectins promote adhesive interactions between leukocytes and endothelial cells, leukocytes and platelets, platelets and endothelial cells, and leukocytes and other leukocytes. These multicellular interactions may amplify the recruitment of leukocytes to lymphoid tissues and inflammatory sites.[58,62,63]

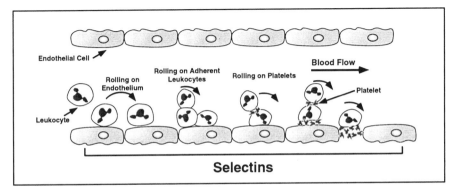

Figure 1-4. Multicellular interactions mediated by selectins under hydrodynamic flow.

Models have been proposed to explain how selectins mediate cell-cell interactions under hydrodynamic flow.[64,65] In these models, selectin-ligand interactions must form rapidly to facilitate tethering, and then dissociate rapidly to facilitate rolling. Furthermore, selectin-ligand bonds must have high tensile strength; that is, shear forces must not significantly accelerate the rate of dissociation. To continue rolling, new selectin-ligand bonds must form at the leading edge of the cell to replace those bonds that dissociate at the trailing edge. The rolling velocity of the cell reflects the balance between the rates of association and dissociation of these bonds. Studies of the transient tethering of neutrophils on very low densities of selectins under flow support the concept that selectin-ligand bonds have fast association and dissociation rates with high tensile strength.[66,67] Significantly, L-selectin-ligand interactions have faster rates of association and dissociation than P-selectin- and E-selectin-ligand interactions. Leukocytes interacting through L-selectin roll faster and require a threshold shear force to support rolling. At low shear stresses, the initial bonds dissociate before new bonds can form; higher shear stresses may rotate the cell faster, bringing L-selectin molecules at the leading edge more quickly in proximity with new ligands on the substrate.[68] Shear-enhanced bond formation may compensate for shear-accelerated bond dissociation, thereby stabilizing selectin-mediated rolling adhesion.[69]

Biochemical assays indicate that selectins bind to their ligands with fast association and dissociation kinetics,[24,70] in agreement with cell tether lifetimes, which were extrapolated to estimate unstressed bond dissociation rates.[66,67] Other factors may also contribute to the kinetics of cell-cell interactions. For example, L-selectin is rapidly shed from the surface of leukocytes after activation[71] or after antibody crosslinking of L-selectin.[72] Inhibitors of this proteolytic event markedly slow the velocity of neutrophils rolling on L-selectin ligands.[73] Thus, interactions of L-selectin with cell-surface ligands may promote shedding of L-selectin, effectively reducing the number of L-selectin-ligand bonds between cells. The orientation of selectins and their ligands on cell surfaces may also enhance their interactions under hydrodynamic flow. On leukocytes, L-selectin is concentrated on the tips of microvilli, the regions most likely to contact the surface of another cell under shear.[74] A chimeric L-selectin molecule containing the transmembrane and cytoplasmic domains of CD44 is randomly distributed on the surface of transfected leukocytes; the chimera is much less effective than wild type L-selectin in mediating tethering of cells on L-selectin-ligand substrates.[75] P-selectin is a highly extended protein, suggesting that projection of the lectin domain well above the cell surface may optimize its ability to interact with ligands on leukocytes under shear forces.[21] Consistent with this hypothesis, flowing neutrophils roll much less effectively on transfected cells that express shorter forms of P-selectin in which some of the SCRs have been deleted than they do on cells expressing wild type P-selectin.[76] Clustering of selectins and their ligands on apposing cell surfaces may also increase the number of interactions under flow conditions.[77,78]

PSGL-1 carries only a small fraction of the total sLe^x-like glycans found on the surfaces of leukocytes,[41] and the other high-affinity glycoprotein ligands for selectins are probably also minor components of the cell surface. Theoretically, large numbers of very low affinity ligands might support selectin-dependent tethering and/or rolling as well as or better than a small number of higher affinity ligands. However, it is clear that PSGL-1 plays an important role in shear-dependent interactions of leukocytes with P- and L-selectin. PL1, a monoclonal antibody (mAb) to human PSGL-1, prevents binding of purified PSGL-1 to purified P-selectin.[79] The mAb also completely blocks tethering and rolling of leukocytes on P-selectin substrates *in vitro*[79,80] and *in vivo*.[81] Flowing neutrophils tether to and roll on purified, immobilized PSGL-1; this interaction is blocked by PL1 or by mAbs to L-selectin.[62] Furthermore, PL1 significantly inhibits the L-selectin-dependent rolling of neutrophils on adherent neutrophils[62] or the L-selectin-dependent aggregation of stirred neutrophils.[82] PL1 only modestly inhibits binding of purified PSGL-1 to purified E-selectin.[80] However, it indirectly inhibits the accumulation of rolling neutrophils on E-selectin substrates by blocking L-selectin-PSGL-1 interactions that mediate neutrophil-neutrophil contacts[80,83] The PL1 epitope is near the N-terminus of PSGL-1; it overlaps the tyrosine sulfation sites and includes a threonine to which a critical O-glycan is attached.[84] This is consistent with the notion that P- and L-selectin identify a specific, composite binding site on the molecule.[53,85] Because PSGL-1 is highly extended, the binding site extends well above the cell surface.[84] PSGL-1 is also concentrated on the tips of leukocyte microvilli.[79,86] Thus, the structure and orientation of PSGL-1 appear to be ideal for efficient interactions with L- and P-selectin under flow. PL1 also blocks adhesion of leukocytes to P-selectin under static conditions.[79] Therefore, PSGL-1-P-selectin interactions are essential for cell-cell contact even when there is no requirement for rapid bond formation. Leukocytes also require PSGL-1 to roll on P-selectin in mice[87,88]

It is not yet known if the other described glycoprotein ligands for selectins mediate leukocyte tethering or rolling under shear conditions. *In vitro*, flowing leukocytes use L-selectin to tether to and roll on immobilized mucosal cell addressin molecule-1 (MAdCAM-1)[89] or CD34[90] Whether these glycoproteins serve as L-selectin ligands on an intact cell is less clear. *In vivo*, mAbs to MAdCAM-1 partially block the rolling of L-selectin-expressing cells on certain endothelial cells, although it is not certain whether this occurs because the mAbs mask the O-glycan-rich domain of MAdCAM-1.[91] Mice rendered genetically deficient in CD34 have no obvious defects in leukocyte trafficking, indicating that CD34 is not essential for the tethering or rolling of leukocytes under the conditions examined.[92] Lymphoid tissues secrete GlyCAM-1 into plasma,[93] where it may activate rolling lymphocytes.[94] The function of ESL-1 in mediating rolling of leukocytes on E-selectin has not been studied. Other glycoprotein ligands for selectins have been reported, but they are less well characterized.[35,44,95,96]

Signaling through Selectins and Modulation of Selectin Function by Signaling

Leukocytes roll at velocities that are significantly slower than those of free-flowing cells, allowing juxtacrine activation by lipid autacoids and chemoattractants that are secreted or presented on the endothelial cell surface.[97] The activation of leukocytes is critical for upregulating integrin function, which stabilizes adhesion and then mediates emigration.[98] There is increasing interest in the possibility that selectin-ligand interactions themselves transmit signals into leukocytes. Most studies suggest that adhesion of leukocytes to purified E- or P-selectin does not cause overt activation, as defined by changes in cell shape, upregulation of integrin function, secretion of granule contents, or release of oxygen-derived radicals, cytokines, or other molecules.[97,99-104] However, such interactions may produce signals that are integrated with those produced by chemokines or lipid autacoids to fully activate leukocytes.[105-107] For example, monocytes mobilize the transcription factor NF-κB and secrete cytokines when incubated with P-selectin and platelet-activating factor together, but not with either molecule individually.[100] Monocytes secrete a different profile of cytokines when incubated with activated platelets; this response is mimicked when monocytes are incubated with P-selectin plus purified RANTES, a chemokine that is secreted by activated platelets.[101] Cross-linking of L-selectin on neutrophils with antibodies or sulfatides triggers an increase in cytosolic free calcium,[108,109] potentiates the oxidative burst in response to the bacterial peptide N-formyl Met-Leu-Phe,[109] and enhances integrin-dependent adhesion in response to low concentrations of chemokines.[110-112] The requirement for co-stimulatory signals may limit the activation of leukocytes to sites where a conventional agonist is co-expressed with a selectin or a selectin ligand.

Leukocyte adhesion or activation may also modulate the functions of selectins or their ligands. Adhesion of neutrophils to activated endothelial cells promotes clustering of E-selectin through cytoskeletal interactions.[113] Activated neutrophils shed L-selectin[71] and redistribute PSGL-1 from the tips of microvilli to the uropods of the polarized cells.[114] Both events may facilitate transfer of adhesive control from selectins to integrins as the leukocyte begins to emigrate into tissues.

Regulation of Expression of Selectins

The expression of selectins is tightly regulated to ensure that leukocytes tether to and roll on the blood vessel wall only at appropriate sites. As mentioned previously, L-selectin is constitutively expressed on most leukocytes, but it is proteolytically shed after leukocyte activation[71] and perhaps after interactions with cell-surface ligands.[73] In contrast, E- and P-selectin are not constitutively expressed on the cell surface, but are transported to the plasma membrane upon cellular activation. As discussed in detail in Chapter 2, inflammatory mediators such as tumor necrosis factor-α, interleukin-1β, or lipopolysaccharide transiently induce endothelial cells to transcribe E-selectin mRNA, which leads to synthesis of E-selectin protein. There is a lag of 1-2 hours before E-selectin appears on the endothelial cell surface, reflecting the time requirement for new protein synthesis. Unlike E-selectin, P-selectin is constitutively synthesized by both endothelial cells and megakaryocytes.[115,116] It is then stored in secretory storage granules: the Weibel-Palade bodies of endothelial cells[116,117] and the α granules of platelets.[118,119] Mediators such as thrombin, histamine, complement components, and oxygen-derived radicals rapidly redistribute P-selectin to the cell surface through fusion of granule membranes with the plasma membrane.[116,118-121] Because new protein synthesis is not required, this translocation step requires only seconds to minutes.

In vitro, P-selectin remains on the surface of activated platelets for many hours.[122] *In vivo*, however, P-selectin is rapidly proteolytically shed from the surface of activated platelets.[123] P- and E-selectin are cleared from the endothelial cell surface by endocytosis.[120,124] The endocytosis of P-selectin, which is particularly rapid, has been shown to occur through clathrin-coated pits in transfected CHO cells.[125] Both P- and E-selectin recycle from early endosomes to the plasma membrane. However, both molecules have short half-lives because they are rapidly degraded in lysosomes.[124,126-128] The short half-life of P-selectin is due to efficient endosomal sorting.[126] Some P-selectin molecules also recycle from endosomes to the trans-Golgi network and enter new Weibel-Palade bodies.[129] The 35-residue cytoplasmic domain of P-selectin carries signals that direct sorting into secretory granules,[130] endocytosis in clathrin-coated pits,[125] and movement from endosomes to lysosomes.[126] Mutational analysis suggests that residues throughout the cytoplasmic domain contribute to sorting.[125,131] Although less well studied, the cytoplasmic domain of E-selectin has signals that regulate the efficiency of internalization and endosomal sorting.[132] In mice, targeted deletion of the cytoplasmic domain of P-selectin prevents sorting of P-selectin to Weibel-Palade bodies in endothelial cells. However, P-selectin is still sorted into α granules of platelets, suggesting that the cytoplasmic domain is not required for sorting in this specialized cell type.[133]

The basal rate of transcription of P-selectin in endothelial cells can be further elevated by certain inflammatory cytokines. An increase in synthesis of P-selectin may saturate the sorting pathway into secretory granules, leading to immediate delivery of P-selectin to the cell surface. Subsequent challenges with thrombin or related agonists may further increase surface levels of P-selectin. In murine endothelial cells, tumor necrosis factor-α or lipopolysaccharide increases transcription of P-selectin both *in vitro* and *in vivo*.[134-136] The kinetics of induction are similar to those of E-selectin, with mRNA levels peaking after 4-6 hours and declining to basal levels after 12-24 hours.

However, these mediators do not increase P-selectin mRNA in cultured human endothelial cells.[137,138] Furthermore, lipopolysaccharide injected into the skin of non-human primates does not increase P-selectin protein in venules,[139] and *Escherichia coli* injected intravenously into baboons does not increase P-selectin mRNA in any tissue examined.[140] Therefore, there are important differences in the transcriptional regulation of P-selectin in primates and rodents. Notably, interleukin-4 or oncostatin M markedly increase transcription of mRNA for P-selectin, but not E-selectin, in cultured human endothelial cells.[138] Transcriptional induction requires new protein synthesis and results in a sustained increase in P-selectin mRNA and protein for at least 72 hours. Interleukin-13 also elevates levels of P-selectin in cultured human endothelial cells.[141] The elaboration of interleukin-4, interleukin-13, and oncostatin M may in part account for the sustained expression of P-selectin, but not E-selectin, on the apical surface of endothelial cells in some human tissues with chronic or allergic inflammation.[142-144]

The transcriptional regulation of the P-selectin gene has been less studied than that of the E-selectin gene. The exon-intron organization of the human P-selectin gene has been determined,[145] and the 5' flanking region has been isolated.[146] Unlike the E-selectin promoter, the P-selectin promoter has no canonical TATA box, and transcription is initiated at multiple sites. The first 249 base pairs upstream of the translational initiation site confer constitutive expression of a reporter gene in transfected endothelial cells. A GATA element is required for optimal constitutive expression. There is also a novel κB element that binds homodimers of p50 or p52, but not heterodimers containing p65.[147] Co-transfection experiments indicate that interactions of Bcl-3 and p52 homodimers with the κB element augment expression of a reporter gene in endothelial cells. In contrast, co-expression of p50 homodimers represses expression, although the repression is prevented by co-expression of Bcl-3[147] These data suggest that differential interactions of Bcl-3 with p50 and p52 homodimers help regulate the expression of P-selectin in humans. Interleukin-4 augments transcription of human P-selectin mRNA in part by binding to Stat6 elements in the 5' flanking region of the gene.[148]

The 5' flanking region of the murine P-selectin gene shares several conserved elements with those found in the human P-selectin gene. However, there are important differences, which may explain why tumor necrosis factor-α, interleukin-1β, and lipopolysaccharide activate the murine but not the human P-selectin gene. The murine gene lacks the unique κB site specific for p50 or p52 homodimers found in the human gene. Instead, it contains unique tandem κB elements and a variant activating transcription factor/cAMP response element.[149] These resemble sites in the E-selectin gene that are required for induction by tumor necrosis factor-α (see Chapter 2). Mutation of these elements in reporter gene constructs confirms their requirement for tumor necrosis factor-α or lipopolysaccharide to optimally induce expression of the murine P-selectin gene.[150]

In Vivo Studies of Selectins

Numerous *in vivo* studies have confirmed the biological importance of the selectins.[151-154] Of particular significance is the discovery of a congenital disorder of fucose metabolism in humans, termed leukocyte adhesion deficiency-2 (LAD-2).[155-157] Patients with LAD-2 lack fucosylated glycoconjugates. As a result, they fail to express functional selectin ligands on leukocytes (and probably also on endothelial cells). Leukocytes from these patients do not attach to and roll on E- or P-selectin. Consistent with these abnormalities, the patients have increased numbers of infections. Mice deficient in Fuc-TVII and Fuc-TIV, the two α1,3 fucosyltransferases that fucosylate selectin ligands on blood and vascular cells, have defective leukocyte recruitment into inflamed tissues and markedly diminished leukocyte rolling on selectins in shear flow.[158,159]

Mice rendered genetically deficient in each of the three selectins are healthy in the absence of specific infectious challenges.[136,160,161] L-selectin-deficient mice have fewer lymphocytes located in peripheral lymph nodes and are less efficient in homing of lymphocytes to lymphatic tissues.[161] Mice lacking L- or P-selectin have impaired rolling of leukocytes in venules of exteriorized tissues.[162] The defect in rolling is observed earlier in P-selectin-deficient mice, as expected if P-selectin is normally rapidly mobilized to the endothelial cell surface after tissue trauma. Mice lacking either L- or P-selectin have impaired leukocyte recruitment in models of both acute and chronic inflammation.[136,163,164] These abnormalities are more evident in E-selectin-deficient mice when P-selectin is also blocked by infusion of a mAb.[160] Mice deficient in both E- and P-selectin have frequent severe infections and shortened survival.[165,166] Mice deficient in all three selectins have a similar or worse phenotype, confirming the importance of these molecules in host defense.[167,168]

In vitro, P- and L-selectin bind to human hematopoietic progenitor cells, probably through PSGL-1.[169-172] These interactions may modulate hematopoiesis. Circulating hematopoietic progenitor cells home to bone marrow, in part, by interacting with P- and E-selectin.[173] *In vitro* studies suggest that E-selectin contributes to angiogenesis.[174,175] However, neither selectin-deficient mice nor LAD-2 patients have obvious abnormalities in angiogenesis. It is possible that selectins normally participate in this process, but that other molecules can substitute in their absence.

Overall, the available data suggest that the selectins have overlapping functions *in vivo*. This is consistent with *in vitro* and *in vivo* studies that demonstrate distinct differences in the regulation of expression of these molecules. Data from a variety of animal models suggest that dysregulated expression of selectins contributes to the pathogenesis of inflammatory and thrombotic diseases.[176-178] The ability of selectin antagonists to reduce tissue damage in these models suggests that inhibitors of selectin function or expression may be effective therapeutics for such diseases in humans.

14

REFERENCES

1. Carlos, T.M. and Harlan, J.M. Leukocyte-endothelial adhesion molecules. *Blood* 1994;84:2068-2101.
2. Springer, T.A. Traffic signals on endothelium for lymphocyte recirculation and leukocyte emigration. *Annu.Rev.Physiol.* 1995;57:827-872.
3. Butcher, E.C. and Picker, L.J. Lymphocyte homing and homeostasis. *Science* 1996;272:60-66.
4. McEver, R.P., Moore, K.L., and Cummings, R.D. Leukocyte trafficking mediated by selectin-carbohydrate interactions. *J.Biol.Chem.* 1995;270:11025-11028.
5. Kansas, G.S. Selectins and their ligands: current concepts and controversies. *Blood* 1996;88:3259-3287.
6. Graves, B.J., Crowther, R.L., Chandran, C., Rumberger, J.M., Li, S., Huang, K.-S., Presky, D.H., Familletti, P.C., Wolitzky, B.A., and Burns, D.K. Insight into E-selectin/ligand interaction from the crystal structure and mutagenesis of the lec/EGF domains. *Nature* 1994;367:532-538.
7. Erbe, D.V., Watson, S.W., Presta, L.G., Wolitzky, B.A., Foxall, C., Brandley, B.K., and Lasky, L.A. P- and E-selectin use common sites for carbohydrate ligand recognition and cell adhesion. *J.Cell Biol.* 1993;120:1227-1235.
8. Erbe, D.V., Wolitzky, B.A., Presta, L.G., Norton, C.R., Ramos, R.J., Burns, D.K., Rumberger, J.M., Rao, B.N.N., Foxall, C., Brandley, B.K. *et al.* Identification of an E-selectin region critical for carbohydrate recognition and cell adhesion. *J.Cell Biol.* 1992;119:215-227.
9. Hollenbaugh, D., Bajorath, J., Stenkamp, R., and Aruffo, A. Interaction of P-selectin (CD62) and its cellular ligand: Analysis of critical residues. *Biochemistry* 1993;32:2960-2966.
10. Bajorath, J., Hollenbaugh, D., King, G., Harte, W.Jr., Eustice, D.C., Darveau, R.P., and Aruffo, A. CD62/P-selectin binding sites for myeloid cells and sulfatides are overlapping. *Biochemistry* 1994;33:1332-1339.
11. Revelle, B.M., Scott, D., Kogan, T.P., Zheng, J.H., and Beck, P.J. Structure-function analysis of P-selectin-sialyl Lewis^x binding interactions - Mutagenic alteration of ligand binding specificity. *J.Biol.Chem.* 1996;271:4289-4297.
12. Pigott, R., Needham, L.A., Edwards, R.M., Walker, C., and Power, C. Structural and functional studies of the endothelial activation antigen endothelial leucocyte adhesion molecule-1 using a panel of monoclonal antibodies. *J.Immunol.* 1991;147:130-135.
13. Bowen, B.R., Fennie, C., and Lasky, L.A. The Mel 14 antibody binds to the lectin domain of the murine peripheral lymph node homing receptor. *J.Cell Biol.* 1990;110:147-153.
14. Kansas, G.S., Saunders, K.B., Ley, K., Zakrzewicz, A., Gibson, R.M., Furie, B.C., Furie, B., and Tedder, T.F. A role for the epidermal growth factor-like domain of P-selectin in ligand recognition and cell adhesion. *J.Cell Biol.* 1994;124:609-618.
15. Gibson, R.M., Kansas, G.S., Tedder, T.F., Furie, B., and Furie, B.C. Lectin and epidermal growth factor domains of P-selectin at physiologic density are the recognition unit for leukocyte binding. *Blood* 1995;85:151-158.

16. Tu, L.L., Chen, A.J., Delahunty, M.D., Moore, K.L., Watson, S.R., McEver, R.P., and Tedder, T.F. L-selectin binds to P-selectin glycoprotein ligand-1 on leukocytes. *J.Immunol.* 1996;157:3995-4004.

17. Kolbinger, F., Patton, J.T., Geisenhoff, G., Aenis, A., Li, X.H., and Katopodis, A.G. The carbohydrate-recognition domain of E-selectin is sufficient for ligand binding under both static and flow conditions. *Biochemistry* 1996;35:6385-6392.

18. Dwir, O., Kansas, G.S., and Alon, R. An activated L-selectin mutant with conserved equilibrium binding properties but enhanced ligand recognition under shear flow. *J.Biol.Chem.* 2000;275:18682-18691.

19. Watson, S.R., Imai, Y., Fennie, C., Geoffrey, J., Singer, M., Rosen, S.D., and Lasky, L.A. The complement binding-like domains of the murine homing receptor facilitate lectin activity. *J.Cell Biol.* 1991;115:235-243.

20. Li, S.H., Burns, D.K., Rumberger, J.M., Presky, D.H., Wilkinson, V.L., Anostario, M.Jr., Wolitzky, B.A., Norton, C.R., Familletti, P.C., Kim, K.J. *et al.* Consensus repeat domains of E-selectin enhance ligand binding. *J.Biol.Chem.* 1994;269:4431-4437.

21. Ushiyama, S., Laue, T.M., Moore, K.L., Erickson, H.P., and McEver, R.P. Structural and functional characterization of monomeric soluble P-selectin and comparison with membrane P-selectin. *J.Biol.Chem.* 1993;268:15229-15237.

22. Moore, K.L., Eaton, S.F., Lyons, D.E., Lichenstein, H.S., Cummings, R.D., and McEver, R.P. The P-selectin glycoprotein ligand from human neutrophils displays sialylated, fucosylated, O-linked poly-N-acetyllactosamine. *J.Biol.Chem.* 1994;269:23318-23327.

23. Hensley, P., McDevitt, P.J., Brooks, I., Trill, J.J., Feild, J.A., McNulty, D.E., Connor, J.R., Griswold, D.E., Kumar, N.V., Kopple, K.D. *et al.* The soluble form of E-selectin is an asymmetric monomer. Expression, purification, and characterization of the recombinant protein. *J.Biol.Chem.* 1994;269:23949-23958.

24. Mehta, P., Cummings, R.D., and McEver, R.P. Affinity and kinetic analysis of P-selectin binding to P-selectin glycoprotein ligand-1. *J.Biol.Chem.* 1998;273:32506-32513.

25. Varki, A. Selectin ligands. *Proc.Natl.Acad.Sci.USA* 1994;91:7390-7397.

26. Jacob, G.S., Kirmaier, C., Abbas, S.Z., Howard, S.C., Steininger, C.N., Welply, J.K., and Scudder, P. Binding of sialyl Lewis x to E-selectin as measured by fluorescence polarization. *Biochemistry* 1995;34:1210-1217.

27. Cooke, R.M., Hale, R.S., Lister, S.G., Shah, G., and Weir, M.P. The conformation of the sialyl Lewis X ligand changes upon binding to E-selectin. *Biochemistry* 1994;33:10591-10596.

28. Aruffo, A., Kolanus, W., Walz, G., Fredman, P., and Seed, B. CD62/P-selectin recognition of myeloid and tumor cell sulfatides. *Cell* 1991;67:35-44.

29. Norgard-Sumnicht, K.E., Varki, N.M., and Varki, A. Calcium-dependent heparin-like ligands for L-selectin in nonlymphoid endothelial cells. *Science* 1993;261:480-483.

30. Moore, K.L., Stults, N.L., Diaz, S., Smith, D.L., Cummings, R.D., Varki, A., and McEver, R.P. Identification of a specific glycoprotein ligand for P-selectin (CD62) on myeloid cells. *J.Cell.Biol.* 1992;118:445-456.

31. Steegmaier, M., Levinovitz, A., Isenmann, S., Borges, E., Lenter, M.,

16

Kocher, H.P., Kleuser, B., and Vestweber, D. The E-selectin-ligand ESL-1 is a variant of an FGF-receptor. *Nature* 1995;373:615-620.

32. Sako, D., Chang, X.-J., Barone, K.M., Vachino, G., White, H.M., Shaw, G., Veldman, G.M., Bean, K.M., Ahern, T.J., Furie, B. *et al.* Expression cloning of a functional glycoprotein ligand for P-selectin. *Cell* 1993;75:1179-1186.

33. Levinovitz, A., Mühlhoff, J., Isenmann, S., and Vestweber, D. Identification of a glycoprotein ligand for E-selectin on mouse myeloid cells. *J.Cell Biol.* 1993;121:449-459.

34. Patel, T.P., Goelz, S.E., Lobb, R.R., and Parekh, R.B. Isolation and characterization of natural protein-associated carbohydrate ligands for E-selectin. *Biochemistry* 1994;33:14815-14824.

35. Lenter, M., Levinovitz, A., Isenmann, S., and Vestweber, D. Monospecific and common glycoprotein ligands for E- and P-selectin on myeloid cells. *J.Cell Biol.* 1994;125:471-481.

36. Mebius, R.E. and Watson, S.R. L- and E-selectin can recognize the same naturally occurring ligands on high endothelial venules. *J.Immunol.* 1993;151:3252-3260.

37. Baumhueter, S., Singer, M.S., Henzel, W., Hemmerich, S., Renz, M., Rosen, S.D., and Lasky, L.A. Binding of L-selectin to the vascular sialomucin CD34. *Science* 1993;262:436-438.

38. Lasky, L.A., Singer, M.S., Dowbenko, D., Imai, Y., Henzel, W.J., Grimley, C., Fennie, C., Gillett, N., Watson, S.R., and Rosen, S.D. An endothelial ligand for L-selectin is a novel mucin-like molecule. *Cell* 1992;69:927-938.

39. Briskin, M.J., McEvoy, L.M., and Butcher, E.C. MAdCAM-1 has homology to immunoglobulin and mucin-like adhesion receptors and to IgA1. *Nature* 1993;363:461-463.

40. Imai, Y., Singer, M.S., Fennie, C., Lasky, L.A., and Rosen, S.D. Identification of a carbohydrate-based endothelial ligand for a lymphocyte homing receptor. *J.Cell Biol.* 1991;113:1213-1222.

41. Norgard, K.E., Moore, K.L., Diaz, S., Stults, N.L., Ushiyama, S., McEver, R.P., Cummings, R.D., and Varki, A. Characterization of a specific ligand for P-selectin on myeloid cells. A minor glycoprotein with sialylated *O*-linked oligosaccharides. *J.Biol.Chem.* 1993;268:12764-12774.

42. Li, F., Wilkins, P.P., Crawley, S., Weinstein, J., Cummings, R.D., and McEver, R.P. Post-translational modifications of recombinant P-selectin glycoprotein ligand-1 required for binding to P- and E-selectin. *J.Biol.Chem.* 1996;271:3255-3264.

43. Imai, Y., Lasky, L.A., and Rosen, S.D. Sulphation requirement for GlyCAM-1, an endothelial ligand for L-selectin. *Nature* 1993;361:555-557.

44. Hemmerich, S., Butcher, E.C., and Rosen, S.D. Sulfation-dependent recognition of high endothelial venules (HEV)-ligands by L-selectin and MECA 79, an adhesion-blocking monoclonal antibody. *J.Exp.Med.* 1994;180:2219-2226.

45. Wilkins, P.P., Moore, K.L., McEver, R.P., and Cummings, R.D. Tyrosine sulfation of P-selectin glycoprotein ligand-1 is required for high affinity binding to P-selectin. *J.Biol.Chem.* 1995;270:22677-22680.

46. Pouyani, T. and Seed, B. PSGL-1 recognition of P-selectin is controlled by a tyrosine sulfation consensus at the PSGL-1 amino terminus. *Cell* 1995;83:333-343.

47. Sako, D., Comess, K.M., Barone, K.M., Camphausen, R.T., Cumming, D.A.,

and Shaw, G.D. A sulfated peptide segment at the amino terminus of PSGL-1 is critical for P-selectin binding. *Cell* 1995;83:323-331.

48.   Hemmerich, S. and Rosen, S.D. 6'-sulfated sialyl Lewis x is a major capping group of GlyCAM-1. *Biochemistry* 1994;33:4830-4835.

49.   Hemmerich, S., Bertozzi, C.R., Leffler, H., and Rosen, S.D. Identification of the sulfated monosaccharides of GlyCAM-1, an endothelial-derived ligand for L-selectin. *Biochemistry* 1994;33:4820-4829.

50.   Hemmerich, S., Leffler, H., and Rosen, S.D. Structure of the *O*-glycans in GlyCAM-1, an endothelial-derived ligand for L-selectin. *J.Biol.Chem.* 1995;270:12035-12047.

51.   Wilkins, P.P., McEver, R.P., and Cummings, R.D. Structures of the O-glycans on P-selectin glycoprotein ligand-1 from HL-60 cells. *J.Biol.Chem.* 1996;271:18732-18742.

52.   Maemura, K. and Fukuda, M. Poly-*N*-acetyllactosaminyl *O*-glycans attached to leukosialin: The presence of sialyl Le^x structures in *O*-glycans. *J.Biol.Chem.* 1992;267:24379-24386.

53.   Leppanen, A., Mehta, P., Ouyang, Y.-B., Ju, T., Helin, J., Moore, K.L., van Die, I., Canfield, W.M., McEver, R.P., and Cummings, R.D. A novel glycosulfopeptide binds to P-selectin and inhibits leukocyte adhesion to P-selectin. *J.Biol.Chem.* 1999;274:24838-24848.

54.   Crottet, P., Kim, Y.J., Varki, A. Subsets of sialylated, sulfated mucins of diverse origins are recognized by L-selectin. *Glycobiology* 1996;6:191-208.

55.   Picker, L.J., Kishimoto, T.K., Smith, C.W., Warnock, R.A., and Butcher, E.C. ELAM-1 is an adhesion molecule for skin-homing T cells. *Nature* 1991;349:796-799.

56.   Moore, K.L. and Thompson, L.F. P-selectin (CD62) binds to subpopulations of human memory T lymphocytes and natural killer cells. *Biochem.Biophys.Res.Commun.* 1992;186:173-181.

57.   Borges, E., Tietz, W., Steegmaier, M., Moll, T., Hallmann, R., Hamann, A., and Vestweber, D. P-selectin glycoprotein ligand-1 (PSGL-1) on T helper 1 but not on T helper 2 cells binds to P-selectin and supports migration into inflamed skin. *J.Exp.Med.* 1997;185:573-578.

58.   Diacovo, T.G., Puri, K.D., Warnock, R.A., Springer, T.A., and Von Andrian, U.H. Platelet-mediated lymphocyte delivery to high endothelial venules. *Science* 1996;273:252-255.

59.   Frenette, P.S., Johnson, R.C., Hynes, R.O., and Wagner, D.D. Platelets roll on stimulated endothelium *in vivo*: An interaction mediated by endothelial P-selectin. *Proc.Natl.Acad.Sci.USA* 1995;92:7450-7454.

60.   Bargatze, R.F., Kurk, S., Butcher, E.C., and Jutila, M.A. Neutrophils roll on adherent neutrophils bound to cytokine-induced endothelial cells via L-selectin on the rolling cells. *J.Exp.Med.* 1994;180:1785-1792.

61.   Simon, S.I., Rochon, Y.P., Lynam, E.B., Smith, C.W., Anderson, D.C., and Sklar, L.A. β2-Integrin and L-selectin are obligatory receptors in neutrophil aggregation. *Blood* 1993;82:1097-1106.

62.   Walcheck, B., Moore, K.L., McEver, R.P., and Kishimoto, T.K. Neutrophil-neutrophil interactions under hydrodynamic shear stress involve L-selectin and PSGL-1: a mechanism that amplifies initial leukocyte accumulation on P-selectin *in vitro*. *J.Clin.Invest.* 1996;98:1081-1087.

63.   Alon, R., Fuhlbrigge, R.C., Finger, E.B., and Springer, T.A. Interactions

through L-selectin between leukocytes and adherent leukocytes nucleate rolling adhesions on selectins and VCAM-1 in shear flow. *J.Cell Biol.* 1996;135:849-865.

64. Hammer, D.A. and Apte, S.M. Simulation of cell rolling and adhesion on surfaces in shear flow: General results and analysis of selectin-mediated neutrophil adhesion. *Biophys.J.* 1992;63:35-57.

65. Tözeren, A. and Ley, K. How do selectins mediate leukocyte rolling in venules? *Biophys.J.* 1992;63:700-709.

66. Alon, R., Hammer, D.A., and Springer, T.A. Lifetime of the P-selectin: carbohydrate bond and its response to tensile force in hydrodynamic flow. *Nature* 1995;374:539-542.

67. Alon, R., Chen, S.Q., Puri, K.D., Finger, E.B., and Springer, T.A. The kinetics of L-selectin tethers and the mechanics of selectin-mediated rolling. *J.Cell Biol.* 1997;138:1169-1180.

68. Finger, E.B., Puri, K.D., Alon, R., Lawrence, M.B., Von Andrian, U.H., and Springer, T.A. Adhesion through L-selectin requires a threshold hydrodynamic shear. *Nature* 1996;379:266-269.

69. Chen, S.Q. and Springer, T.A. An automatic braking system that stabilizes leukocyte rolling by an increase in selectin bond number with shear. *J.Cell Biol.* 1999;144:185-200.

70. Nicholson, M.W., Barclay, A.N., Singer, M.S., Rosen, S.D., and Van der Merwe, P.A. Affinity and kinetic analysis of L-selectin (CD62L) binding to glycosylation-dependent cell-adhesion molecule-1. *J.Biol.Chem.* 1998;273:763-770.

71. Kishimoto, T.K., Jutila, M.A., Berg, E.L., and Butcher, E.C. Neutrophil Mac-1 and MEL-14 adhesion proteins inversely regulated by chemotactic factors. *Science* 1989;245:1238-1241.

72. Patecanda, A., Walcheck, B., Bishop, D.K., and Jutila, M.A. Rapid activation-independent shedding of leukocyte L-selectin induced by cross-linking of the surface antigen. *Eur.J.Immunol.* 1992;22:1279-1286.

73. Walcheck, B., Kahn, J., Fisher, J.M., Wang, B.B., Fisk, R.S., Payan, D.G., Feehan, C., Betageri, R., Darlak, K., Spatola, A.F. *et al.* Neutrophil rolling altered by inhibition of L-selectin shedding *in vitro. Nature* 1996;380:720-723.

74. Picker, L.J., Warnock, R.A., Burns, A.R., Doerschuk, C.M., Berg, E.L., and Butcher, E.C. The neutrophil selectin LECAM-1 presents carbohydrate ligands to the vascular selectins ELAM-1 and GMP-140. *Cell* 1991;66:921-933.

75. Von Andrian, U.H., Hasslen, S.R., Nelson, R.D., Erlandsen, S.L., and Butcher, E.C. A central role for microvillous receptor presentation in leukocyte adhesion under flow. *Cell* 1995;82:989-999.

76. Patel, K.D., Nollert, M.U., and McEver, R.P. P-selectin must extend a sufficient length from the plasma membrane to mediate rolling of neutrophils. *J.Cell Biol.* 1995;131:1893-1902.

77. Setiadi, H., Sedgewick, G., Erlandsen, S.L., and McEver, R.P. Interactions of the cytoplasmic domain of P-selectin with clathrin-coated pits enhance leukocyte adhesion under flow. *J.Cell Biol.* 1998;142:859-871.

78. Li, X., Steeber, D.A., Tang, M.L.K., Farrar, M.A., Perlmutter, R.M., and Tedder, T.F. Regulation of L-selectin-mediated rolling through receptor dimerization. *J.Exp.Med.* 1998;188:1385-1390.

79. Moore, K.L., Patel, K.D., Bruehl, R.E., Fugang, L., Johnson, D.A., Lichenstein, H.S., Cummings, R.D., Bainton, D.F., and McEver, R.P. P-selectin glycoprotein ligand-1 mediates rolling of human neutrophils on P-selectin. *J.Cell Biol.* 1995;128:661-671.

80. Patel, K.D., Moore, K.L., Nollert, M.U., and McEver, R.P. Neutrophils use both shared and distinct mechanisms to adhere to selectins under static and flow conditions. *J.Clin.Invest.* 1995;96:1887-1896.

81. Norman, K.E., Moore, K.L., McEver, R.P., and Ley, K. Leukocyte rolling *in vivo* is mediated by P-selectin glycoprotein ligand-1. *Blood* 1995;86:4417-4421.

82. Guyer, D.A., Moore, K.L., Lynam, E., Schammel, C.M.G., Rogelj, S., McEver, R.P., and Sklar, L.A. P-selectin glycoprotein ligand-1 (PSGL-1) is a ligand for L-selectin in neutrophil aggregation. *Blood* 1996;88:2415-2421.

83. Patel, K.D. and McEver, R.P. Comparison of tethering and rolling of eosinophils and neutrophils through selectins and P-selectin glycoprotein ligand-1. *J.Immunol.* 1997;159:4555-4565.

84. Li, F., Erickson, H.P., James, J.A., Moore, K.L., Cummings, R.D., and McEver, R.P. Visualization of P-selectin glycoprotein ligand-1 as a highly extended molecule and mapping of protein epitopes for monoclonal antibodies. *J.Biol.Chem.* 1996;271:6342-6348.

85. Ramachandran, V., Nollert, M.U., Qiu, H., Liu, W., Cummings, R.D., Zhu, C., and McEver, R.P. Tyrosine replacement in P-selectin glycoprotein ligand-1 affects distinct kinetic and mechanical properties of bonds with P- and L-selectin. *Proc.Natl.Acad.Sci.USA* 1999;96:13771-13776.

86. Bruehl, R.E., Moore, K.L., Lorant, D.E., Borregaard, N., Zimmerman, G.A., McEver, R.P., and Bainton, D.F. Leukocyte activation induces surface redistribution of P-selectin glycoprotein ligand-1. *J.Leukoc.Biol.* 1997;61:489-499.

87. Borges, E., Eytner, R., Moll, T., Steegmaier, M., Campbell, M.A., Ley, K., Mossman, H., and Vestweber, D. The P-selectin glycoprotein ligand-1 is important for recruitment of neutrophils into inflamed mouse peritoneum. *Blood* 1997;90:1934-1942.

88. Yang, J., Hirata, T., Croce, K., Merrill-Skoloff, G., Tchernychev, B., Williams, E., Flaumenhaft, R., Furie, B.C., and Furie, B. Targeted gene disruption demonstrates that P-selectin glycoprotein ligand 1 (PSGL-1) is required for P-selectin-mediated but not E-selectin-mediated neutrophil rolling and migration. *J.Exp.Med.* 1999;190:1769-1782.

89. Berg, E.L., McEvoy, L.M., Berlin, C., Bargatze, R.F., and Butcher, E.C. L-selectin-mediated lymphocyte rolling on MAdCAM-1. *Nature* 1993;366:695-698.

90. Puri, K.D., Finger, E.B., Gaudernack, G., and Springer, T.A. Sialomucin CD34 is the major L-selectin ligand in human tonsil high endothelial venules. *J.Cell Biol.* 1995;131:261-270.

91. Bargatze, R.F., Jutila, M.A., and Butcher, E.C. Distinct roles of L-selectin and integrins α4β7 and LFA-1 in lymphocyte homing to Peyer's patch-HEV in situ: the multistep model confirmed and refined. *Immunity* 1995;3:99-108.

92. Cheng, J., Baumhueter, S., Cacalano, G., Carver-Moore, K., Thibodeaux, H., Thomas, R., Broxmeyer, H.E., Cooper, S., Hague, N., Moore, M. *et al.* Hematopoietic defects in mice lacking the sialomucin CD34. *Blood*

20

1996;87:479-490.

93. Brustein, M., Kraal, G., Mebius, R.E., and Watson, S.R. Identification of a soluble form of a ligand for the lymphocyte homing receptor. *J.Exp.Med.* 1992;176:1415-1419.

94. Hwang, S.T., Singer, M.S., Giblin, P.A., Yednock, T.A., Bacon, K.B., Simon, S.I., and Rosen, S.D. GlyCAM-1, a physiologic ligand for L-selectin, activates β2 integrins on naive peripheral lymphocytes. *J.Exp.Med.* 1996;184:1343-1348.

95. Walcheck, B., Watts, G., and Jutila, M.A. Bovine γ/δ T cells bind E-selectin via a novel glycoprotein receptor: First characterization of a lymphocyte/E-selectin interaction in an animal model. *J.Exp.Med.* 1993;178:853-863.

96. Sassetti, C., Tangemann, K., Singer, M.S., Kershaw, D.B., and Rosen, S.D. Identification of podocalyxin-like protein as a high endothelial venule ligand for L-selectin: Parallels to CD34. *J.Exp.Med.* 1998;187:1965-1975.

97. Lorant, D.E., Patel, K.D., McIntyre, T.M., McEver, R.P., Prescott, S.M., and Zimmerman, G.A. Coexpression of GMP-140 and PAF by endothelium stimulated by histamine or thrombin: A juxtacrine system for adhesion and activation of neutrophils. *J.Cell Biol.* 1991;115:223-234.

98. Zimmerman, G.A., McIntyre, T.M., and Prescott, S.M. Adhesion and signaling in vascular cell-cell interactions. *J.Clin.Invest.* 1996;98:1699-1702.

99. Lorant, D.E., Topham, M.K., Whatley, R.E., McEver, R.P., McIntyre, T.M., Prescott, S.M., and Zimmerman, G.A. Inflammatory roles of P-selectin. *J.Clin.Invest.* 1993;92:559-570.

100. Weyrich, A.S., McIntyre, T.M., McEver, R.P., Prescott, S.M., and Zimmerman, G.A. Monocyte tethering by P-selectin regulates monocyte chemotactic protein-1 and tumor necrosis factor-α secretion. *J.Clin.Invest.* 1995;95:2297-2303.

101. Weyrich, A.S., Elstad, M.R., McEver, R.P., McIntyre, T.M., Moore, K.L., Morrissey, J.H., Prescott, S.M., and Zimmerman, G.A. Activated platelets signal chemokine synthesis by human monocytes. *J.Clin.Invest.* 1996;97:1525-1534.

102. Abbassi, O., Kishimoto, T.K., McIntire, L.V., Anderson, D.C., and Smith, C.W. E-selectin supports neutrophil rolling *in vitro* under conditions of flow. *J.Clin.Invest.* 1993;92:2719-2730.

103. Lawrence, M.B. and Springer, T.A. Leukocytes roll on a selectin at physiologic flow rates: Distinction from and prerequisite for adhesion through integrins. *Cell* 1991;65:859-873.

104. Lawrence, M.B. and Springer, T.A. Neutrophils roll on E-selectin. *J.Immunol.* 1993;151:6338-6346.

105. Hidari, K.I.P.J., Weyrich, A.S., Zimmerman, G.A., and McEver, R.P. Engagement of P-selectin glycoprotein ligand-1 enhances tyrosine phosphorylation and activates mitogen-activated protein kinases in human neutrophils. *J.Biol.Chem.* 1997;272:28750-28756.

106. Evangelista, V., Manarini, S., Sideri, R., Rotondo, S., Martelli, N., Piccoli, A., Totani, L., Piccardoni, P., Vestweber, D., de Gaetano, G. *et al.* Platelet/polymorphonuclear leukocyte interaction: P-selectin triggers protein-tyrosine phosphorylation-dependent CD11b/CD18 adhesion: Role of PSGL-1 as a signaling molecule. *Blood* 1999;93:876-885.

107. Simon, S.I., Hu, Y., Vestweber, D., and Smith, C.W. Neutrophil tethering on E-selectin activates β_2 integrin binding to ICAM-1 through a mitogen-activated protein kinase signal transduction pathway. *J.Immunol.* 2000;164:4348-4358.

108. Laudanna, C., Constantin, G., Baron, P., Scarpini, E., Scarlato, G., Cabrini, G., Dechecchi, C., Rossi, F., Cassatella, M.A., and Berton, G. Sulfatides trigger increase of cytosolic free calcium and enhanced expression of tumor necrosis factor-α and interleukin-8 mRNA in human neutrophils. Evidence for a role of L-selectin as a signaling molecule. *J.Biol.Chem.* 1994;269:4021-4026.

109. Waddell, T.K., Fialkow, L., Chan, C.K., Kishimoto, T.K., and Downey, G.P. Potentiation of the oxidative burst of human neutrophils. A signaling role for L-selectin. *J.Biol.Chem.* 1994;269:18485-18491.

110. Simon, S.I., Cherapanov, V., Nadra, I., Waddell, T.K., Seo, S.M., Wang, Q., Doerschuk, C.M., and Downey, G.P. Signaling functions of L-selectin in neutrophils: Alterations in the cytoskeleton and colocalization with CD18. *J.Immunol.* 1999;163:2891-2901.

111. Smolen, J.E., Petersen, T.K., Koch, C., O'Keefe, S.J., Hanlon, W.A., Seo, S., Pearson, D., Fossett, M.C., and Simon, S.I. L-selectin signaling of neutrophil adhesion and degranulation involves p38 mitogen-activated protein kinase. *J.Biol.Chem.* 2000;275:15876-15884.

112. Tsang, Y.T.M., Neelamegham, S., Hu, Y., Berg, E.L., Burns, A.R., Smith, C.W., and Simon, S.I. Synergy between L-selectin signaling and chemotactic activation during neutrophil adhesion and transmigration. *J.Immunol.* 1997;159:4566-4577.

113. Yoshida, M., Westlin, W.F., Wang, N., Ingber, D.E., Rosenzweig, A., Resnick, N., and Gimbrone, M.A.Jr. Leukocyte adhesion to vascular endothelium induces E-selectin linkage to the actin cytoskeleton. *J.Cell Biol.* 1996;133:445-455.

114. Lorant, D.E., McEver, R.P., McIntyre, T.M., Moore, K.L., Prescott, S.M., and Zimmerman, G.A. Activation of polymorphonuclear leukocytes reduces their adhesion to P-selectin and causes redistribution of ligands for P-selectin on their surfaces. *J.Clin.Invest.* 1995;96:171-182.

115. Johnston, G.I., Kurosky, A., and McEver, R.P. Structural and biosynthetic studies of the granule membrane protein, GMP-140, from human platelets and endothelial cells. *J.Biol.Chem.* 1989;264:1816-1823.

116. McEver, R.P., Beckstead, J.H., Moore, K.L., Marshall-Carlson, L., and Bainton, D.F. GMP-140, a platelet alpha-granule membrane protein, is also synthesized by vascular endothelial cells and is localized in Weibel-Palade bodies. *J.Clin.Invest.* 1989;84:92-99.

117. Bonfanti, R., Furie, B.C., Furie, B., and Wagner, D.D. PADGEM (GMP 140) is a component of Weibel-Palade bodies of human endothelial cells. *Blood* 1989;73:1109-1112.

118. Stenberg, P.E., McEver, R.P., Shuman, M.A., Jacques, Y.V., and Bainton, D.F. A platelet alpha-granule membrane protein (GMP-140) is expressed on the plasma membrane after activation. *J.Cell Biol.* 1985;101:880-886.

119. Berman, C.L., Yeo, E.L., Wencel-Drake, J.D., Furie, B.C., Ginsberg, M.H., and Furie, B. A platelet alpha granule membrane protein that is associated with the plasma membrane after activation. *J.Clin.Invest.* 1986;78:130-137.

22

120.   Hattori, R., Hamilton, K.K., Fugate, R.D., McEver, R.P., and Sims, P.J. Stimulated secretion of endothelial von Willebrand factor is accompanied by rapid redistribution to the cell surface of the intracellular granule membrane protein GMP-140. *J.Biol.Chem.* 1989;264:7768-7771.

121.   Hattori, R., Hamilton, K.K., McEver, R.P., and Sims, P.J. Complement proteins C5b-9 induce secretion of high molecular weight multimers of endothelial von Willebrand factor and translocation of granule membrane protein GMP-140 to the cell surface. *J.Biol.Chem.* 1989;264:9053-9060.

122.   George, J.N., Pickett, E.B., Saucerman, S., McEver, R.P., Kunicki, T.J., Kieffer, N., and Newman, P.J. Platelet surface glycoproteins. Studies on resting and activated platelets and platelet membrane microparticles in normal subjects, and observations in patients during adult respiratory distress syndrome and cardiac surgery. *J.Clin.Invest.* 1986;78:340-348.

123.   Michelson, A.D., Barnard, M.R., Hechtman, H.B., MacGregor, H., Connolly, R.J., Loscalzo, J., and Valeri, C.R. *In vivo* tracking of platelets: circulating degranulated platelets rapidly lose surface P-selectin but continue to circulate and function. *Proc.Natl.Acad.Sci.USA* 1996;93:11877-11882.

124.   Kuijpers, T.W., Raleigh, M., Kavanagh, T., Janssen, H., Calafat, J., Roos, D., and Harlan, J.M. Cytokine-activated endothelial cells internalize E-selectin into a lysosomal compartment of vesiculotubular shape: A tubulin-driven process. *J.Immunol.* 1994;152:5060-5069.

125.   Setiadi, H., Disdier, M., Green, S.A., Canfield, W.M., and McEver, R.P. Residues throughout the cytoplasmic domain affect the internalization efficiency of P-selectin. *J.Biol.Chem.* 1995;270:26818-26826.

126.   Green, S.A., Setiadi, H., McEver, R.P., and Kelly, R.B. The cytoplasmic domain of P-selectin contains a sorting determinant that mediates rapid degradation in lysosomes. *J.Cell Biol.* 1994;124:435-448.

127.   Bevilacqua, M.P., Stengelin, S., Gimbrone, M.A.Jr., and Seed, B. Endothelial leukocyte adhesion molecule 1: an inducible receptor for neutrophils related to complement regulatory proteins and lectins. *Science* 1989;243:1160-1165.

128.   Smeets, E.F., de Vries, T., Leeuwenberg, J.F.M., Van den Eijnden, D.H., Buurman, W.A., and Neefjes, J.J. Phosphorylation of surface E-selectin and the effect of soluble ligand (Sialyl Lewis^x) on the half-life of E-selectin. *Eur.J.Immunol.* 1993;23:147-151.

129.   Subramaniam, M., Koedam, J.A., and Wagner, D.D. Divergent fates of P- and E-selectins after their expression on the plasma membrane. *Mol.Biol.Cell* 1993;4:791-801.

130.   Disdier, M., Morrissey, J.H., Fugate, R.D., Bainton, D.F., and McEver, R.P. Cytoplasmic domain of P-selectin (CD62) contains the signal for sorting into the regulated secretory pathway. *Mol.Biol.Cell* 1992;3:309-321.

131.   Straley, K.S., Daugherty, B.L., Aeder, S.E., Hockenson, A.L., Kim, K., and Green, S.A. An atypical sorting determinant in the cytoplasmic domain of P-selectin mediates endosomal sorting. *Mol.Biol.Cell* 1998;9:1683-1694.

132.   Chuang, P.I., Young, B.A., Thiagarajan, R.R., Cornejo, C., Winn, R.K., and Harlan, J.M. Cytoplasmic domain of E-selectin contains a non-tyrosine endocytosis signal. *J.Biol.Chem.* 1997;272:24813-24818.

133.   Hartwell, D.M., Mayadas, T.N., Berger, G., Frenette, P.S., Rayburn, H., Hynes, R.O., and Wagner, D.D. Role of P-selectin cytoplasmic domain in granular targeting *in vivo* and in early inflammatory responses. *J.Cell Biol.*

1998;143:1129-1141.

134.   Weller, A., Isenmann, S., and Vestweber, D. Cloning of the mouse endothelial selectins. Expression of both E- and P-selectin is inducible by tumor necrosis factor. *J.Biol.Chem.* 199;267:15176-15183.

135.   Sanders, W.E., Wilson, R.W., Ballantyne, C.M., and Beaudet, A.L. Molecular cloning and analysis of *in vivo* expression of murine P-selectin. *Blood* 199;80:795-800.

136.   Mayadas, T.N., Johnson, R.C., Rayburn, H., Hynes, R.O., and Wagner, D.D. Leukocyte rolling and extravasation are severely compromised in P selectin-deficient mice. *Cell* 1993;74:541-554.

137.   Burns, S.A., DeGuzman, B.J., Newburger, J.W., Mayer, J.E.Jr., Neufeld, E.J., and Briscoe, D.M. P-selectin expression in myocardium of children undergoing cardiopulmonary bypass. *J.Thorac.Cardiovasc.Surg.* 1995;110:924-933.

138.   Yao, L., Pan, J., Setiadi, H., Patel, K.D., and McEver, R.P. Interleukin 4 or oncostatin M induces a prolonged increase in P-selectin mRNA and protein in human endothelial cells. *J.Exp.Med.* 1996;184:81-92.

139.   Silber, A., Newman, W., Reimann, K.A., Hendricks, E., Walsh, D., and Ringler, D.J. Kinetic expression of endothelial adhesion molecules and relationship to leukocyte recruitment in two cutaneous models of inflammation. *Lab.Invest.* 1994;70:163-175.

140.   Yao, L., Setiadi, H., Xia, L., Laszik, Z., Taylor, F.B., and McEver, R.P. Divergent inducible expression of P-selectin and E-selectin in mice and primates. *Blood* 1999;94:3820-3828.

141.   Woltmann, G., McNulty, C. A., Dewson, G., Symon, F. A., and Wardlaw, A. J. Interleukin-13 induces PSGL-1/P-selectin-dependent adhesion of eosinophils, but not neutrophils, to human umbilical vein endothelial cells under flow. *Blood* 2000;95:3146-3152.

142.   Symon, F.A., Walsh, G.M., Watson, S.R., and Wardlaw, A.J. Eosinophil adhesion to nasal polyp endothelium is P-selectin-dependent. *J.Exp.Med.* 1994;180:371-376.

143.   Grober, J.S., Bowen, B.L., Ebling, H., Athey, B., Thompson, C.B., Fox, D.A., and Stoolman, L.M. Monocyte-endothelial adhesion in chronic rheumatoid arthritis: in situ detection of selectin and integrin-dependent interactions. *J.Clin.Invest.* 1993;91:2609-2619.

144.   Johnson-Tidey, R.R., McGregor, J.L., Taylor, P.R., and Poston, R.N. Increase in the adhesion molecule P-selectin in endothelium overlying atherosclerotic plaques. Coexpression with intercellular adhesion molecule-1. *Am.J.Pathol.* 1994;144:952-961.

145.   Johnston, G.I., Bliss, G.A., Newman, P.J., and McEver, R.P. Structure of the human gene encoding granule membrane protein-140, a member of the selectin family of adhesion receptors for leukocytes. *J.Biol.Chem.* 1990;265:21381-21385.

146.   Pan, J. and McEver, R.P. Characterization of the promoter for the human P-selectin gene. *J.Biol.Chem.* 1993;268:22600-22608.

147.   Pan, J. and McEver, R.P. Regulation of the human P-selectin promoter by Bcl-3 and specific homodimeric members of the NF-κB/Rel family. *J.Biol.Chem.* 1995;270:23077-23083.

148.   Khew-Goodall, Y., Wadham, C., Stein, B.N., Gamble, J.R., and Vadas, M.A.

24

Stat6 activation is essential for interleukin-4 induction of P-selectin transcription in human umbilical vein endothelial cells. *Arterioscler.Thromb.Vasc.Biol.* 1999;19:1421-1429.

149. Pan, J., Xia, L., and McEver, R.P. Comparison of promoters for the murine and human P-selectin genes suggests species-specific and conserved mechanisms for transcriptional regulation in endothelial cells. *J.Biol.Chem.* 1998;273:10058-10067.

150. Pan, J., Xia, L., Yao, L., and McEver, R.P. Tumor necrosis factor-α or lipopolysaccharide-induced expression of the murine P-selectin gene in endothelial cells involves novel κB sites and a variant ATF/CRE element. *J.Biol.Chem.* 1998;273:10068-10077.

151. Granger, D.N. and Kubes, P. The microcirculation and inflammation: modulation of leukocyte-endothelial cell adhesion. *J.Leukoc.Biol.* 1994;55:662-675.

152. Ley, K. and Tedder, T.F. Leukocyte interactions with vascular endothelium: New insights into selectin-mediated attachment and rolling. *J.Immunol.* 1995;155:525-528.

153. Hynes, R.O. and Wagner, D.D. Genetic manipulation of vascular adhesion molecules in mice. *J.Clin.Invest.* 1996;98:2193-2195.

154. Frenette, P.S. and Wagner, D.D. Insights into selectin function from knockout mice. *Thromb.Haemost.* 1997;78:60-64.

155. Etzioni, A., Frydman, M., Pollack, S., Avidor, I., Phillips, M.L., Paulson, J.C., and Gershoni-Baruch, R. Brief report: recurrent severe infections caused by a novel leukocyte adhesion deficiency. *N.Engl.J.Med.* 1992;327:1789-1792.

156. Von Andrian, U.H., Berger, E.M., Ramezani, L., Chambers, J.D., Ochs, H.D., Harlan, J.M., Paulson, J.C., Etzioni, A., and Arfors, K.-E. *In vivo* behavior of neutrophils from two patients with distinct inherited leukocyte adhesion deficiency syndromes. *J.Clin.Invest.* 1993;91:2893-2897.

157. Philips, M.L., Schwartz, B.R., Etzioni, A., Bayer, R., Ochs, H.D., Paulson, J.C., and Harlan, J.M. Neutrophil adhesion in leukocyte adhesion deficiency syndrome type 2. *J.Clin.Invest.* 1995;96:2898-2906.

158. Maly, P., Thall, A.D., Petryniak, B., Rogers, G.E., Smith, P.L., Marks, R.M., Kelly, R.J., Gersten, K.M., Cheng, G.Y., Saunders, T.L. *et al.* The α(1,3)Fucosyltransferase Fuc-TVII controls leukocyte trafficking through an essential role in L-, E-, and P-selectin ligand biosynthesis. *Cell* 1996;86:643-653.

159. Weninger, W., Ulfman, L. H., Cheng, G., Souchkova, N., Quackenbush, E. J., Lowe, J. B., and von Andrian, U. H. Specialized contributions by alpha(1,3)-fucosyltransferase-IV and FucT-VII during leukocyte rolling in dermal microvessels [In Process Citation]. *Immunity* 2000;12:665-676.

160. Labow, M.A., Norton, C.R., Rumberger, J.M., Lombard-Gillooly, K.M., Shuster, D.J., Hubbard, J., Bertko, R., Knaack, P.A., Terry, R.W., Harbison, M.L. *et al.* Characterization of E-selectin-deficient mice: demonstration of overlapping function of the endothelial selectins. *Immunity* 1994;1:709-720.

161. Arbones, M.L., Ord, D.C., Ley, K., Ratech, H., Maynard-Curry, C., Otten, G., Capon, D.J., and Tedder, T.F. Lymphocyte homing and leukocyte rolling and migration are impaired in L-selectin-deficient mice. *Immunity* 1994;1:247-260.

162.   Ley, K., Bullard, D.C., Arbonés, M.L., Bosse, R., Vestweber, D., Tedder, T.F., and Beaudet, A.L. Sequential contribution of L- and P-selectin to leukocyte rolling *in vivo. J.Exp.Med.* 1995;181:669-675.

163.   Subramaniam, M., Saffaripour, S., Watson, S.R., Mayadas, T.N., Hynes, R.O., and Wagner, D.D. Reduced recruitment of inflammatory cells in a contact hypersensitivity response in P-selectin-deficient mice. *J.Exp.Med.* 1995;181:2277-2282.

164.   Tedder, T.F., Steeber, D.A., and Pizcueta, P. L-selectin-deficient mice have impaired leukocyte recruitment into inflammatory sites. *J.Exp.Med.* 1995;181:2259-2264.

165.   Bullard, D.C., Kunkel, E.J., Kubo, H., Hicks, M.J., Lorenzo, I., Doyle, N.A., Doerschuk, C.M., Ley, K., and Beaudet, A.L. Infectious susceptibility and severe deficiency of leukocyte rolling and recruitment in E-selectin and P-selectin double mutant mice. *J.Exp.Med.* 1996;183:2329-2336.

166.   Frenette, P.S., Mayadas, T.N., Rayburn, H., Hynes, R.O., and Wagner, D.D. Susceptibility to infection and altered hematopoiesis and mice deficient in both P- and E-selectin. *Cell* 1996;84:563-574.

167.   Robinson, S.D., Frenette, P.S., Rayburn, H., Cummiskey, M., Ullman-Cullere, M., Wagner, D.D., and Hynes, R.O. Multiple, targeted deficiencies in selectins reveal a predominant role for P-selectin in leukocyte recruitment. *Proc Natl Acad Sci U S A* 1999;96:11452-7.

168.   Jung, U. and Ley, K. Mice lacking two or all three selectins demonstrate overlapping and distinct functions for each selectin. *J Immunol* 1999;162:6755-62.

169.   Zannettino, A.C.W., Berndt, M.C., Butcher, C., Butcher, E.C., Vadas, M.A., and Simmons, P.J. Primitive human hematopoietic progenitors adhere to P-selectin (CD62P). *Blood* 1995;85:3466-3477.

170.   Dercksen, M.W., Weimar, I.S., Richel, D.J., Breton-Gorius, J., Vainchenker, W., Slaper-Cortenbach, I.C.M., Pinedo, H.M., von dem Borne, A.E.G.Kr., Gerritsen, W.R., and Van der Schoot, C.E. The value of flow cytometric analysis of platelet glycoprotein expression of CD34⁺ cells measured under conditions that prevent P-selectin-mediated binding of platelets. *Blood* 1995;86:3771-3782.

171.   Laszik, Z., Jansen, P.J., Cummings, R.D., Tedder, T.F., McEver, R.P., and Moore, K.L. P-selectin glycoprotein ligand-1 is broadly expressed in cells of myeloid, lymphoid, and dendritic lineage and in some nonhematopoietic cells. *Blood* 1996;88:3010-3021.

172.   Spertini, O., Cordey, A.-S., Monai, N., Giuffre, L., and Schapira, M. P-selectin glycoprotein ligand-1 (PSGL-1) is a ligand for L-selectin on neutrophils, monocytes and CD34⁺ hematopoietic progenitor cells. *J.Cell Biol.* 1996;135:523-531.

173.   Frenette, P.S., Subbarao, S., Mazo, I.B., Von Andrian, U.H., and Wagner, D.D. Endothelial selectins and vascular cell adhesion molecule-1 promote hematopoietic progenitor homing to bone marrow. *Proc.Natl.Acad.Sci.USA* 1998;95:14423-14428.

174.   Nguyen, M., Strubel, N.A., and Bischoff, J. A role for sialyl Lewis-X/A glycoconjugates in capillary morphogenesis. *Nature* 1993;365:267-269.

175.   Koch, A.E., Halloran, M.M., Haskell, C.J., Shah, M.R., and Polverini, P.J. Angiogenesis mediated by soluble forms of E-selectin and vascular cell

adhesion molecule-1. *Nature* 1995;376:517-519.

176.   Lefer, A.M., Weyrich, A.S., and Buerke, M. Role of selectins, a new family of adhesion molecules, in ischaemia-reperfusion injury. *Cardiovasc.Res.* 1994;28:289-294.

177.   Sharar, S.R., Winn, R.K., and Harlan, J.M. The adhesion cascade and anti-adhesion therapy: An overview. *Springer Semin.Immunopathol.* 1995;16:359-378.

178.   Lefer, D.J. Pharmacology of selectin inhibitors in ischemia/reperfusion states. *Annu.Rev.Pharmacol.Toxicol.* 2000;40:283-294.

179.   Cyster, J.G., Shotton, D.M., and Williams, A.F. The dimensions of the T lymphocyte glycoprotein leukosialin and identification of linear protein epitopes that can be modified by glycosylation. *EMBO J.* 1991;10:893-902.

Chapter 2

TRANSCRIPTIONAL REGULATION OF THE E-SELECTIN GENE

Tucker Collins,[+] Kelly-Ann Sheppard,[+] Simos Simeonidis,[+] and David W. Rose[*]

[+]Vascular Research Division, Department of Pathology,
Brigham and Women's Hospital and Harvard Medical School,
221 Longwood Avenue, Boston, MA 02115
[*]Department of Medicine and Whittier Diabetes Program
University of California San Diego
9500 Gilman Drive, La Jolla, CA 92093

INTRODUCTION

Recruitment of circulating leukocytes by cytokine-activated endothelium is important in both acute and chronic inflammation and is controlled, in part, by the selectin family of adhesion molecules. As outlined in the previous chapter, one of the members of this family, E-selectin, plays a key role in the initial stages of leukocyte recruitment and is tightly regulated in the vasculature. In this chapter we will focus on the molecular mechanisms that regulate the expression of the E-selectin gene.

E-selectin (CD62E), also known as endothelial-leukocyte adhesion molecule-1 (ELAM-1), mediates the adhesion of leukocytes to endothelium and is thought to play a role in recruiting circulating blood cells to sites of inflammatory responses. E-selectin is a member of the selectin family of proteins which consists of three members: E-selectin, P-selectin, and L-selectin (reviewed in 1; see Chapter 1 of this volume). Human E-selectin has an N-terminal lectin domain, an epidermal growth factor-like domain, six domains homologous to complement regulatory domains, a single transmembrane, and a short cytoplasmic domain. It recognizes specific proteins that have sialyl Lewis X (or related carbohydrates) and are found on leukocytes. Several leukocyte glycoproteins have been identified that bind to E-selectin, including E-selectin ligand-1 (ESL-1) and P-selectin glycoprotein ligand (PSGL-1). Interaction of E-selectin with its ligand contributes to the initial tethering of leukocytes to the endothelial surface.

Establishing the physiologic role of E-selectin has been a challenge. The process of leukocyte recruitment is a dynamic one and involves multiple steps. Selectins mediate the initial tethering and rolling of leukocytes along the endothelium of the vessel wall (reviewed in Chapter 1). Analysis of mice with targeted mutations in the selectins suggests the presence of redundant mechanisms controlling leukocyte recruitment. Studies with these animal models *in vivo* suggest that the initial leukocyte rolling is P-selectin-dependent, whereas the subsequent leukocyte-endothelial interaction is L-selectin-dependent. Initial studies in mice with a targeted mutation in the E-selectin gene found no defects in leukocyte recruitment unless P-selectin function was also blocked.[2] Subsequent work demonstrates that rolling of leukocytes *in vivo* in E-selectin null animals may be faster than in normal animals.[3] E-selectin is also required *in vivo* for normal levels of leukocyte firm adhesion to cytokine-activated microvascular endothelium.[4,5] Decreased leukocyte transit time in the E-selectin null animals may diminish exposure time to chemoattractants and decrease firm adhesion and subsequent transmigration. Collectively, these observations support the concept that E-selectin serves an important function during leukocyte recruitment.

E-SELECTIN GENE EXPRESSION

Although the members of the selectin family are structurally and functionally related, the mechanisms controlling their expression are quite distinct. Expression of E-selectin is normally not detected in resting endothelium, but is strongly and rapidly induced by the inflammatory cytokines, such as tumor necrosis factor-α (TNF-α) and interleukin-1 (IL-1), as well as by lipopolysaccharide (LPS).[6,7] E-selectin appears on the endothelial cell surface within 1 to 2 hours of cytokine treatment, is expressed maximally at 4 to 6 hours, and then rapidly declines even in the continuous presence of cytokine. The kinetics of accumulation of E-selectin mRNA parallels the rate of transcription of the E-selectin gene.[8] As an inflammatory or immune reaction evolves, expression of some endothelial adhesion molecules diminishes. This exchange of cell surface molecules correlates with a change in leukocyte traffic. Decreased expression of E-selectin correlates with the switch from neutrophil to mononuclear cell recruitment during some inflammatory responses (reviewed in Chapters 1 and 3).

E-selectin gene expression can be induced by a number of agonists, including ionizing radiation, oxygen radicals, complement components, thrombin, physical forces, and phorbol esters (reviewed in 1 and 9). In addition to the inflammatory cytokines and LPS, E-selectin is also induced by other cytokines, such as interleukins-3 and −10.[10,11] In contrast, induction of E-selectin by inflammatory cytokines is inhibited by transforming growth factor-β[12] or interleukin-4[13,14] both of which have anti-inflammatory properties *in vivo*. E-selectin expression is also inhibited by specific anti-inflammatory agents (reviewed in Chapter 8). For example, dexamethasone has been reported to inhibit expression of this adhesion molecule in some endothelial cell types.[15]

In addition to soluble factors, interactions between the endothelial surface and leukocytes can induce the expression of E-selectin. For example, CD40 is expressed on a variety of cells, including vascular endothelium.[16,17] CD40 interacts with CD40L, an activation-induced CD4+ T cell surface molecule. These interactions are known to play a role in B cell activation and differentiation. Interactions between endothelial cell CD40 and a recombinant soluble form of CD40 ligand lead to endothelial cell activation, E-selectin expression, and leukocyte adhesion. This could be a mechanism by which activated T cells augment inflammatory responses *in vivo* by upregulating the expression of endothelial cell surface adhesion molecules[18] (reviewed in 19).

Changes in the level of E-selectin gene expression are primarily regulated at the transcriptional level. A comparison of the kinetics of accumulation of E-selectin mRNA, with changes in the rate of transcription of the E-selectin gene, indicate that the transient induction of the gene is due to a dramatic transcriptional activation followed by postinduction repression of transcription.[8,20] The kinetics of induction and decay of E-selectin mRNA can be changed by the protein synthesis inhibitor cycloheximide. In the presence of the inhibitor, cytokine induces the E-selectin gene to higher levels, and the level of E-selectin remains high after induction. This phenomenon, called superinduction, is seen with many highly inducible genes. Nuclear run-off studies demonstrate that cycloheximide blocks the transcriptional repression of the gene, since in the presence of the inhibitor the rate of transcription remains high after induction.[20] This inhibitor may therefore block the synthesis of a transcriptional repressor involved in post-induction shutdown of the E-selectin gene. The molecular events in this process will be discussed in more detail in a subsequent section of this chapter. Cycloheximide may also inhibit the rapid degradation of the E-selectin transcript, but nuclear run-off experiments suggest that this effect is modest. The decrease in cell surface expression of E-selectin also involves internalization and degradation of the protein in the lysosomes.[21]

ENDOTHELIAL CELL-SPECIFIC GENE EXPRESSION OF E-SELECTIN

E-selectin expression is largely restricted to venular and capillary endothelial cells, although it is not uniformly expressed in the vasculature.[22,23] It can be localized to physiologically relevant vascular structures, such as post-capillary venules. This spatial heterogeneity may restrict leukocyte recruitment to appropriate vascular locations. Although there has been some interest in the mechanism controlling cell-type specific expression of the E-selectin gene, little insight has been gained into the factors that regulate this process. Additionally, virtually nothing is known about the factors that further restrict expression of E-selectin to particular endothelial subpopulations.

The promoter region of the E-selectin gene can mediate inflammatory cytokine induction of reporter constructs in non-endothelial cells (e.g., HeLa or COS cells). This suggests that at least some of the element(s) controlling endothelial-specific expression are likely to be found outside of the promoter region of the gene. There is some evidence that methylation of an element in the promoter might repress transcription of the E-selectin gene in non-endothelial cells.[24] Additionally, there is evidence for a tissue-specific repressor that inhibits E-selectin expression.[25] As expected, E-selectin was not detected in cultured human aortic smooth muscle cells after treatment with TNF-α. However, E-selectin mRNA was detected after the same cells were pretreated with a protein synthesis inhibitor and then challenged with inflammatory cytokine. These findings suggest that a labile repressor protein(s) plays a role in inhibiting E-selectin expression in aortic smooth muscle cells at the level of gene transcription. To date, the nature of the inhibitor and the mechanism of repression are not known. We will focus below on the cytokine-induced expression of the E-selectin gene.

ORGANIZATION OF THE CYTOKINE RESPONSE REGION OF THE E-SELECTIN GENE

The 5' flanking sequence of the E-selectin gene is capable of conferring cytokine responsiveness on heterologous promoter-reporter genes. Deletion analysis of the human E-selectin promoter in several types of endothelial cells revealed that all of the activity was contained in the first 160 base pairs of the promoter.[26-28] This regulatory section of the gene has been defined as the cytokine response region, and confers cytokine responsiveness in both endothelial and non-endothelial cells.

Partial DNase I-hypersensitive site mapping of the E-selectin gene revealed that a single site was present in the proximal promoter only following TNF-α stimulation.[27] DNase I-hypersensitive sites are often associated with regulatory DNA sequences involved in transcriptional control. Additionally, the appearance of DNase I-hypersensitive sites correlates reasonably well with the binding of transcriptional regulatory proteins. Remarkably, a more detailed DNase I-hypersensitive site map of the E-selectin gene has not been completed. Such a map might provide insight into regulatory elements located within the gene that are unique to endothelial cells. Collectively, the deletion analysis and the existing hypersensitive site mapping studies suggest that the proximal promoter region of the E-selectin gene is necessary for cytokine-induced gene activation.

The organization of the regulatory elements required for cytokine-induced expression of the E-selectin gene has been defined and reveals a surprisingly complex arrangement of positive regulatory domains (PDs).[26,29] In our opinion, the PD nomenclature is useful because multiple regulatory proteins may interact with specific sites contained within each of these regulatory regions. Determining which of these interactions is actually involved in cytokine-induced expression of E-selectin remains a key challenge. We will describe below our current understanding of E-selectin regulatory sequences and summarize progress in the identification of proteins that bind to the cytokine response region.

The E-selectin cytokine response region contains four positive regulatory elements (Fig. 2-1). PDI (GGGATTTCCTC) is located at the 3' end of the cytokine response region in the E-selectin promoter. PDI contains a consensus NF-κB site and was historically the first regulatory element to be identified in the E-selectin promoter. Deletion and point mutation analysis establish that this site is necessary but not sufficient for TNF-α-induced expression of E-selectin.[8] A single copy of this element does not confer IL-1-induced gene expression when fused to a reporter, although a multimerized version of this element will transfer cytokine-induced gene expression to a reporter gene (Whitley and Collins, unpublished observations).

PDII (TGACATCA) is located at the 5' end of the cytokine response region. This domain contains an ATF-like element that is important for cytokine-induced expression of the E-selectin gene. Deletion and site-directed mutagenesis studies identify the CRE/ATF site as a functional element for maximal cytokine-induced expression of the gene.[30]

PDIII and PDIV (GGGAAAGTTTT and GGATATTCCC), which are located in the central portion of the cytokine response region, have been mapped by deletion and point mutation analysis, as well as by systematic site-directed mutagenesis.[26-28] Conservation of the DNA sequence of all four of the positive regulatory domains in the human,[8,31] mouse,[32] and rabbit[33] E-selectin promoters is consistent with functional roles for these sites in the regulation of E-selectin expression. The transcriptional activators that can interact positively with these elements in the E-selectin promoter are outlined below.

PROTEINS THAT BIND TO PDI, III, AND IV

Several proteins bind specifically to these elements in the E-selectin cytokine response region, including nuclear factor-κB,[8,26-28,34,35] high mobility group I(Y) (HMG I(Y)),[26,28] nuclear factor of activated T cells (NFAT) (M. Whitley and T. Collins, unpublished), and STAT 6,[14] although only NF-κB and HMG I(Y) have been implicated in the positive regulation of E-selectin expression.

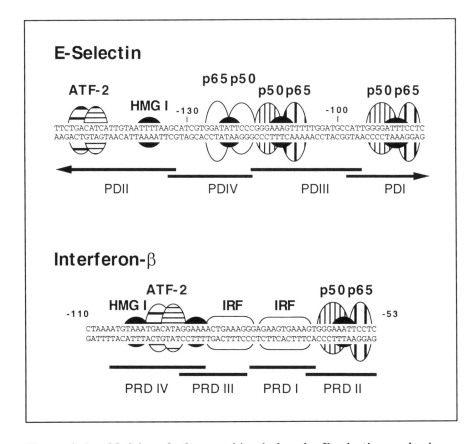

Figure 2-1. Models of the cytokine-induced E-selectin and virus-induced IFN-β enhancers. The double-stranded DNA sequences are shown of the E-selectin promoter from -156 to -83 and the IFN-β promoter from -110 to -53, relative to the transcriptional start sites. The inducible enhancers in IFN-β and E-selectin contain four regulatory domains, designated PRDI-PRDIV in the human IFN-β gene and PDI-IV in the E-selectin gene. Transcription factors that bind to each of the elements are shown for both genes. Transcription factors that bind to the E-selectin cytokine response region are as follows: NF-κB binds to PDI, III and IV; ATF-2 or ATF-a homodimers or the corresponding c-JUN heterodimers bind to PDII. HMG I(Y) binds to a region within PDII, as well as to the AT-rich regions in the κB sites contained in PDI, III and IV. Those for the IFN-β gene are as follows (reviewed in 37): NF-κB binds to PRDII, interferon regulatory factors bind to PRDI and III, and an activating transcription factor-2 (ATF-2)/c-JUN heterodimer binds to PRDIV. Binding sites for HMG I(Y) are found in PRDs II and IV and are indicated. The binding of HMG I(Y) at multiple sites increases the binding NF-κB and ATF2/c-JUN heterodimer for their respective sites and bends DNA in a way that facilitates the formation of a high order complex necessary for transcriptional activation. The transcription factors make extensive protein-protein contacts and the complex interacts as a unit with both the co-activator complexes containing CBP and P/CAF, as well as with the basal transcription machinery.

NF-κB Binds to PDI, III, and IV

NF-κB binding to PDI, III, and IV is required for cytokine inducibility of the E-selectin gene. Since these regulatory domains bind similar factors, the elements will be discussed together. Two observations indicate that NF-κB may play an important role in cytokine induction of the E-selectin gene. First, mutations that decrease the binding of NF-κB to any of the NF-κB binding sites *in vitro* results in diminished cytokine-induced E-selectin expression. Second, Rel family proteins found in nuclear extracts from cytokine−activated endothelial cells specifically interact with these elements. Supershift and UV cross-linking studies demonstrate that heterodimeric p50/p65 is the NF-κB species binding to these elements.[26-28] The three κB elements contained in PDIV, PDIII, and PDI have also been designated κB-1, κB-2 and κB-3, respectively.[27]

The presence of three closely spaced functionally important NF-κB sites raises the question of whether one site is preferentially occupied. DNase I footprinting studies reveal that when exposed to increasing amounts of p50/p65 heterodimer, PDI and PDIII (κB3 and κB2) are occupied, whereas PDIV (κB-1) is not. Additionally, footprinting studies using recombinant forms of NF-κB failed to detect any cooperative interactions between NF-κB proteins bound to the three sites within the E-selectin promoter.[26,27] This is interesting because the Rel homology domain of NF-κB participated in cooperative binding to κB sites present in reconstituted nucleosome cores.[36] Competition studies were used to measure the relative binding affinities of the NF-κB proteins to the three κB sites independently. These results suggest that p50/p65 and p50 bind more tightly with PDIII and PDI than with PDIV, and that PDIV preferentially binds p65.[27]

Several models may explain the preferential occupancy of two NF-κB elements and the apparent functional requirement for three NF-κB elements for cytokine inducibility. First, other members of the Rel family may bind PDIV. We have noted that recombinant c-Rel homodimers can simultaneously bind to all three κB elements in the cytokine response region of the intact promoter.[26] Additionally, members of the extended Rel family, such as NFAT, may bind to elements within PDIII. Binding of NFAT to PDIII may permit NF-κB to bind to PDI and PDIV. Second, other factors interacting with the E-selectin promoter may alter the conformation of the protein-DNA complex, allowing NF-κB to interact with PDIV. HMG I(Y) is required for cytokine-induced E-selectin expression, binds to both PDIII and IV, and enhances the binding of NF-κB to these elements.[26] As discussed in subsequent sections, HMG I(Y) may function as an architectural element in the assembly of the cytokine-induced enhancer complex. HMG I(Y) could facilitate occupancy of all three κB elements in the cytokine response region. However, initial footprinting studies done with recombinant forms of NF-κB and HMG I(Y) do not support this model (D. Thanos and T. Collins, unpublished). Finally, an authentic chromatin environment may facilitate binding of NF-κB to all three κB elements in the promoter. NF-κB cannot bind simultaneously to PDI, PDIII, and PDIV in the E-selectin cytokine response region in the form of linear DNA, but may be able to bind when the DNA is presented on a nucleosomal surface. The identification of DNA-binding proteins capable of interacting with the E-selectin cytokine response region when it is found in chromatin is an important area for future investigation.

HMG I(Y)

HMG I(Y) appears to be required for cytokine induction of the human E-selectin gene. The HMG proteins are low molecular weight non-histone chromosomal proteins that have been placed into three groups: HMG 1/2, HMG 14/17, and HMG I(Y). These proteins do not have transcriptional activating capacity, but facilitate the assembly of transcriptionally competent complexes.

HMG I(Y) binds to a region within the E-selectin PDII, as well as to the AT-rich regions in the κB sites contained in PDI, III, and IV.[26,28] The binding sites for HMG I(Y) found in PDs III and IV in the E-selectin gene promoter are indicated in Figure 2-1. The relationship between HMG I(Y) and NF-κB binding at PDIII and PDIV was investigated by comparing the effects of single base mutations on HMG I(Y) and on NF-κB binding with the effects of the same mutations on TNF-α-induced E-selectin promoter reporter gene expression. In general, mutations in the AT-rich region in the middle of PDIII and PDIV have the strongest negative effect on HMG I(Y) binding, whereas mutations in the GC-rich flanking sequence primarily decrease binding of NF-κB. Significantly, both types of mutations decrease the level of TNF-α-induced expression of the reporter gene. HMG I(Y) probably functions as an important architectural component in the assembly of the cytokine-induced transcription complex on the E-selectin promoter, as it does in the IFN-β gene promoter (reviewed in 37).

PROTEINS THAT BIND TO PDII

The sequence of PDII (TGACATCA) is similar to the DNA sequences recognized by the activating transcription factor (ATF) family of proteins. This family of transcriptional regulators shares a basic leucine zipper motif and can selectively heterodimerize with members of the c-FOS and c-JUN groups of factors.

This E-selectin site has been shown to be bound by recombinant ATF–a, ATF-2, ATF-3, c-JUN. Additionally, changes in the level of induced E-selectin expression can be correlated with alterations in the composition of the proteins binding to this element.[30]

Occupancy of PDII by ATF2/c-JUN confers a significantly stronger response to cytokines. Moreover, E-selectin induction is diminished in mice lacking ATF-2.[38] Despite the potential for overlap in function among the members of the ATF family, ATF-2 is probably important in cytokine-induced expression of the E-selectin gene.

SIGNALING PATHWAYS ACTIVATING E-SELECTIN TRANSCRIPTION

Two parallel signaling pathways are required for activation of E-selectin gene transcription in response to TNF-α. In endothelial cells, TNF-α interaction with a TNF-α receptor (TNFR1) induces coupling with receptor-associated proteins and generation of various TNF-induced signals (reviewed in 39). One signaling pathway results in phosphorylation, ubiquitination, and degradation of IκBα by the proteasome with nuclear accumulation of NF-κB (reviewed in 40-42). Concomitantly, the second set of TNF-α-induced events leads to activation of the c-JUN N-terminal kinase (JNK) kinase, resulting in phosphorylation of ATF-2 and c-JUN. These two pathways are rapidly activated and converge on the E-selectin promoter to result in cytokine responsiveness of this gene.[43,44]

Support for the involvement of JNK kinase in the cytokine-induced expression of E-selectin is provided by studies defining signaling pathways in endothelial cells. Inflammatory cytokines (TNF-α or IL-1) and CD40 ligand activate similar responses in endothelial cells. Quantitatively, however, TNF-α is more potent that IL-1, which is much more potent than CD40 ligand.[45] The same hierarchy is observed for transcriptional activation of an E-selectin promoter-reporter construct. Although these agents had a similar effect on NF-κB, activation of JNK showed the same hierarchy of potency. This direct correlation between the level of JNK activation and the activity of the promoter-reporter construct is consistent with a role for this kinase in E-selectin gene expression. JNK may also be involved in the phenomenon known as homologous desensitization, or the inability of individual cytokines to re-induce E-selectin expression. Stimulation of E-selectin either by inflammatory cytokine or by CD40 ligand renders endothelial cells transiently unresponsive to restimulation by the same, but not by a different, agonist. Interestingly, such homologous desensitization is paralleled by changes in JNK activation, suggesting that JNK may regulate this process.[45]

THE E-SELECTIN CYTOKINE-INDUCED ENHANCEOSOME

Analysis of the E-selectin promoter reveals a small cytokine response region with multiple NF-κB sites and a CRE/ATF-like element.

This cytokine-induced enhancer may involve a specific spatial arrangement of transcription factors, in association with other protein components functioning as architectural elements, to generate a unique higher order complex.[46] This complex may serve as a platform to recruit coactivators and interact with components of the basal complex. Similar findings from the analysis of other inducible promoters suggest that assembly of unique enhancer complexes from similar sets of transcription factors may be a common theme and may provide the specificity required for the regulation of complex patterns of gene expression. The general concept of enhanceosomes will be introduced below, followed by a more detailed discussion of the evidence that this type of transcriptional regulatory element assembles on the E-selectin promoter in response to inflammatory cytokine.

ENHANCEOSOMES AND TRANSCRIPTIONAL SYNERGY

A diverse group of inducible genes is regulated by a limited number of transcriptional activators. While this minimizes the complexity necessary to link different signaling systems, it raises the problem of how specificity is achieved in response to environmental cues. This problem is resolved by employing the principles of transcriptional synergy, whereby small combinations of ubiquitous as well as cell- and signal-specific activators collectively interact to generate a greater-than-additive transcriptional effect (reviewed in 47). This collection of activators has been called an enhanceosome. More specifically, an enhanceosome is a specific arrangement of transcription factor recognition sites and a precise collection of bound activators which together generate a network of protein-protein and protein-DNA interactions that are unique to a given enhancer. The sequence-specific activators and architectural or DNA-bending proteins engage in cooperative protein-protein interactions to create a stable platform. The enhanceosome then presents a distinct surface that is recognized both by coactivator complexes and by the basal transcriptional machinery. This leads to cooperative recruitment of RNA polymerase II and its associated factors and results in synergistic activation of transcription.

Perhaps the best characterized enhanceosome is that in the virus-inducible enhancer of the human interferon-β (IFN-β) gene (reviewed in 37). The human IFN-β gene is transiently induced by virus. The positive regulatory elements required for viral activation have been localized to the first 104 base pairs immediately upstream of the start site of transcription. This regulatory region consists of four overlapping positive regulatory domains (PRDI-PRDIV). NF-κB binds to PRDII, members of the interferon regulator factor (IRF) family bind to PRDI and III, and a ATF-2/c-JUN heterodimer binds to PRDIV (Fig. 2-1). HMG I(Y) is required for viral induction of the human IFN-β gene.[48] Binding of HMG I(Y) to the enhancers alters the structure of the DNA, allowing cooperative recruitment of the IFN-β gene activators that, together with HMG I(Y), assemble into a remarkably stable higher order nucleoprotein complex termed the IFN-β enhanceosome.[49,50] Mutations that alter the binding or positioning of either the transcriptional activators or HMG I(Y), or alterations that diminish the interactions between any of these components, decrease enhanceosome stability and transcription activation. Important for this discussion is the observation that the cytokine response region of the E-selectin and virus response region of the IFN-β gene show remarkable similarities[26,29] (Fig. 2-1).

A Model of the Assembly of the Cytokine-Induced E-Selectin Transcription Enhancer

After induction by cytokine, heterodimers of NF-κB bind to the promoter, which is constitutively occupied by an ATF-2/c-JUN heterodimer. In parallel with nuclear accumulation of NF-κB, ATF-2 and c-JUN are phosphorylated by c-JUN N-terminal kinase (JNK).[43] The binding of HMG I(Y) at multiple sites increases the binding of NF-κB and bends DNA[51] in a way that facilitates the formation of high order complex. Collectively these events place multiple transcriptional activators in a favorable architecture to compete for transcriptional coactivators. The transcriptional activators in the enhancer then make extensive protein-protein contacts with both coactivator complexes and possibly with the basal transcriptional machinery. We will develop some of the key aspects of this model below.

Physical Constraints and Architectural Proteins

An important component of this model of the cytokine-induced enhancer complex is the spatial arrangement of the elements. The relative helical orientation or phasing of PDI and PDII in the E-selectin enhancer is more important than the distance between these elements for cytokine responsiveness.[51] Using a panel of promoter mutants, it was demonstrated that moving these elements 5 bp (half-helix turns) up- or downstream resulted in a strong reduction in promoter activity, whereas shifts of 10 bp (a full helix turn) in either direction minimally affected IL-1 inducibility. This spacing dependence suggests that the transcription factors need to bind at the same face of the DNA helix for optimal interactions.

Heterodimeric transcription factors, such as the basic region-leucine zipper proteins ATF-2/c-JUN, are components of the both the E-selectin and IFN-β enhanceosomes. In recent studies it was demonstrated that within the interferon-β enhanceosome, the

ATF-2/c-JUN heterodimer binds in a specific orientation, which is required for assembly of a complex between ATF-2-c-JUN and IRF-3.[52] The DNA-bound ATF-2/c-JUN heterodimer interacts with the DNA-binding domain of IRF-3, bound to an adjacent site in the IFN-β enhancer. Additionally, *in vivo* chromatin immunoprecipitation studies reveal that recruitment of IRF-3 to the IFN-β promoter upon viral infection is dependent on the orientation of the ATF-2/c-JUN heterodimer binding site. Interestingly, the E-selectin ATF-2/c-JUN binding site is functional in the context of the IFN-β enhancer, resulting in wild type levels of virus induction.[52] It is possible that the assembly of a functional E-selectin enhanceosome is also dependent on the orientation of the heterodimer and its ability to interact with an adjacent transcription factor.

Protein-induced DNA bending also plays an important role in E-selectin gene expression, as has been suggested for the IFN-β gene enhancer.[49] Direct interaction between proteins bound to separate sites on DNA requires a minimum length of 130 base pairs for looping of the intervening DNA to occur. This suggests that interaction between the functional elements in the E-selectin promoter would require a bend or twist in the promoter. The energy required for juxtaposition of adjacent functional elements can be provided by proteins that bend the DNA helix. All of the transcription factors that are known to bind the E-selectin promoter can bend DNA. Circular permutation assays, which detect protein-induced and sequence-specific deformation of the DNA helix, indicate that both ATF-2 and NF-κB can bend the corresponding E-selectin elements.[51] These sequence-specific DNA-bending proteins may direct deformation of the DNA helix and allow for a precise alignment of functional elements, thus stabilizing weak protein-protein associations between the transcriptional activators.

Another important aspect of the model is the presence of a protein that functions primarily as an architectural component of the complex. As noted above, we have demonstrated that HMG I(Y) is involved in the regulation of the E-selectin gene through both κB elements in PDIII and PDIV.[26] HMG I(Y) may also play a role in ATF-2 interactions. HMG I(Y) binding apparently alters the structure of DNA, thereby increasing the affinity of NF-κB and ATF-2 for their respective recognition sequences. When the HMG I(Y) binding sites are placed on a model of a DNA double helix, they can be positioned on the same face of the helix. Thus, if HMG I(Y) bound to each of these sites simultaneously and induced a DNA bend in the same direction at each site, a loop containing the E-selectin enhancer would be formed. This loop would facilitate the interactions of NF-κB and ATF-2 and subsequent stabilization of the complex. Collectively, the DNA-bending proteins and the other architectural elements may create a scaffold which could provide a supporting structure for the assembly of a higher order complex.

Consistent with this model is the crystal structure of a large fragment of the p50 subunit, as well as the structure of a p50/p65 heterodimer bound to DNA.[53,54] The p50 dimer envelops about two-thirds of the cylindrical surface of the DNA helix, making specific contacts along the 10-base-pair κB recognition site. Only the DNA's minor groove was exposed. This is the site where minor groove binding proteins such as HMG I(Y) may interact with DNA along with NF-κB.[48] Understanding the structure of the individual transcriptional activators bound to DNA is the first step in generating an authentic model of the way in which the transcription factors assemble into the cytokine-induced enhancer.

PROTEIN-PROTEIN INTERACTIONS

The higher order structural features of the cytokine-induced enhancer would facilitate interactions between the proteins in the assembled complex. Direct protein-protein associations are important mechanisms by which transcription factors synergistically cooperate. NF-κB, a necessary element in these complexes, has been shown to physically interact with ATF-2, ATF-a, c-JUN, TFIID, and HMG I(Y). Additionally, direct protein-protein associations have been demonstrated between ATF-2 and HMG I(Y). These interactions may stabilize the cytokine-induced transcriptional activator complexes and provide a unique interface between the assembled enhancer and the coactivators or basal factors in the transcription machinery.

NF-κB DEPENDENT GENE EXPRESSION AND THE RECRUITMENT OF COACTIVATORS

NF-κB-dependent gene expression involves the recruitment of complex(es) of nuclear factors termed transcriptional coactivators that probably function by facilitating or bridging the sequence-specific activators to the basal transcriptional machinery and altering chromatin structure (Fig. 2-2).

THE CREB-BINDING PROTEIN (CBP) COMPLEX

CBP was originally identified as a coactivator that interacts with the cAMP-response element binding protein (CREB), a process dependent on the cAMP-dependent protein kinase A and its phosphorylation of CREB (reviewed in 55-57). A closely related cofactor, p300, was independently isolated on the basis of its interaction with adenovirus E1A. Both CBP and p300 interact with a variety of transcriptional activators. For example, work from our own group as well as others demonstrated that the NF-κB component p65 interacts with this class of coactivators.[46,58] Further work revealed that phosphorylation of p65 by protein kinase A modulates the interaction with CBP.[59] In recent studies, we have shown that inactivation of CBP by nuclear antibody microinjection prevents NF-κB-dependent transactivation.[60] These studies establish the *in vivo* relevance of CBP to NF-κB-dependent gene expression.

CBP and p300 facilitate gene expression through multiple mechanisms. First, CBP and p300 function as molecular scaffolds coupling a variety of transcription factors to the basal transcriptional machinery. The constitutive association of CBP with RNA polymerase II[61] could facilitate recruitment of the polymerase as part of a holoenzyme complex.[62] Interactions between CBP and TFIIB and the TBP component of TFIID also suggest a bridging function for the coactivator. Second, the histone acetyltransferase (HAT) activity inherent in CBP and p300 may facilitate transcription through chromatin remodeling.[63,64] Third, CBP-associated proteins, such as the p300/CBP associated factor P/CAF (see below), or the steroid receptor-coactivator–1 (SRC-1), a member of the p160 family of coactivators,[65,66] also have HAT activity and could change chromatin structure. These HATs may function by acetylating the N-terminal tails of core histones. This is an intriguing possibility, since histone hyperacetylation has been correlated with transcription of active genes, whereas hypoacetylated histones have been associated with transcriptionally inactive chromatin (reviewed in 67). Fourth, CBP/p300 can acetylate interacting transcription factors, such as p53[68] and the hematopoietic differentiation factor GATA-1.[69] Fifth, CBP can also acetylate architectural proteins, such as HMG I(Y), and facilitate gene expression.[70] Finally, recruitment of CBP by inducible enhancers, such as the IFN-β enhanceosome, is required for synergistic activation of transcription.[71] Taken together, these results suggest that CBP is a versatile coactivator with multiple roles in inducible gene expression.

Coactivator-mediated localized hyperacetylation probably plays an important role in inducible gene expression. An attractive model for the importance of these HAT activities was developed from work on the IFN-β enhanceosome. In these studies, viral infection led to localized hyperacetylation of histones H3 and H4 at the IFN-β promoter.[72] Alterations in the transcriptional activators found in the enhanceosome which diminish CBP recruitment, suppress both histone hyperacetylation and activation of the IFN-β gene. A similar set of events may occur in the induction of the E-selectin gene. Inflammatory cytokines would induce the formation of the E-selectin enhanceosome, which in turn would recruit a series of coactivators with HAT activity. Collectively, these events would result in the localized hyperacetylation of histones in the region of the E-selectin promoter, facilitating the recruitment of the transcriptional machinery to the promoter, or promoter clearance by RNA polymerase II.

CBP interacts with a remarkably diverse group of signal-dependent transcriptional activators. This has led to proposals that the coactivator may function as a signal integrator by coordinating diverse signal transduction events at the transcriptional level.[73] This concept is consistent with observations that levels of the CBP homolog, p300, are limiting relative to those of p65,[74] and that competition for CBP may regulate p65 transactivation.[75,76] Additionally, in mice lacking one allele of CBP, it was shown that the gene dosage of CBP is selectively required for the induction of TNF-α transcription by the T cell receptor, but not by virus infection.[77] This suggests a stimulus-specific requirement for levels of CBP in some forms of inducible gene expression.

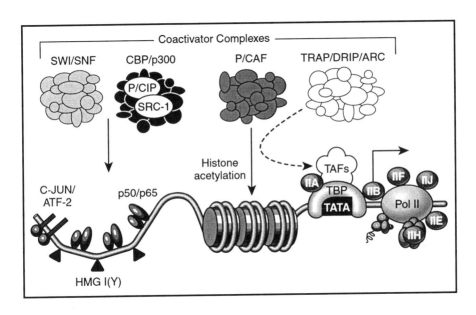

Figure 2-2. A model for the role of coactivation complexes in the cytokine-induced expression of the E-selectin gene. Following induction by TNA-α, heterodimers of NF-κB (p50/p65) bind to the cytokine response region of the E-selectin promoter which is constitutively occupied by c-JUN/ATF-2 heterodimer. The binding of HMG I(Y) at multiple sites increases the binding of NF-κB and bends DNA in a manner that facilitates protein-protein contacts and the formation of a higher-order complex. Collectively, these events place the transcriptional activators in a favorable position to interact with coactivators, including the SWI/SNF, CBP, P/CAF, and the ARC/DRIP/TRAP complexes. These coactivator complexes may act to relieve chromatin-mediate repression and facilitate the recruitment of the basal transcriptional machinery. Modified and redrawn from Glass and Rosenfeld, ref. 89.

THE P/CAF COMPLEX

P/CAF was originally identified as a CBP-associated factor and was cloned based on its sequence similarity to the yeast histone acetyltransferase, yGCN5.[78] In its native state, P/CAF is found in a complex with some 20 distinct polypeptides. Interestingly, some of these polypeptides are identical to the TBP-associated factors (TAFs) that form the TFIID complex.[79] P/CAF has been shown to interact with multiple coactivators including activator of the thyroid and retinoic acid receptors (ACTR) and SRC-1, which is also known as nuclear receptor coactivator-1 (NCoA-1).[65,80] In addition, P/CAF has been implicated in the coactivation of numerous transcription factors including the nuclear hormone receptors MyoD and p53, as well as NF-κB.[60,81-83]

The manner in which P/CAF further activates transcription following recruitment to DNA is not fully resolved. However, like CBP/p300, the P/CAF structure reveals a potent HAT domain. Histone acetylation may play an important role in the modulation of chromatin structure, thereby allowing for the formation of a transcriptionally active complex on the DNA. P/CAF's ability to associate with numerous coactivators that possess intrinsic HAT activity suggests the formation of large coactivator complexes by some transcription factors. However, the use of HAT activity appears to be selective. For example, the HAT activity of P/CAF appears to be important for nuclear receptor and MyoD function, but that of CBP does not. In the case of CREB, CBP HAT activity is required, whereas P/CAF HAT activity is not. In the case of p53, direct acetylation by both of these coactivators results in activation of the transcription factor following DNA damage.[84]

The adenoviral oncoprotein E1A inhibits NF-κB stimulated E-selectin promoter-reporter gene expression.[46] E1A inhibits p300/CBP activity in part by inhibiting the CBP-P/CAF interaction.[85] However, the inhibition observed may not merely reflect displacement of P/CAF from p300/CBP. More recently, E1A has been demonstrated to directly interact with p/CAF and inhibit its ability to acetylate nucleosomal histones and p53.[86,87] This targeting of P/CAF by the transforming viral protein E1A suggests a role for P/CAF in cellular processes that may be independent from that requiring interaction with p300/CBP.

Recent work by our laboratory suggests a role for P/CAF in mediating NF-κB-dependent gene expression of E-selectin. Exogenous expression of P/CAF potentiates p65-dependent activation of an E-selectin reporter construct and co-injection of anti-P/CAF antibodies into cells acts to block p65-mediated transcription of an E-selectin-LacZ reporter construct.[60] Transfection and nuclear microinjection studies demonstrate that the HAT activity of P/CAF is required for co-activation of NF-κB-dependent gene, showing the *in vivo* relevance of the P/CAF coactivator complex in NF-κB dependent gene expression.

THE P160 FAMILY OF PROTEINS

NF-κB-dependent gene expression involves a third class of transcriptional coactivators. SRC-1 or NCoA-1 (SRC-1/NCoA-1) is believed to interact with p50 to potentiate NF-κB-mediated transactivation.[88] This coactivator is a member of a group of three related coactivators (the p160 family) that includes: first, SRC-1/NCoA-1; second, NCoA-2 (also known as transcriptional intermediate factor-2 (TIF-2), or glucocorticoid receptor interaction protein (GRIP-1)); and a third member, p300/CBP co-integrator associated protein (p/CIP) (also known as receptor–associated coactivator-3 (RAC-3)), amplified in breast carcinoma (AIB1), activator of the thyroid and retinoic acid receptors (ACTR), and thyroid hormone activator molecule (TRAM-1)). Many of these coactivators were initially identified as ligand-dependent nuclear-receptor-interacting factors (reviewed in 89). The p160 coactivators interact with the nuclear receptors through a series of helical domains that contain a core LXXLL consensus sequence. Each of these domains is sufficient for ligand-dependent interaction with the nuclear receptors. Additionally, this family of coactivators also interacts with CBP through a distinct LXXLL motif. To date, the p160 family of coactivators is largely restricted to regulating the nuclear receptors, although most of the coactivators will potentiate the transcriptional activity of several types of nuclear hormone receptors.

Recent work by our laboratory and others has suggested a role for members of the p160 family in mediating NF-κB-dependent gene expression of E-selectin. Single-cell nuclear microinjection studies were used to demonstrate the *in vivo* relevance of the p160 family to NF-κB-dependent gene expression.[60] Microinjection of anti-SRC-1/NCoA-1 antibody inhibited transcriptional activation of an E-selectin promoter-reporter construct. Co-injection of an SRC-1/NCoA-1 expression vector reversed the blocking effect of the anti-SRC-1/NCoA-1 antibody, suggesting that the effect was specific. Consistent with the microinjection studies, overexpression of SRC-1/NCoA-1 augmented p65-dependent transcriptional activation and overexpression of a CBP mutant lacking the SRC-1/NCoA-1 interacting domain inhibited p65-dependent gene expression. Additionally, LXD4 of SRC-1/NCoA-1 (the CBP interaction motif) is required for both nuclear receptor and NF-κB-dependent gene expression. TIF-2 and GRIP-1 (or NCoA-2) are functional homologs of SRC-1/NCoA-1 which have been shown to enhance the ligand-dependent transcriptional activity of some members of the nuclear receptor family. Overexpression studies demonstrated that either GRIP-1 or TIF-2 augmented p65 activation of an E-selectin promoter reporter gene. Interestingly, rescue studies done using the microinjection approach suggest that SRC-1/NCoA-1 and TIF-2/GRIP-1/NCoA-2 may have distinct roles in NF-κB-mediated transcription.[60] Taken together, these studies suggest a role for the p160 family members in NF-κB dependent gene expression.

ARC/DRIP/TRAP COMPLEX

Another class of multisubunit coactivator complex capable of modifying chromatin has been designated activator-recruited cofactors (ARCs), thyroid hormone receptor-associated proteins (TRAPs), or vitamin D receptor-interacting proteins (DRIPs) (reviewed in 90). In recent studies, this cofactor complex was shown to interact directly with several different activators, including the p65 component of NF-κB.[91] This association strongly enhances transcription directed by p65 *in vitro* with chromatin-assembled DNA templates. The ARC/DRIP complex consists of 16 or more subunits, including some novel proteins, as well as components present in other multisubunit assemblies. The complex does not appear to contain HAT activity. This large composite coactivator is recruited by different types of signal-dependent transcription factors to increase gene expression.

Collectively, these findings demonstrate that NF-κB-dependent gene expression probably requires multiple coactivators. This raises several intriguing questions about how these complexes are utilized. It is possible that the coactivation complexes are recruited to NF-κB-dependent promoters in a sequential fashion. In this model, one complex is recruited to perform a specific function, such as relieving the repressive actions of chromatin. For example, the SWI/SNF complex stimulated nucleosome binding by the Rel homology domain of NF-κB.[92] Upon completion of this process, the complex leaves the promoter and is replaced by a second complex that performs a distinct process required for transactivation, perhaps the recruitment of coactivator complexes with HAT activity. In a second, combinatorial model, a transcription factor present in an enhanceosome may recruit different coactivator complexes that synergize to achieve physiologic levels of gene expression. In a third possibility, the same promoter may recruit different coactivator complexes in response to specific signaling pathways. Future studies should provide new insights into the relative importance of the sequential or combinatorial actions of the coactivator complexes.

NF-κB-DEPENDENT GENE EXPRESSION AND INTERACTIONS WITH BASAL TRANSCRIPTION FACTORS

NF-κB components have been shown to interact with factors in the basal transcription machinery.[93-95] For example, p65 and c-Rel bind to TATA box binding protein (TBP) and TFIIB *in vitro*, but the relevance of these isolated interactions to NF-κB-dependent gene expression has not been established. It is thought that once assembled, enhanceosomes present distinct surfaces that are recognized by the basal transcriptional machinery and by coactivator complexes. This leads to cooperative recruitment of RNA polymerase II and its associated factors and results in synergistic activation of transcription.

Although there is no evidence to date suggesting that the cytokine-induced E-selectin enhanceosome interacts with the basal machinery, the concept that the enhanceosome engages in multiple specific contacts with the general machinery is supported by work with the IFN-β promoter.[96,97] In reconstitution studies, the formation of a stable enhanceosome-dependent pre-initiation complex requires cooperative interactions between the enhancesome and some of the components of the basal transcriptional apparatus (reviewed in Chapter 7). Additionally, correct assembly of the IFN-β enhanceosome is required for the efficient recruitment of TFIID into a template-committed TFIIB-TFIA-USA complex, and for the subsequent recruitment of the RNA polymerase II holoenzyme complex. Taken together, these studies and those from other systems suggest that the enhanceosome recruits both coactivators and general factors, which collectively add to the stability of the final complex.

TRANSCRIPTIONAL REPRESSION OF THE E-SELECTIN GENE

Tumor necrosis factor-alpha (TNF-α) induction of the E-selectin gene leads to transient accumulation of high levels of mRNA in endothelial cells. The increase in mRNA after induction is due to an increase in the rate of gene transcription that is maintained for several hours in the continuous presence of cytokine. Following removal of TNF-α, there is a dramatic post-induction transcriptional repression that is protein synthesis-dependent. This post-induction transcriptional repression mechanism may be one component of a program that prevents inappropriate and prolonged expression of the adhesion molecule. In contrast to the study of the activation of the E-selectin cytokine response region, relatively little is known about the mechanism of repression of the gene. Some of the initial findings about these mechanisms are outlined below.

The endothelial NF-κB/IκBα system is necessary for induction of the E-selectin gene and probably plays a role in the repression of the gene.[20] Nuclear localization of the p50 and p65 subunits of NF-κB in endothelial cells requires the continuous presence of TNF-α. Following removal of TNF-α, the E-selectin gene is actively repressed. The repression correlates with the protein synthesis-dependent loss of both the p50 and p65 subunits of NF-κB from the nucleus. IκBα, an inhibitor of NF-κB, can be detected in the nucleus, and its nuclear localization is increased following removal of TNF-α. IκBα is capable of specifically displacing heterodimeric p50/p65 from the E-selectin κB elements, while having no effect on the binding of p50 homodimer. These results suggest that repression of E-selectin transcription following cytokine removal requires the loss of nuclear p50 and p65, and involves IκBα.[20]

The function of the cytokine response region of the E-selectin promoter requires NF-κB binding to PDI, PDIII, and PDIV, as discussed previously. During continuous incubation with TNF-α, gel shift analysis revealed persistent binding activity of NF-κB proteins to two of the E-selectin NF-κB binding sites (PDI and PDIII). However, binding activity of proteins that recognize a third NF-κB element (PDIV) was lost following prolonged inflammatory cytokine stimulation. These findings demonstrate that termination of E-selectin expression occurs with loss of protein-DNA interactions at only one of the three NF-κB binding sites in the E-selectin promoter.[98] These results suggest that in addition to IκBα displacement of NF-κB, other mechanisms may exist to decrease expression of the E-selectin gene.

The post-induction repression of the E-selectin gene could involve some or all of the following possibilities. First, NF-κB components may still occupy some of the PDs, but during the post-induction repression period they may undergo a change (e.g., dephosphorylation) such that one or more of the coactivation complexes (e.g., CBP or P/CAF) is displaced by a repressive complex. This exchange of coregulators is well documented to occur with the nuclear receptors and may be a common theme in transcriptional regulation (reviewed in 89). Second, other known transcription factors in the E-selectin cytokine response region could be inactivated. As discussed above, levels of JNK activity parallel that of the E-selectin promoter. It is possible that the activity of this kinase may decrease during the post-induction repression period, resulting in lower levels of phosphorylated c-JUN and ATF-2. Additionally, members of the ATF family may heterodimerize with inhibitors, such as ATF-a0, that function like dominant negatives and suppress transactivation.[99] Third, alterations in the architectural components of the E-selectin enhanceosome may decrease gene expression. It is interesting that acetylation of HMG I(Y) by CBP extinguishes expression of the IFN-β gene by disrupting the enhanceosome.[68] Fourth, transcriptional repressors may interact with the E-selectin cytokine response region. These repressors may displace activating factors, as has been suggested for IL-4-mediated suppression of TNF-α-activated E-selectin expression.[14] In this instance, STAT 6 antagonizes NF-κB. Additionally, the transcriptional repressors could recruit co-repressor complexes that alters chromatin structure and decrease gene expression. If the mechanistic similarities between the E-selectin cytokine response region and the viral response region of the IFN-β promoter extend to post-induction repression, it is likely that negative regulatory regions exist in the E-selectin promoter that interact with DNA-binding proteins. Given the potential consequences of inappropriate expression of E-selectin, it would not be surprising if multiple mechanisms were in place to control the expression of the gene.

SUMMARY AND FUTURE DIRECTIONS

Expression of the human E-selectin gene is transiently induced by inflammatory cytokines. The regulatory sequences required for this on and off regulation are located within a region of 160 base pairs upstream of the start site of transcription. This cytokine response region contains at least four distinct positive regulatory domains. The mechanism of TNF-α induction involves at least two endothelial signaling pathways, NF-κB/IκBα and JNK/p38 MAP kinases. Future studies may define additional signaling events that are important in the induction of E-selectin.[100] The NF-κB/IκBα signaling pathway results in phosphorylation, ubiquitination, and degradation of IκBα by the proteasome with nuclear accumulation of the p50 and p65 components of NF-κB. Concomitantly, a second set of TNF-α-induced events leads to activation of the c-Jun N-terminal kinase (JNK1), but not extracellular signal-regulated kinase (ERK1), resulting in phosphorylation of ATF-2 and c-Jun. These two pathways rapidly converge on the E-selectin promoter and place multiple transcription factors in a favorable architectural configuration to assemble into a composite structure, or enhanceosome. This platform competes for limiting amounts of several different types of coactivators, including both the CBP and P/CAF complexes. Understanding the nature of these complexes and how they function is a key direction for future work. Additionally, the E-selectin enhanceosome probably interacts with the basal transcriptional machinery to stimulate gene expression. The E-selectin gene undergoes an active transcriptional repression following cytokine induction which is mediated by IκBα-dependent displacement of NF-κB and by other mechanisms. Future studies will be needed to fully define the mechanisms involved in post-induction repression.

ACKNOWLEDGMENTS

We wish to thank the former members of the Collins laboratory who have worked on E-selectin transcription for their valuable insights and contributions to our understanding of this gene. These include Drs. Mary Gerritsen (Genentech, Inc.), Matthew Frosch (Harvard University), Andrew Neish (Emory University), Jacqueline Pierce (Millennium), Margaret Read (Millennium), Maryann Whitley (Genetics Institute), and Amy Williams (Millennium), as well as insights from our collaborators Drs. Christopher Glass, Tom Maniatis, Geoff Rosenfeld, and Dimitris Thanos. Additionally, we would like to acknowledge the technical support of Zaffar Haque, Sarah Moore, and Kathy Phelps. This work is supported by NIH grant RO1 DK54802 to D.R., and grants RO1 HL45462, and PO1 HL36028 to T.C.

48

REFERENCES

1. Vestweber D, Blanks JE. Mechanisms that regulate the function of the selectins and their ligands. Physiol Rev 1999; 79:181-213.
2. Labow MA, Norton CR, Rumberger JM, Lombard-Gillooly KM, Shuster DJ, Hubbard J, Bertko R, Knaack PA, Terry RW, Harbison ML. Characterization of E-selectin-deficient mice: demonstration of overlapping function of the endothelial selectins. Immunity 1994; 1:709-720.
3. Kunkel EJ, Ley K. Distinct phenotype of E-selectin-deficient mice. E-selectin is required for slow leukocyte rolling in vivo. Circ Res 1996; 79:1196-1204.
4. Ley K, Allietta M, Bullard DC, Morgan S. Importance of E-selectin for firm leukocyte adhesion in vivo. Circ Res 1998; 83:287-294.
5. Milstone DS, Fukumura D, Padgett RC, O'Donnell PE, Davis VM, Benavidez OJ, Monsky WL, Melder RJ, Jain RK, Gimbrone MA. Mice lacking E-selectin show normal numbers of rolling leukocytes but reduced leukocyte stable arrest on cytokine-activated microvascular endothelium. Microcirculation 1998; 5:153-171.
6. Bevilacqua MP, Pober JS, Mendric DL, Cotran RS, Gimbrone MA. Identification of an inducible endothelial-leukocyte adhesion molecule. Proc Natl Acad Sci 1987; 84:9238-9242.
7. Bevilacqua MP, Stengelin S, Gimbrone Jr MA, Seed B. Endothelial leukocyte adhesion molecule 1: An inducible receptor for neutrophils related to complement regulatory proteins and lectins. Science 1989; 243:1160-1165.
8. Whelan J, Ghersa P, Hooft van Huijsduijnen R, Gray J, Chandra G, Talabot F, DeLamarter J F. An NF-κB-like factor is essential but not sufficient for cytokine induction of endothelial adhesion molecule 1 (ELAM-1) gene transcription. Nucleic Acids Res 1991; 19:2645-2653.
9. Pober JS, Cotran RC. (1990) Cytokines and endothelial cell biology. Physiol Rev 1990; 70:427-451.
10. Brizzi MF, Garbarino G, Rossi PR, Pagliardi GL, Aruino C, Avanzi GC, Pegoraro L. Interleukin 3 stimulates proliferation and triggers endothelial-leukocyte adhesion molecule 1 gene activation of human endothelial cells. J Clin Invest 1993; 91:2887-2892.
11. Vora M, Romero LI, Karasek MA. Interleukin-10 induces E-selectin on small and large blood vessel endothelial cells. J Exp Med 1996; 184:821-829.
12. Gamble JR, Khew-Goodall Y, Vadas MA. Transforming growth factor-beta inhibits E-selectin expression on human endothelial cells. J Immunol 1993; 150:4495-4503.
13. Thornhill MH, Haskard DO. IL-4 regulates endothelial cell activation by IL-1 tumor necrosis factor or IFN-gamma. J Immunol 1990; 145:865-872.
14. Bennett BL, Cruz R, Lacson RG, Manning AM. Interleukin-4 suppression of tumor necrosis factor alpha-stimulated E-selectin gene transcription is mediated by STAT6 antagonism of NF-kappaB. J Biol Chem 1997; 272:10202-10209.
15. Brostjan C, Anrather J, Csizmadia V, Natarajan G, Winkler H. Glucocorticoids inhibit E-selectin expression by targeting NF-kappaB and not ATF/c-Jun. J Immunol 1997; 158:3836-3844.
16. Karmann K, Hughes CC, Schechner J, Fanslow WC, Pober JS. CD40 on human endothelial cells: inducibility by cytokines and functional regulation of adhesion molecule expression. Proc Natl Acad Sci USA 1995; 92:4342-4346.

17.   Hollenbaugh D, Mischel-Petty N, Edwards CP, Simon JC, Denfeld RW, Kiener PA, Aruffo A. Expression of functional CD40 by vascular endothelial cells. J Exp Med 1995; 182:33-40.

18.   Yellin MJ, Brett J, Baum D, Matsushima A, Szabolics M, Stern D, Chess L. Functional interactions of T cells with endothelial cells: the role of CD40L-CD40-mediate signals. J Exp Med 1995; 182:1857-1864.

19.   Pober JS. Immunobiology of human vaascular endothelium. Immunol. Res 1999; 19:225-232.

20.   Read MA, Neish AS, Gerritsen ME, Collins T. Postinduction transcriptional repression of E-selectin and vascular cell adhesion molecule-1. J Immunol 1996; 157, 3472-3479.

21.   Subramaniam M, Koedam JA, Wagner DD (1993) Divergent fates of P- and E-selectins after their expression on the plasma membrane. Mol Biol Cell 1993; 8:791-801.

22.   Eppihimer MJ, Wolitsky B, Anderson DC, Labow MA, Ganger DN. Heterogeneity of expression of E- and P-selectins in vivo. Circ Res 1996; 79:560-569.

23.   Hickey MJ, Kanwar S, McCafferty DM, Granger DN, Eppihimer MJ, Kubes P. Varying roles of E-selectin and P-selectin in different microvascular beds in response to antigen. J. Immunol 1999; 162:11137-1143.

24.   Smith GM, Whelan J, Pescini R, Ghersa P, DeLamarter JF, Hooft van Huijsduijnen R. DNA-methylation of the E-selectin promoter represses NF-kappaB transactivation. Biochem Biophys Res Commun 1993; 194:215-21.

25.   Chen XL, Tummala PE, Olliff L, Medford RM. E-selectin gene expression in vascular smooth muscle cells. Evidence for a tissue-specific repressor protein. Circ Res. 1997; 80, 305-311.

26.   Whitley MZ, Dimitris T, Read MA, Maniatis T, Collins, T. A striking similarity in the organization of the E-selectin and beta interferon gene promoters. Mol. Cell. Biol. 1994; 14, 6464-6475.

27.   Schindler U, Baichwal VR. Three NF-κB binding sites in the human E-selectin gene required for maximal tumor necrosis factor alpha-induced expression. Mol Cell Biol 1994; 14, 5820-5831.

28.   Lewis H, Kaszubska W, DeLamarter JF, Whelan J. Cooperativity between two NF-κB complexes, mediated by high-mobility-group protein I(Y), is essential for cytokine-induced expression of the E-selectin promoter. Mol Cell Biol 1994; 14, 5701-5709.

29.   Collins T, Read MA, Neish AS, Whitley MZ, Thanos D, Maniatis T. Transcriptional regulation of endothelial cell adhesion molecules: NF-κB and cytokine-inducible enhancers. FASEB J. 1995; 9:899-909.

30.   Kaszubska W, Hooft van Huijsduijnen R, Ghersa P, DeRaemy-Schenk A-M, Chen BPC, Hai T, DeLamarter JF, Whelan J. Cyclic AMP-independent ATF family members interact with NF-κB and function in the activation of the E-selectin promoter in response to cytokines. Mol Cell Biol 1993; 13:7180-7190.

31.   Collins T, Williams A, Johnson GI, Kim J, Eddy R, Shows T, Gimbrone Jr MA, Bevilaqua MP. Structure and chromosomal location of the gene for endothelial-leukocyte adhesion molecule 1. J. Biol. Chem. 1991; 266:2466-2473.

32.   Becker-André M, Hooft van Huijsduijnen R, Losberger C, Whelan J, DeLamarter JF. Murine endothelial leukocyte-adhesion molecule 1 is a close structural and functional homologue of the human protein. Eur J Biochem 1992; 206:401-411.

50

33.   Larigan JD, Tsang TC, Rumberger JM, Burns DK. Characterization of cDNA and genomic sequences encoding rabbit ELAM-1: conservation of structure and functional interactions with leukocytes. DNA Cell Biol 1992; 11:149-162.

34.   Montgomery KF, Osborn L, Hession C, Tizard R, Goff D, Vassallo C, Tarr PI, Bomsztyk K, Lobb R, Harlan JM, Pohlman TH. Activation of endothelial leukocyte adhesion molecule 1 (ELAM-1) gene transcription. Proc Natl Acad Sci USA 1991; 88:6523-6527.

35.   Read MA, Whitley MZ, Williams AJ, Collins T. NF-κB and IκB-α: An inducible regulatory system in endothelial activation. J Exp Med 1994; 179:503-512.

36.   Adams CC, Workman JL. Binding of disparate transcriptional activators to nucleosomal DNA is inherently cooperative. Mol Cell Biol 1995; 15:1405-1421.

37.   Maniatis T, Falvo JV, Kim TH, Dim TK, Lin CH, Parekh BS, Wathelet MG. Structure and funciton of the interferon-beta enhanceosome. Cold Spring Harb Symp Quant Biol 1998; 63:609-620.

38.   Reimold AM, Grusby MJ, Kosaras B, Fries JW, Mori R, Maniwa S, Clauss IM, Collins T, Sidman RL, Glimcher MJ, Glimcher LH. Chondrodysplasia and neurological abnormalities in ATF-2-deficine mice. Nature 1996; 379:262-265.

39.   Ashkenazi A, Dixit VM. Death receptors: signaling and modulation. Science 1998; 281:1305-1308.

40.   Maniatis T. A ubiquitin ligase complex essential for the NF-kappaB, Wnt/Wingless, and Hedgehog signaling pathways. Genes Dev 1999; 13:505-510.

41.   Karin, M. The beginning of the end: IkappaB kinase (IKK) and NF-kappaB activation. J Biol Chem 1999; 274:27339-27342.

42.   Israel A. The IKK complex: an integrator of all signals that activate NF-kappaB? Trends Cell Biol 2000; 10:129-133.

43.   Read MA, Whitley MZ, Gupta S, Pierce JW, Best J, Davis RJ, Collins T. Tumor necrosis factor alpha-induced E-selectin expression is activated by the nuclear factor-kappaB and c-JUN N-terminal kinase/p38 mitogen-activated protein pathways. J Biol Chem 1997; 272:2753-2761.

44.   Min W, Pober JS. TNF initiates E-selectin transcription through parallel TRAF-NF-kappa B and TRAF-RAC/CDC42-JNK-c-Jun/ATF2 pathways. J Immunol 1997; 159:3508-3518.

45.   Karmann K, Min W, Fanslow WC, Pober JS. Activation and homologous densitization of human endothelial cells by CD40 ligand, tumor necrosis factor, and interleukin 1. J Exp Med 1996; 1154:173-182.

46.   Gerritsen ME, Williams AJ, Neish AS, Moore S, Shi Y, Collins T. CREB-binding protein/p300 are transcriptional coactivators of p65. Proc Natl Acad Sci 1997; 94:2927-2932.

47.   Carey M. The enhanceosome and transcriptional synergy. Cell 1998; 92:5-8.

48.   Thanos D, Maniatis T. The high mobility group protein HMG I(Y) is required for NF-κB-dependent virus induction of the human IFN-ß gene. Cell 1992; 71:777-789.

49.   Falvo JV, Thanos D, Maniatis T. Reversal of intrinsic DNA bends in the IFN beta gene enhancer by transcription factors and the architectural protein HMG I(Y). Cell 1995; 83:1101-1111.

50.   Thanos D, Maniatis T. Virus induction of human IFN beta gene expression requires the assembly of an enhanceosome. Cell 1995; 83:1091-1100.

51.   Meacock S, Pescini-Gobert R, Delmarter JF, Hooft van Huijsduijnen R. Transcription factor-induced, phased bending of the E-selectin promoter. J Biol Chem   1994; 269:31756-31762.

52.   Falvo JV, Parekh BS, Lin CH, Fraenkel E, Maniatis T.   Assembly of a functional beta interferon enhanceosome is dependent on ATF-2-c-jun heterodimer orientation.   Mol Cell Biol   2000; 20:4814-4825.

53.   Ghosh G, van Duyne G, Ghosh S, Sigler PB.   Structure of NF-κB p50 homodimer bound to a κB site. Nature   1995; 373:303-310.

54.   Chen FE, Huang DB, Chen YQ, Ghosh G.   Crystal structure of p50/p65 heterodimer of transcription factor NF-κB bound to DNA.   Nature   1998; 391:410-413.

55.   Janknecht R, Hunter T. A growing coactivator network.   Nature   1996; 383:22-23.

56.   Shikama N, Lyon J, Thangue NBL.   The p300/CBP family: Integrating signals with transcription factors and chromatin.   Trends Cell Biol   1997; 7:230-236.

57.   Goodman RH, Smolik S.   CBP/p300 in cell growth, transformation and development.   Genes & Dev   2000; 14:1553-1577.

58.   Perkins ND, Felzien LK, Betts JC, Leung K, Beach DH, Nabel GJ.   Regulation of NF-κB by cyclin-dependent kinases associated with the p300 coactivator. Science   1997; 275:523-527.

59.   Zhong H, Voll RE, Ghosh S.   Phosphorylation of NF-kappa B p65 by PKA stimulates transcriptional activity by promoting a novel bivalent interaction with the coactivator CBP/p300.   Mol Cell   1998; 1:661-671.

60.   Sheppard KA, Rose DW, Haque ZK, Kurokawa R, McInerney E, Westin S, Thanos D, Rosenfeld MG, Glass CK, Collins T.   Transcriptional activation by NF-κB requires multiple coactivators.   Mol Cell Biol   1999; 19:6367-6378.

61.   Kee BL, Arias J, Montminy MR.   Adaptor-mediated recruitment of RNA polymerase II to a signal-dependent activator.   J Biol Chem   1996; 271:2373-2375.

62.   Ossipow V, Tassan JP, Nigg EA, Schibler U.   A mammalian RNA polymerase II holoenzyme containing all components required for promoter-specific transcription initiation.   Cell   1995; 83:137-146.

63.   Bannister AJ, Kouzarides T.   The CBP coactivator is a histone acetyltransferase.   Nature   1996; 384:641-643.

64.   Ogryzko VV, Schiltz RL, Russanova V, Howard BH, Nakatani Y.   The transcriptional coactivators p300 and CBP are histone acetyltransferases. Cell   1996; 87:953-959.

65.   Yao TP, Ku G, Zhou N, Scully R, Livingston DM.   The nuclear hormone receptor coactivator SRC-1 is a specific target of p300.   Proc Natl Acad Sci USA   1996; 93:10626-10631.

66.   Spencer TE, Jenster G, Burcin MM, Allis CD, Zhou J, Mizzen CA, McKenna NJH, Onate   SA, Tsai SY, Tsai MJ, O'Malley BW.   Steroid receptor coactivator-1 is a histone acetyltransferase.   1997; 389:194-198.

67.   Kuo M-H, Allis CD.   Roles of histone aceytltransferases and deacetylases in gene regulation.   BioEssays   1998; 20:615-626.

68.   Gu W, Shi XL, Roeder RG.   Synergistic activation of transcription by CBP and p53.   Nature   1997; 387:819-823.

69.   Boyes J, Byfield P, Nakatani Y, Ogryzko V.   Regulation of activity of the transcription factor GATA-1 by acetylation.   Nature   1998; 396:594-598.

70.   Munshi N, Merika M, Yie J, Senger K, Chen G, Thanos D.   Acetylation of HMG I(Y) by CBP turns off IFN beta expression by disrupting the enhanceosome.   Mol Cell   1998; 4:457-467.

52

71. Merika M, Williams AJ, Chen G, Collins T, Thanos D. Recruitment of CBP/p300 by the IFN-β enhanceosome is required for synergistic activation of transcription. Mol. Cell 1998; 1:277-287.

72. Parekh BS, Maniatis T. Virus infection leads to localized hyperacetylation of histone H3 and H4 at the IFN-β promoter. Mol Cell 1999; 3:125-129.

73. Kamei Y, Xu L, Heinzel T, Torchia J, Kurokawa R, Gloss B, Lin SC, Heyman RA, Rose DW, Glass CK, Rosenfeld MG. A CBP integrator complex mediates transcriptional activation and AP-1 inhibition by nuclear receptors. Cell 1996; 85:403-414.

74. Hottiger MO, Felzien LK, Nabel GJ. Modulation of cytokine-induced HIV gene expression by competitive binding of transcription factors to the coactivator p300. EMBO J. 1998; 17:3124-3134.

75. Sheppard KA, Phelps KM, Williams AJ, Thanos D, Glass CK, Rosenfeld MG, Gerritsen ME, Collins T. Nuclear integration of glucocorticoid receptor and nuclear factorkappaB signaling by CREB-binding protein and steroid receptor coactivator-1. J Biol Chem 1998; 273:29291-29294.

76. Webster GA, Perkins ND. Transcriptional cross talk between NF-κB and p53. Mol Cell Biol 1999; 19:3485-95.

77. Falvo JV, Brinkman BM, Tsytsykova AV, Tsai EY, Yao TP, Kung AL, Goldfeld AE. A stimulus-specific role for CREB-binding protein (CBP) in T cell receptor-activated tumor necrosis factor alpha gene expression. Proc Natl Acad Sci USA 2000; 97:3925-3929.

78. Yang XJ, Ogryzko VV, Nishikawa J, Howard BH, Nakatani Y. A p300/CBP-associated factor that competes with the adenoviral oncoprotein E1A. Nature 1996; 382:319-324.

79. Ogryzko VV, Kotani T, Zhang X, Schlitz RL, Howard T, Yang XJ, Howard BH, Qin J, Nakatani Y. Histone-like TAFs within the PCAF histone acetylase complex. Cell 1998; 94:35-44.

80. Chen H, Lin RJ, Schiltz RL, Chakravarti D, Nash A, Nagy L, Privalsky ML, Nakatani Y, Evans RM. Nuclear receptor coactivator ACTR is a novel histone acetyltransferase and forms a multimeric activation complex with P/CAF and CBP/p300. Cell 1997; 90:569-580.

81. Korzus E, Torchia J, Rose DW, Xu L, Kurokawa R, McInerney EM, Mullen TM, Glass Ck, Rosenfeld MG. Transcription factor-specific requirements for coactivators and their acetyltransferase functions. Science 1998; 279:703-707.

82. Puri PL, Sartorelli V, Yang XJ, Hamamori Y, Ogryzko VV, Howard BH, Kedes L, Wang JY, Graessmann, A, Nakatani Y, Levrero M. Differential roles of p300 and PCAF acetyltransferases in muscle differentiation. Mol Cell 1997; 1:35-45.

83. Scolnick DM, Chehab NH, Stavridi ES, Lien MC, Caruso L, Moran E, Berger SL, Halazonetis TD. CREB-binding protein and p300/CBP-associated factor are transcriptional coactivators of the p53 tumor suppressor protein. Cancer Res 1997; 57:3693-3696.

84. Sakaguchi K, Herrera JE, Saito S, Miki T, Bustin M, Vassilev A, Anderson CW, Apella E. DNA damage activates p53 through a phosphorylation-acetylation cascade. Genes Dev 1998;12:2831-2841.

85. Yang XJ, Ogryzko VV, Nishikawa J, Howard BH, Nakatani Y. A p300/CBP-associated factor that competes with the adenoviral oncoprotein E1A. Nature 1996; 382:319-324.

86. Reid JL, Bannister AJ, Zegerman P, Martinez-Balbas MA, Douzarides T. E1A directly binds and regulates the P/CAF acetyltransferase. EMBO J 1998; 17:4469-4477.

87.   Chakravarti D, Ogryzko V, Kao HY, Nash A, Chen H, Nakatani Y, Evans RM. A viral mechanism for inhibition of p300 and PCAF acetyltransferase activity. Cell 1999; 96:393-403.

88.   Na SY, Lee SK, Han SJ, Choi HS, Im SY, Lee JW. Steroid receptor coactivator-1 interacts with the p50 subunit and coactivates nuclear factor kappaB-mediated transcription. J Biol Chem 1998; 273:10831-10834.

89.   Glass CK, Rosenfeld MG. The coregulator exchange in transcriptional functions of nuclear receptors. Genes & Dev 2000; 14:121-141.

90.   Kingston RE. A shared but complex bridge. Nature 1999; 399:199-200.

91.   Naar AM, Beaurang PA, Zhou S, Abraham S, Solomaon W, Tjian. Composite coactivator ARC mediates chromatin-directed trancriptional activation. Nature 1999; 398, 828-832.

92.   Utley RT, Cote J, Owen-Hughes T, Workman JL. SWI/SNF stimulates the formation of disparate activator-nucleosome complexes but is partially redundant with cooperative binding. J Biol Chem 1997; 272:12642-12649.

93.   Kerr LD, Ransone LJ, Wamsley P, et al. Association between proto-oncoprotein Rel and TATA-binding protein mediates transcriptional activation by NF-κB. Nature 1993; 365:412-419.

94.   Xu X, Prorock C, Ishikawa H, et al. Functional interaction of the v-Rel and C-Rel oncoproteins with the TATA-binding protein and association with transcription factor IIB. Mol Cell Biol 1993; 13:6733-6741.

95.   Schmitz ML, Stelzer G, Altmann H, et al. Interaction of the COOH-terminal transactivation domains of p65 NF-κB with TATA-binding protein transcription factor IIB, and coactivators. J Biol Chem 1995; 270:7219-7226.

96.   Kim TK, Maniatis T. The mechanism of transcriptional synergy of an in vitro assembled interferon-beta enhanceosome. Mol. Cell 1997; 1:119-129.

97.   Kim TK, Kim TH, Maniatis T. Efficient recruitment of TFIIB and CBP-RNA polymerase II holoenzyme by an interferon-beta enhanceosome in vitro. Proc Natl Acad Sci 1998; 95:12191-12196.

98.   Boyle EM, Sato TT, Noel RF Verrier ED Pohlman TH. Transcriptional arrest of the human E-selectin gene. J Surg Res 1999; 82:194-200.

99.   Pescini R, Kaszubska W, Whelan J, DeLamarter JF, Hooft van Huijsduijnen R. ATF-a0, a novel variant of the ATF/CREB transcription factor family, forms a dominant transcription inhibitor in ATF-a heterodimers. J Biol Chem 1994; 269:1159-1165.

100.   Bandyopadhyay SK, de La Motte CA, Williams BR. Induction of E-selectin expression by double stranded RNA and TNF-α is attenuated in murine aortic endothelial cells derived from double-stranded RNA-activated kinase (PKR)-null mice. J Immunol 2000; 164:2077-2083.

Chapter 3

THE IMMUNOGLOBULIN SUPERFAMILY IN LEUKOCYTE RECRUITMENT

Ulrich H. von Andrian

The Center for Blood Research and
Department of Pathology, Harvard Medical School
200 Longwood Avenue
Boston, MA 02115

INTRODUCTION

Leukocyte Adhesion in Inflammation and Homing

In order to function as effector cells of the innate or specific immune system, leukocytes must leave the intravascular compartment. In the course of this process, circulating cells are exposed to significant hydrodynamic forces exerted by rapidly moving blood components.[1] Consequently, an intricate system of specialized surface molecules has evolved on leukocytes and endothelial cells (EC) that confers the mechanical stability necessary for interactions between blood cells and the vessel wall in the presence of flow. These receptors include a group of endothelial immunoglobulin superfamily (IgSF) members, several leukocyte integrins that function as IgSF counter-receptors, the three selectins and their sialomucin ligands, and a number of other molecules on leukocytes, platelets, EC, and within the extracellular matrix (reviewed in refs. 2-5).

It has been well established that many adhesion molecules are preferentially expressed or upregulated on distinct leukocyte subsets or are restricted to macro- or microvascular beds exposed to specific inflammatory signals or specialized environments. Specificity in adhesion receptor expression correlates often with the preferential recruitment of leukocyte subsets *in vivo*. Typical examples include the predominant influx of neutrophils and eosinophils in acute and allergic inflammation, respectively, or the selective migration (homing) of lymphocyte subsets to lymphoid organs or chronically inflamed tissues (reviewed in refs. 6, 7). It is now widely accepted that this multitude of

56

migratory specificities cannot be fully explained by the differential expression or function of a single adhesion receptor. This was first demonstrated by studies on neutrophils in inflamed venules and in a flow apparatus[8-10] that indicated that adhesion in the presence of flow follows a multi-step cascade involving distinct primary and secondary adhesion molecules (Fig. 3-1). Accordingly, most adhesion receptors can be assigned to one of two (or three) groups that differ with regard to their predominant function in the adhesion cascade.

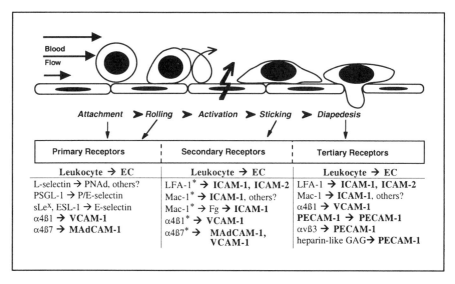

Figure 3-1 Schematic diagram of the multi-step adhesion cascade required for leukocyte accumulation in tissues. Known molecular pathways for individual adhesion steps are shown. Endothelial IgSF members are in bold print (* indicates requirement for enhanced ligand binding activity). Some interactions are instrumental in tissue-specific homing events and/or the trafficking of specialized leukocyte subsets, others have a broader role. Several surface molecules on leukocytes and EC that have not been included here can also mediate leukocyte adhesion. However, their role in leukocyte trafficking has not been determined.

Primary adhesion receptors are specialized to initiate interactions between free-flowing cells and the endothelium. Consequently, their binding kinetics are characterized by a fast on-rate and high tensile strength.[11,12] Most primary adhesion molecules on leukocytes are concentrated on the tips of microvilli (Fig. 3-2), a feature that optimizes their availability for cell tethering under high shear flow.[13] Interactions between primary receptors and their endothelial ligands are mostly short-lived. Thus, as the blood stream exerts continuous pressure on a tethered cell, adhesive bonds at the cell's trailing end break while new bonds are formed downstream. This results in a characteristic slow rolling motion, a hallmark of interactions involving primary adhesion receptors. However, rolling alone is not sufficient for emigration; the cell must first come to a complete stop through engagement of secondary adhesion molecules that are distinct from primary receptors in most cases.

Interactions between receptor-ligand pairs that mediate secondary adhesion events require that the adhesion partner on the leukocyte is in a state of enhanced binding activity. Thus, a leukocyte begins to roll via primary interactions and, subsequently, encounters an activating stimulus that triggers intracellular signaling cascades and induces rapid conformational changes within secondary receptors (i.e., $\beta 2$ or $\alpha 4$ integrins, or both). When activated, these molecules bind more strongly and with a longer bond lifetime to their endothelial counterparts than primary receptors. Their engagement can thus arrest a rolling cell even under high shear conditions. Once a cell has firmly attached itself to the lumenal endothelial surface, it must diapedese into the extravascular compartment and migrate through the interstitium toward its target. Emigration involves secondary adhesion molecules, but also requires additional (tertiary) adhesion pathways.

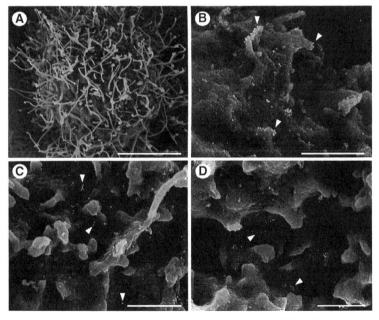

Figure 3-2. Scanning electron micrographs of immunogold-labeled lymphocytes. **(A)** A L1-2 pre-B lymphoid cell is shown at low magnification (scale bar represents 3 µm). Note numerous filamentous surface extensions (microvilli), a typical topographic feature of lymphoid cells. **(B)** Immunogold staining (the bright white dots indicated by arrowheads) localizes clusters of human L-selectin on L1-2 cells preferentially to the tips of microvilli. In contrast, immunogold staining of TK-1 T lymphoma cells with monoclonal antibodies to ICAM-1 **(C)** and PECAM-1 **(D)** reveals that both IgSF members are preferentially expressed on the planar cell body. This distribution of ICAM-1 and PECAM-1 is not random, but appears to reflect a preferential recruitment of these receptors away from the most distal aspects of the cell surface. Scale bars in panels B-D reflect 500 nm.

After a leukocyte has migrated into an area of inflammation or a lymphoid organ, it must detect, identify, and respond to specific signals within the local environment. Such clues may be in the form of soluble factors, microorganisms, tissue debris, extracellular matrix molecules, professional antigen-presenting cells (APC), or other cells in the tissue. Surface adhesion molecules are central for these interstitial events as well. Their engagement helps to transmit or amplify signals to and from leukocytes, is essential for phagocytosis, and is one of the main effects through which cytokines and chemokines affect leukocyte behavior. Thus, adhesion molecules are not only needed for tissue-specific leukocyte migration, but are also involved in the subsequent activation, differentiation, proliferation, and even death of cells. This chapter will provide an overview of the role of the endothelial IgSF adhesion molecules in these processes. To this end, one must understand the principles of IgSF distribution and expression and the structural aspects that allow these molecules to interact with their counter-receptors. The evidence implicating individual IgSF members in primary, secondary, and tertiary adhesion events will be reviewed and a (necessarily limited) selection of studies will be discussed to summarize the current knowledge of endothelial IgSF molecules in the recruitment and homeostasis of leukocyte subsets in health and disease.

Endothelial Leukocyte Adhesion Receptors of the Immunoglobulin Gene Superfamily

The immunoglobulin gene superfamily consists of a diverse and widespread group of soluble and membrane-bound glycoproteins that share a characteristic structural motif, the Ig domain, which can be present as one or several tandem copies within an IgSF member.[14] The primary sequence of the Ig domain consists of 70 - 110 amino acids, usually with two cysteine residues that are linked by a disulfide bond. Two sheets of anti-parallel β strands form a characteristic tertiary conformation, the so-called antibody fold. Many Ig family members combine Ig domains with other structural motifs such as hinge regions and mucin-like domains. Only antibodies and MHC molecules contain hypervariable regions that mediate the recognition or presentation of antigens.

Phylogenetically, IgSF members are thought to have evolved by gene duplication and exon shuffling from a single ancestral gene. This is still evident in the organization of modern IgSF members with multiple Ig domains; most Ig domains are encoded by a single exon, and essentially all are separated by phase I introns of variable length. The earliest IgSF probably consisted of molecules that were involved in pattern recognition and/or conferred mechanical stability to interactions between two or more cells. Derivatives of these IgSF members are found today in both vertebrates and insects.[15] They possess characteristic homologous Ig domains that belong to the so-called C2 subclass.[16] Another (presumably younger) subfamily of IgSF members that are restricted to vertebrates features specialized regions involved in immune recognition of antigens. This family, which includes soluble and surface-bound antibodies, T cell receptors, and MHC molecules, shares C1 and/or V-type Ig domains that are structurally distinct from C2 motifs. V-like domains are also found in many other proteins.

This chapter will focus on a subgroup of C2-type IgSF molecules that share three characteristic features: they are endothelial-expressed; they can function as ligands for leukocyte integrins; and they play a role in leukocyte adhesion and migration. Specifically, this chapter will summarize the biology and pathophysiologic significance of intercellular adhesion molecule (ICAM)-1 (CD54), ICAM-2 (CD102), platelet endothelial cell adhesion molecule (PECAM)-1 (CD31), vascular cell adhesion molecule (VCAM)-1 (CD106), and mucosal vascular addressin cell adhesion molecule (MAdCAM)-1. A schematic representation of these receptors and their known counter-receptor(s) is shown in Fig. 3-3.

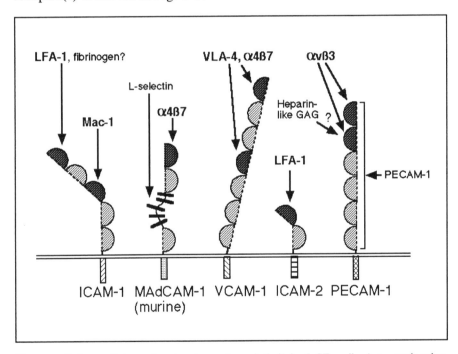

Figure 3-3. Schematic drawing of endothelial IgSF adhesion molecules. Interactions with known counter-receptors are indicated by arrows. Individual Ig domains are represented by semicircles (not drawn to scale). Bends in extracellular domains that have been detected by electron or rotary shadowing microscopy or from crystal structures are shown. Ig domains with binding sites for integrins (bold print) are shown in dark gray. MAdCAM-1 contains a serine- and proline-rich mucin-like domain that can present O-linked carbohydrate ligands to L-selectin. The major murine isoform of MAdCAM-1 is depicted here. Human and primate MAdCAM-1 possess only two N-terminal Ig domains and a longer mucin domain which is not separated from the cell surface by the IgA-like membrane-proximal domain of the murine isoform.

CURRENT RESEARCH

Physiologic Distribution and Regulation of Endothelial IgSF Expression

This section (summarized in Table 3-1) will highlight the regulatory mechanisms of IgSF expression and distribution in different organs, macro- and microvascular segments, and topographic regions on single EC. Most IgSF molecules are also found on other cell types including leukocytes, where they have additional regulatory functions during leukocyte migration and proliferation.

ICAM-1: ICAM-1 has the broadest distribution of all endothelial IgSF members. Many cell types including EC and lymphocytes constitutively express moderate levels of ICAM-1 and increase expression up to 40-fold upon activation.[17,18] ICAM-1 transcription by EC is strongly induced by exposure to cytokines such as tumor necrosis factor (TNF)-α, interferon (IFN)-γ, interleukin (IL)-1 and IL-6, and by a number of other inflammatory agents including bacterial endotoxin (LPS), activators of protein kinase c, retinoic acid, and lysophosphatidylcholine.[17,19-21] ICAM-1 transcription by cultured EC was also found to be inducible by CD40 crosslinking[22] and by exposure to hydrodynamic shear, a process that is thought to involve a cis-acting shear stress response element in the ICAM-1 promoter.[23] Conversely, a number of conditions were identified that decrease constitutive expression or block cytokine-induced upregulation of ICAM-1 on EC. These include factors with angiogenic properties,[24] prostaglandin E1,[25] and certain flavones such as apigenin, an ingredient of Chinese herbal medicines.[26,27] Transcriptional regulation of ICAM-1 expression will be reviewed in detail in Chapter 4.

Several observations indicate that ICAM-1 function and availability is not only transcriptionally regulated, but is also controlled by its distribution on the cell surface. Immunoelectron microscopy studies on cultured EC have shown that the molecule is normally present on all surface membranes that can interact with adherent or migrating leukocytes.[28,29] However, exposure of EC to hydrogen peroxide was found to induce redistribution to the abluminal surface.[30] Since this process is likely to limit further recruitment of circulating leukocytes, it is conceivable that it may reflect a feedback mechanism through which EC are able to avoid excess damage by leukocyte-derived oxygen radicals.

Fluorescence microscopy studies of non-endothelial cells have shown that ICAM-1 accumulates in uropods of polarized B cells and is concentrated in microvillous membrane protrusions of adherent COS transfectants.[31] In contrast, ICAM-1 on peripheral blood lymphocytes is preferentially expressed on the planar cell body (Fig. 3-2).

Table 3-1. Endothelial IgSF distribution and regulation

	ICAM-1	ICAM-2	MAdCAM-1	PECAM-1	VCAM-1
Other names	CD54	CD102		CD31, EndoCAM	CD106, INCAM-110
Distribution in adults	Vascular EC; leukocytes; macrophages, microglia; some neural cells; fibroblasts; epithelial cells; keratinocytes; some tumors; others	Vascular EC; platelets; some lymphocytes; monocytes; DC; basophils; microglia	Venular EC in non-pulmonary mucosa, esp. in HEV of PP and MLN; lamina propria; spleen; chronically inflamed EC; inflamed choroid plexus	All EC (blood > lymph vessels); platelets; all myeloid cells; $CD34^+$ myeloid and B-cell progenitors; lymphocytes (~50%, highest on CD45RA$^+$ CD8$^+$); NK cells	Inflamed EC and vascular SMC; BM EC and fibroblasts; FDC; some macrophages; synoviocytes; osteoblasts; some (inflamed) epithelial, neural and glia cells; some tumors; mesothelium

Table 3-1 (continued). Endothelial IgSF distribution and regulation

	ICAM-1	ICAM-2	MAdCAM-1	PECAM-1	VCAM-1
Regulation in embryogenesis and development	Knockout mice have normal lymphoid tissues; upregulated in fetal and adult thymocytes	Knockout mice are viable and fertile	Expressed on PLN HEV before and several days after birth; absent in spleen of TNFrp55 deficient mice	Multiple isoforms on EC precursors in yolk sac and embryo	Knockout is developmentally lethal; required for development of placenta and heart; role in myogenesis
Regulation on EC	Low baseline expression; strongly induced by cytokines, LPS, activators of PKc and retinoic acid	Constitutively expressed; no change during inflammation, but expression can vary in tumor vessels	Cytokine-inducible on murine endothelioma (NF-κB-dependent); induced in insulitis, thymic hyperplasia, and some forms of arthritis	Concentrated in cell-cell junctions; no change or decrease on activated EC; cytokines induce redistri-bution, phos-phorylation and cytoskeletal association	Absent on most resting EC, except in BM; induction by cytokines is stimulus- and EC-dependent; contact-induced by CD4+ T memory cells

Table 3-1 (continued). Endothelial IgSF distribution and regulation

	ICAM-1	ICAM-2	MAdCAM-1	PECAM-1	VCAM-1
Regulation on hematopoietic cells	Low on myeloid cells; high on NK and monocytes; strong increase on activated lymphocytes	Constitutive on platelets and some lymphocytes; little change upon activation	Not reported on hematopoietic cells	Phosphorylation, redistribution and cytoskeletal association upon activation	Absent on most circulating cells; found on γδ T cell clones and macrophages in arterial plaques

Abbreviations: BM-bone marrow; DC-dendritic cell; EC-endothelial cell; FDC-follicular dendritic cell; LPS-lipopolysaccharide; MLN-mesenteric lymph node; NF-κB-nuclear factor κB; PKc-protein kinase c; PLN-peripheral lymph node; PP-Peyer's patch; SMC-smooth muscle cells.

Exclusion of ICAM-1 from microvilli on resting lymphocytes mirrors that of its counter-receptors, the β2 integrins,[32] and may limit engagement of this pathway to situations that allow close contact between opposing lymphocytes. This may help avoid inadvertent aggregation and/or signaling through receptor crosslinking on blood-borne lymphocytes. The intracellular mechanisms that regulate ICAM-1 surface distribution have not been elucidated. It has been shown that a cationic region within the cytoplasmic tail of ICAM-1 associates with the cytoskeletal molecule α-actinin.[31] However, α-actinin also interacts with microvillous-expressed L-selectin,[33] suggesting that it may serve as a cytoskeletal anchor but not as a mediator of topographic specificity. An additional dimension of ICAM-1 regulation was recently proposed based on the observation that ICAM-1 is expressed as a non-covalent homodimer and that dimerization is necessary for high avidity interactions with its counter-receptors.[34,35] It is not known, however, whether ICAM-1 dimerization constitutes a process that can be actively regulated.

ICAM-2: ICAM-2 is constitutively expressed on vascular EC throughout the body. It has also been detected on lymphocytes, monocytes, basophils, platelets, megakaryocytes, and microglia.[36-38] Low levels of ICAM-2 have been found on thymocytes and some dendritic cells. It is absent from neutrophils. ICAM-2 expression on EC, unlike that of the other IgSF members discussed here, is not altered by exposure to cytokines or other proinflammatory stimuli.[37,39] However, vascular beds in malignant lymphomas were noted to express increased levels of ICAM-2, suggesting that it too can be quantitatively modified.[40] The molecule is also redistributed from the lumenal to the basal surface of EC exposed to hydrogen peroxide.[30] Redistribution of ICAM-2 has also been observed on cell lines where this process involves ezrin, a cytoskeleton-associated molecule. Ezrin can interact with the cytoplasmic tail of ICAM-2 and plays a role in ICAM-2-dependent tumor cell killing.

PECAM-1: PECAM-1 is ubiquitously and constitutively expressed on vascular and lymphatic EC throughout the body. Cultured human EC were found to express as many as 930,000 molecules on their surface.[41] Significant levels of PECAM-1 are also expressed on platelets (~5,000 copies/cell), monocytes and some monocytoid cell lines (~85,000 copies/cell), neutrophils, CD34$^+$ precursor cells, megakaryocytes, and about half of all circulating lymphocytes (reviewed in ref. 42). The majority of PECAM-1$^+$ human lymphocytes are CD45RA$^+$ CD29^{lo} (naive) CD8$^+$ T cells. In addition, NK cells and a subset of CD4$^+$ T cells also express PECAM-1.[43-45]

The regulation of PECAM-1 expression on activated leukocytes is highly complex and appears to depend on the cell type and stimulus.

Activated monocytes do not change PECAM-1 expression, whereas neutrophils were found to downregulate PECAM-1 following formyl-peptide challenge.[46] Phytohemagglutinin (PHA) activation of peripheral blood T cells was shown to suppress PECAM-1 transcription and protein expression.[47] In contrast, expression was increased on lymphokine-activated killer cells after exposure to IL-2 and on the PECAM-1[+] CD4[+] subset following short-term treatment with phorbol ester.[43,44] Repeated cycles of activation, however, induce a gradual loss of PECAM-1 from CD4[+] lymphocytes, and it has been proposed that the reduction in PECAM-1 expression is a measure of maturation of this T cell subset.[48] Cell activation also induces intracellular modifications of PECAM-1. The cytoplasmic tail contains several potential serine, threonine, and tyrosine phosphorylation sites, and rapid phosphorylation was reported in activated platelets, lymphocytes, and EC.[47] However, the significance of phosphorylation remains to be determined.

Immunolocalization studies on the surface distribution of PECAM-1 on confluent bovine and human endothelial monolayers and in blood vessels *in situ* have revealed that PECAM-1 is concentrated at the junctions between adjacent EC.[49a,49b,50] This distribution toward cell-cell boundaries was also observed in transfected cell lines and aggregated platelets.[51,52] *In vitro*, exposure of EC to cytokines alters both PECAM-1 topography and expression levels. Treatment of EC from human umbilical veins (HUVEC) with TNF-α and/or IFN-γ for 12 hours or longer was found to cause redistribution of PECAM-1 away from cell-cell contacts and disassociation from the cytoskeleton.[53] The combination of TNF-α and IFN-γ (but not either cytokine alone) also induced PECAM-1 internalization, a reduction in PECAM-1 mRNA and surface protein levels, and concomitant inhibition of PECAM-1-mediated neutrophil transmigration. This effect was not observed when IL-1, IL-4, IL-6, or endotoxin were used.[54] A similar study with bovine aortic EC has reported that TNF-α plus IFN-γ treatment induced destabilization of PECAM-1 mRNA, but did not affect transcription.[55]

VCAM-1: Resting EC express minimal or undetectable levels of VCAM-1. A notable exception is microvascular EC in the bone marrow, which have been shown to express significant levels of constitutive VCAM-1 on their lumenal surface.[56,57] Constitutive expression has also been found on a variety of extravascular cells in humans including follicular dendritic cells,[58] bone marrow stroma cells,[59] synoviocytes,[60] myoblasts and myotubes in developing muscle,[61] osteoblasts,[62] mesothelial cells,[63] some neural cells,[64] microglia,[65] benign nevi,[66] and epithelial cells in kidney[67] and choroid plexus.[68] Some tumors, especially melanomas and sarcomas, can also express VCAM-1.[66]

A number of cytokines and other proinflammatory agents transcriptionally induce the expression of VCAM-1 not only in cell types with basal expression but also in cells that are normally devoid of VCAM-1, including vascular smooth muscle cells and most EC.[69,70]

Most studies report detectable surface protein between two and four hours after cytokine exposure, reaching maximum levels after twelve hours and lasting for up to 72 hours.[69] A number of studies suggest that the inducibility of endothelial VCAM-1 varies between different vascular beds and also depends on the stimulus. HUVEC monolayers were shown to express VCAM-1 in response to TNF-α, IL-1α, IL-1β, IL-4, IL-6, IL-13, IFN-γ, LPS, and scavenger receptor ligands.[69,71,72] In contrast, microvascular EC from human foreskin were reported to respond to TNF-α but not IL-1α, presumably because an upstream VCAM-1 gene regulatory region inhibits IL-1-induced transcription in this type of EC.[68,73] Human EC from iliac arteries and veins were also found to react differentially to TNF-α treatment; only venous EC could be induced to express VCAM-1, whereas ICAM-1 was upregulated in both.[74] The transcriptional regulation of VCAM-1 will be reviewed in Chapter 5.

Endothelial VCAM-1 expression can also be regulated by microenvironmental factors other than cytokines. For instance, cultured rabbit EC were found to increase both VCAM-1 and ICAM-1 when exposed to the atherogenic phospholipid lysolecithin.[20] HUVEC grown on the basement membrane component type IV collagen expressed much lower levels of VCAM-1 than cells cultured on type I collagen.[75] Further, VCAM-1 expression is influenced by biophysical parameters: a five-day reduction in shear stress in rabbit carotid arteries increased VCAM-1 *in situ*;[76] conversely, VCAM-1 expression decreased on cultured murine EC upon exposure to fluid shear.[77] Finally, direct contact between EC and CD4+CD45RO+ (memory phenotype) T cells can induce VCAM-1.[78] Similar observations have been reported in cocultures of T cells with synovial fibroblasts.[79] It appears reasonable to speculate that this contact induction may involve CD40, a ligand for CD40L on activated CD4+ T cells. Low levels of CD40 are expressed on EC[22] and synoviocytes,[80] and CD40 ligation by multivalent CD40L can induce both VCAM-1 and ICAM-1, suggesting a positive feedback mechanism for the induction of these molecules during T cell recruitment.

MAdCAM-1: MAdCAM-1 was first described as a glycoprotein that is constitutively expressed on high endothelial venules (HEV) of Peyer's patches and mesenteric lymph nodes of mice.[81,82] It has also been detected at lower density in microvessels of the intestinal lamina propria and in lactating mammary glands, and it is induced by inflammatory stimuli in the brain and other sites. More recently, MAdCAM-1 has been identified on marginal zone sinus-lining spleen cells adjacent to the white pulp[83] and on follicular dendritic cells in lymph nodes and Peyer's patches (E.C. Butcher, personal communication). MAdCAM-1 is not expressed in spleens of mice that are either lacking the 55 kD TNF receptor or are doubly deficient in TNF-α and lymphotoxin, suggesting that splenic expression depends on tissue-selective mechanisms that depend on constitutive stimulation via

TNF-α or lymphotoxin.[84,85] This view is consistent with recent findings in transgenic mice expressing lymphotoxin under the rat insulin promoter. These animals were found to develop lymph node-like features including MAdCAM-1⁺ HEV at sites of transgene expression.[86] Proinflammatory cytokines in conjunction with unidentified local factors have also been implicated in MAdCAM-1 induction in chronically inflamed non-lymphoid tissues.

Structural and Functional Properties of Endothelial Immunoglobulins

A summary of the structural features of endothelial IgSF molecules is given in Table 3-2.

Each of the endothelial IgSF molecules discussed here can act as a ligand for leukocyte integrins. ICAM-1, PECAM-1 and MAdCAM-1 can also interact with other molecules (Fig. 3-3 and Table 3-3), but integrin binding is thought to be critical for many of their physiologic functions. IgSF/integrin interactions can contribute to primary, secondary, and tertiary steps in the adhesion cascade (Fig. 3-1). VCAM-1 and MAdCAM-1 support attachment and rolling of mononuclear cells under physiologic shear and in the absence of contributions by other adhesion receptors.[87-89] Both molecules can also contribute to firm adhesion and transendothelial migration when their counter-receptors, the α4 integrins, are in a high affinity state. ICAM-1 cannot support rolling by itself. However, ICAM-1 and -2 play a key role during β2 integrin-mediated sticking[10,90] and each molecule can participate, at least in some settings, in transendothelial migration.[28,91] PECAM-1 was reported to interact with αvβ3 integrin and heparin-like glycosaminoglycans (GAG). It can also undergo homophilic interactions. Neither homophilic nor heterotypic PECAM-1-mediated adhesion events were found to play a role during primary or secondary adhesion. Instead, PECAM-1 is thought to function as a tertiary receptor during transendothelial migration.[92]

There are several common features of integrin-mediated adhesion to IgSF.[4,93] Firstly, integrins must be in an activated state for high affinity binding. This may involve changes in their conformation or surface distribution, or both. Secondly, integrin adhesion to IgSF molecules is temperature- and divalent cation-dependent and is greatly increased upon exposure to extracellular Mn^{2+}. Thirdly, integrin binding sites are restricted to one (mostly the N-terminal) or in some cases two Ig domains. Adjacent domains can often provide structural support to optimize binding, whereas more remote Ig domains do not seem to have a direct role in integrin adhesion. Finally, all integrin binding Ig domains in endothelial IgSF feature a tripeptide motif consisting of an acidic amino acid framed by two residues with aliphatic

Table 3-2. Structural aspects of endothelial IgSF members.

	ICAM-1	ICAM-2	MAdCAM-1	PECAM-1	VCAM-1
Ref. for cDNA sequence[a]	hu: 264,265 mu: 266	hu: 36 mu: 267	hu: 171 mu: 268	hu: 50,123 mu:122	hu: 69,144,145 mu: 269
M_r in kD (protein core)	76-114 (55)	46 (28)	58 - 66 (41)	120 - 130 (80)	110 (81)
No. of Ig domains[b]	5	2	2 (human) or 3 (mouse)	6	7 (6); gpi-linked 3D isoform in rodents
Other structural motifs of ectodomain	8 N-glycosyl-ation sites; bend between D3 and D4	6 N-glycosyl-ation sites (5 in murine ICAM-2)	mu: mucin between D2 and D3 (IgA1-like); hu: no D3, longer mucin	9 N-glycosyl-ation sites	7 N-glycosyl-ation sites in 7D form
Genomic organization, alternatively spliced variants	Single gene on chr. 19 (hu); 7 exons; at least three alt.spliced variants in mice	Single gene on chr. 17 (hu); 4 exons	Single gene on chr. 10 (mu); 5 exons in mice; alt. spliced mRNA lacking exon 4 (mucin + D3) in mice	Single gene on chr. 17 (hu; 2 alleles); 16 exons; alt. splicing of IC domain; one secreted variant	Single gene on chr. 1 (hu); 9 exons (mu: 10); predominant 7D form, a minor 6D form lacks D4 (exon 5);

[a] hu: human, mu: murine.

[b] D1 indicates the N-terminal (most distal extracellular) Ig domain, D2 is immediately C-terminal to D1, and so on.

side chains (Leu/Ile-Glu/Asp-Val/Thr/Ser). This motif is essential for binding of ICAM-1 (D1 and D3: Ile-Glu-Thr) and ICAM-2 (D1: Leu-Glu-Thr) to β2 integrins and of VCAM-1 (D1 and D4: Ile-Asp-Ser) and MAdCAM-1 (D1: Leu-Asp-Thr) to α4 integrins. PECAM-1 contains a similar motif (Leu-Glu-Leu) in the predicted C/D loop of D2. However, it has not been determined whether this region is involved in PECAM-1 binding to αvβ3 integrin which, unlike interactions mediated by β2 and α4 integrins, is inhibited by Arg-Gly-Asp-containing peptides.[94]

The conserved acidic residue in the binding face of the ICAMs has been predicted to ligate a Mg^{2+} ion associated with the metal ion-dependent adhesion site (MIDAS) in the I domain of the LFA-1 α chain.[95] Although it is known that the conserved Asp in VCAM-1 and MAdCAM-1 are equally important, it is unclear whether the corresponding motifs in these IgSF members or in PECAM-1 have a similar ion binding function, especially since neither αvβ3 nor the α4 integrins contains an I domain. Studies on integrin adhesion to synthetic peptides suggest that metal ions are necessary for initial docking, but they may be subsequently displaced from the binding interface.[96] Based on these observations, it has been speculated that integrin association with IgSF could involve two sequential steps. Initially, a cation-dependent low affinity tether is formed via rapid binding to the tripeptide; subsequently, other residues in the binding interface participate in higher affinity binding and confer specificity to interactions between distinct integrin/IgSF pairs.[97] However, this concept has not been tested so far and alternative models have been proposed.[95,98]

ICAM-1: The adhesive function of ICAM-1 was initially recognized by the ability of an anti-ICAM-1 monoclonal antibody to block phorbol ester-induced leukocyte aggregation.[99] Leukocyte aggregation was also found to be inhibited by monoclonal antibodies to lymphocyte function associated antigen (LFA)-1 (CD11a/CD18), and it was later concluded that ICAM-1 and LFA-1 are a receptor-ligand pair. Today, at least four endogenous ligands for ICAM-1 have been identified and partly characterized on a molecular level: the β2 integrins LFA-1,[99] Mac-1 (CD11b/CD18),[100] and p150,95 (CD11c/CD18);[101] and fibrinogen (Fg).[102,103] Two of the 5 Ig domains of ICAM-1 appear to mediate all interactions with these different counter-receptors. Mac-1 was shown to bind to D3.[100] The binding sites for p150/95 and LFA-1 are both located in D1, but the latter interaction is much stronger than the former.[101] D1 is also the binding site for the γ chain of Fg. Consequently, a synthetic peptide that mimics this region of Fg can block Fg-mediated leukocyte adhesion to ICAM-1 *in vitro*[103] and *in vivo*.[104] Interactive sites within the γ chain of soluble Fg may be obscured, but they become exposed when Fg is bound by a surface adhesion molecule such as gpIIbIIIa.[105] It has been speculated that an analogous mechanism also permits Fg-mediated bridging between Mac-

1 and ICAM-1.[104] The interactive domains within ICAM-1 itself are presented in a manner that may promote accessibility to leukocyte receptors; single-molecule electron microscopy of ICAM-1

Table 3-3. Endothelial IgSF counter-receptors and interacting domains.

	ICAM-1	ICAM-2	MAdCAM-1
Counter-receptor(s) → interacting domain(s)	LFA-1→ D1; Mac-1→ D3 Fibrinogen→ D1	LFA-1→ D1	α4β7→D1; L-selectin→ mucin

	PECAM-1	VCAM-1
Counter-receptor(s) → interacting domain(s)	PECAM-1 → D1-6 αVβ3 → D1/2; Heparin-like GAG → D2	α4β1, α4β7 → D1 and D4

show a kink between D3 and D4 which may help expose the distally located integrin binding sites.[106] In addition, cells can probably control the interactivity of ICAM-1 by posttranslational modifications, since Mac-1 binding is negatively regulated by N-glycation of D3.[100]

ICAM-1 was also identified as a target for certain pathogens. Structural motifs within the N-terminal domain that are distinct from the binding site for LFA-1 are recognized by the major group of rhinovirus[107] and by erythrocytes infected with a strain of *Plasmodium falciparum*.[108] One study has used affinity chromatography to suggest that ICAM-1 may interact with the ECM molecule hyaluronan,[109] and another report found evidence that the sialomucin CD43 on Daudi T cells can bind ICAM-1.[110] However, others have been unable to detect interactions between CD43 and ICAM-1.[111,112] Moreover, T cells from CD43-deficient mice have an increased capacity to interact with EC *in vivo* and *in vitro*, suggesting that CD43 plays a major role as an "anti-adhesin" under physiologic conditions.[112a,112b] Thus, the relevance of interactions between ICAM-1 and CD43 or hyaluronan remains to be established.

Under physiologic conditions (i.e., in the presence of flow), ICAM-1 cannot support primary adhesion of untethered cells, but it is highly efficient in mediating activation-induced sticking of rolling neutrophils, monocytes, and lymphocytes.[10,90,113,114] ICAM-1 interactions with β2 integrins also contribute to leukocyte flattening on the endothelial surface and to transendothelial migration, since monoclonal antibodies to both are inhibitory.[28] Also, leukocytes from patients with a genetic defect in β2 integrins have a reduced capacity for spreading and diapedesis.[115] *In situ* studies confirm that ICAM-1 does not play a role in primary rolling adhesions, at least not via interaction with integrins. Accordingly, leukocytes in cremaster muscle venules of ICAM-1-deficient mice have no apparent defect in rolling. However, leukocyte rolling was found to be completely absent in animals that are deficient in both ICAM-1 and P-selectin.[116] In contrast, P-selectin single mutant mice displayed low levels of rolling within one hour after tissue preparation that was inhibited by monoclonal antibodies to L-selectin, but not by a monoclonal antibody to ICAM-1 that neutralizes integrin binding.[116] These observations suggest that ICAM-1 can directly or indirectly contribute to L-selectin-dependent leukocyte adhesion to EC. One possible mechanism for this effect could be inflammation-induced differential glycosylation of ICAM-1 with sugar molecules that are recognized by L-selectin. Alternatively, ICAM-1 may permit a slow accumulation of sticking leukocytes in venules of wild type and P-selectin deficient animals. Adherent neutrophils can express ligands for L-selectin on other leukocytes in the blood stream.[117] Thus, the increased availability of L-selectin ligands on adherent cells may account, at least in part, for the L-selectin-dependent rolling adhesions seen in P-selectin-deficient animals. Since P-selectin/ICAM-1 double knockout mice are likely to permit little or no sticking, rolling via L-selectin may be much less frequent in these animals. At present, it is unclear which of these hypothesized pathways – direct L-selectin ligand presentation or recruitment of adherent leukocytes expressing L-selectin ligands or some other mechanism – is responsible for the absence of L-selectin-dependent rolling in ICAM-1/P-selectin double mutant mice.

ICAM-2: ICAM-2 has two Ig domains that are homologous to the two N-terminal domains of ICAM-1. It has the smallest protein core (28.4 kD) of all endothelial IgSF members.[36] However, posttranslational glycosylation increases its apparent M_r to ~46 kD, accounting for almost half of the mature protein. Recently, analysis of the crystal structure of ICAM-2 revealed that glycosylation has an unprecedented role in ICAM-2; N-linked sugars associated with D2 of ICAM-2 induce a bend in the protein at the junction between D1 and D2 that appears necessary to expose the face of D1 that contains the LFA-1 binding site.[118]

A peptide derived from ICAM-2 was found to bind and activate Mac-1.[119,120] However, the biological implications of this observation are not clear, since full-length ICAM-2 does not appear to interact with Mac-1.[101] Since LFA-1 is expressed on all myeloid and lymphoid cells, ICAM-2 may be able to participate in the intravascular adhesion of any leukocyte subset. ICAM-2 on immobilized platelets was also found to support T cell sticking (but not rolling) when the cells were allowed to interact with platelets under static conditions and shear flow was applied subsequently,[38] but ICAM-2 was much less important in neutrophil adhesion to platelets.[121] ICAM-2 was also shown to support transendothelial migration of T cells.[91] Moreover, ICAM-2 on mononuclear cells or platelets participates in LFA-1-mediated aggregation with leukocytes.[37,38]

PECAM-1: Both murine[122] and human PECAM-1[50,123] possess 6 Ig domains that are homologous to other C2 type Ig motifs, except D5 which is predicted to have an atypical secondary structure. The cytoplasmic tail is relatively large (118 aa) and, unlike corresponding regions in other endothelial IgSF members, is encoded by eight exons (exons 9-16) that give rise to a number of distinct splice variants, including a secreted form that lacks a membrane spanning domain and can be found in human plasma at concentrations ranging from 10 to 25 ng/ml.[124]

In 1985, Ohto et al. reported that two monoclonal antibodies to a human surface glycoprotein (gp120), later identified as PECAM-1, blocked monocyte and neutrophil chemotaxis in response to bacterial endotoxin and suggested a role for this antigen in leukocyte migration.[125] The first direct demonstration that PECAM-1 mediates cell adhesion came from experiments in which cell lines acquired the ability to undergo adhesive interactions after transfection with PECAM-1 cDNA.[51] PECAM-1-mediated adhesion can involve both homophilic and heterophilic interactions.[126] Homophilic adhesion was found to be calcium- and temperature-independent and was abolished in the absence of D1 or D2. Full strength homophilic interactions required all six domains, and binding could be further enhanced by a monoclonal antibody to D6, suggesting that multiple regions throughout PECAM-1 are involved.[127-129]

There is also a role for D6 in heterophilic interactions of monocyte and lymphocyte PECAM-1 with unidentified ligand(s).[127,130] Most PECAM-1 interactions with known heterophilic counter-receptors are calcium- and temperature-sensitive.[126] Such ligands include heparin-like GAG that are thought to interact with a GAG recognition motif in D2 [131] and the integrin αvβ3. The latter interaction was reported in two studies employing murine and human PECAM-1/Fc fusion proteins.[94,132] One report showed that deletion of D2 from mouse PECAM-1 nearly abrogated αvβ3-mediated adhesion of murine lymphoid cells and concluded that D2 contained the αvβ3 binding site.[94] In contrast, a chimeric protein consisting of D1 of human PECAM-1 linked to Ig heavy chain supported significant αvβ3-mediated binding of U937 myelomonocytic cells, and adhesion to PECAM-1/Fc chimera containing all six Ig domains was inhibited by monoclonal antibodies to either D1 or D2, suggesting that both domains may be involved in αvβ3 binding in humans.[132] In contrast to the aforementioned studies in which αvβ3 was shown to interact with surface-bound recombinant PECAM-1, others were unable to detect binding of αvβ3 to PECAM-1 expressed by EC or transfected cell lines (W.A. Muller, personal communication). Recently, a divalent cation-independent 120 kD single chain ligand has been reported on activated CD4⁺CD31⁻ helper cell clones.[133] These studies indicate that PECAM-1-dependent leukocyte/EC interactions can be mediated by several molecularly distinct but partly overlapping pathways that may be differentially regulated by host cell-specific mechanisms and/or distinct microenvironments.

The different adhesive modalities of PECAM-1 are regulated in part by events involving the cytoplasmic domain. For instance, different murine isoforms that vary in their cytoplasmic tail have distinct abilities to mediate heterophilic adhesion and homotypic binding (those containing and lacking exon 14, respectively).[134] Messenger RNA for these isoforms has been detected in developing embryos, and it was proposed that they may have a role in cardiovascular embryogenesis.[135] The aggregation of transfectants via homotypic PECAM-1 binding also requires cytoplasmic sequences that are encoded by exons 9 and 10 and are modulated by structures encoded by exon 14.[128] Intracellular determinants are also required for the conspicuous concentration of PECAM-1 at cell-cell boundaries.[136]

The most prominent function of PECAM-1 is its involvement in transendothelial migration of monocytes and neutrophils across resting or cytokine-activated EC.[92,137-139] Studies with monoclonal antibodies and soluble PECAM-1 showed that both equivalently inhibited transmigration whether they were added to leukocytes. to EC, or to both. Leukocyte adhesion to EC or chemotaxis *per se* was not affected, but cells remained stuck over interendothelial junctions on the apical surface.[92] Recently it has been reported that two different domains are involved in distinct steps of the transmigration process. Monoclonal antibodies mapping to D1 and D2 block monocyte movement across EC junctions, whereas monoclonal antibodies to D6 did not interfere with migration across the endothelial layer, but blocked subsequent penetration of the basal lamina and interstitial migration.[127] Meanwhile, a number of independent studies point to PECAM-1 as an important mediator of transendothelial migration of myeloid cells both *in vitro* and *in vivo*.[137,139,140] In contrast, the role of PECAM-1 in lymphocyte migration is much less clear.

In vitro studies suggest that PECAM-1 contributes to the barrier formation of endothelial monolayers, since bovine EC that were grown in the presence of polyclonal anti-PECAM-1 adhered normally to the culture substrate but failed to establish cell-cell contacts.[49b] In contrast, anti-PECAM-1 did not affect endothelial integrity when added to confluent monolayers. Since PECAM-1 is preferentially located in more basal areas of monolayer junctions,[141] it is possible that it may not be accessible to inhibitory antibodies after tight junctions have formed. Alternatively, PECAM-1 may not be critical for the maintenance of EC junctions once they have been established.

PECAM-1 also has a role in cell-cell communication and signaling. Monoclonal antibody-induced ligation of PECAM-1 on monocytes enhances β2 integrin-dependent adhesion to EC and stimulates an oxidative burst in these cells, but not in neutrophils.[46,142] Binding of whole monoclonal antibodies or Fab fragments to PECAM-1 also increases Mac-1-dependent adhesion of neutrophils to EC and to C3bi-coated erythrocytes.[142] In contrast, PECAM-1 crosslinking on lymphocytes or CD34[+] hematopoietic cells was only found to increase binding of $\alpha4\beta1$ to VCAM-1, whereas LFA-1 was not activated.[143] Recent studies have also demonstrated that PECAM-1 plays a role in lymphocyte activation during mixed lymphocyte reactions (MLR).[130] Thus, PECAM-1-mediated signaling can trigger distinct adhesion and activation cascades on different host cells.

VCAM-1: The first human VCAM-1 cDNA was isolated by expression cloning of a single chain type 1 transmembrane molecule. The clone encoded a molecule with 6 N-terminal Ig domains.[69] Subsequent studies determined that this 6D form was a minor alternatively spliced variant that lacked D4 of the predominant isoform in EC which has 7 Ig domains.[144,145] The genes for rat and mouse VCAM-1 contain an additional exon encoding a unique C-terminus. This variant gives rise to a truncated gpi-linked molecule that contains the three N-terminal domains.[146]

Rotary shadowing and electron microscopy of recombinant 7D VCAM-1 revealed that the Ig domains are extended in a slightly bent linear array.[147] Thus, the N-terminus of VCAM-1 is highly exposed on the EC surface. This may facilitate VCAM-1 interactions with its two leukocyte counter-receptors, the $\alpha4$ integrins $\alpha4\beta1$ (VLA-4) and $\alpha4\beta7$. Rapid adhesion to VCAM-1 *in vivo* is probably further improved by the clustered expression of these integrins on the tips of microvilli, a topographic feature that greatly enhances contact formation under flow.[13,88] Both $\alpha4$ integrins bind regions in D1 and D4 of the 7D form of VCAM-1.[148-152] However, $\alpha4\beta1$ is more potent in binding VCAM-1 than $\alpha4\beta7$ in most settings.[151] The binding sites for both integrins appear to overlap, but they may not be identical.[153] The crystal structure of a VCAM-1 fragment containing D1 and D2 showed that the $\alpha4\beta1$-binding motif (Gln_{38}-Ile-Asp-Ser-Pro-Leu) in D1 forms a distinctive loop between beta sheets C and D that protrudes from the domain surface.[154,155] Mutagenesis studies revealed a second region in the EF loop adjacent to the primary binding site that is also involved in integrin binding.[147] It is not known whether this region provides structural support or participates directly in adhesive interactions.

The $\alpha4$ integrin counter-receptors for VCAM-1 are expressed by lymphocytes, monocytes, eosinophils and basophils, but not by circulating neutrophils. Consequently, immobilized VCAM-1 or transfectants supported adhesion of the former, but not the latter.[69,156] Intravital microscopy of fluorescently labeled human eosinophils in inflamed rabbit mesentery venules revealed that an $\alpha4$ integrin-dependent pathway participated in eosinophil rolling.[157] Subsequent *in vitro* studies on VCAM-1 interactions with lymphocytes in the presence of hydrodynamic shear showed that VCAM-1 could support $\alpha4\beta1$-mediated primary attachment and rolling as well as secondary arrest.[87-89]

Intriguingly, the α4β1/VCAM-1 pathway does not participate in primary adhesion in all settings. A recent study found that T cells rolled on VCAM-1 expressed on transfected L cells, but when T cells were tested on IL-1-treated HUVEC monolayers, endothelial VCAM-1 mediated only sticking, but not rolling.[114] Similarly, human lymphocytes and monocytes that flowed over cytokine-activated HUVEC did not utilize VCAM-1 during initial attachment, but VCAM-1 was found to contribute to sticking and slowing of rolling cells.[158,159] VCAM-1 is also involved in the migration of monocytes and eosinophils across cytokine-activated EC.[160-162] In contrast, transendothelial migration of normal or β2 integrin-deficient lymphocytes was minimally affected by monoclonal antibodies to α4β1 or VCAM-1, suggesting that VCAM-1 is a minor contributor to lymphocyte diapedesis.[28,160,163] However, lymphocytes can utilize immobilized VCAM-1/Ig fusion proteins as a substrate for α4β1-dependent migration across polycarbonate membranes.[164] In addition, VCAM-1 binding to encephalitogenic murine Th1 cells was found to induce transcription and surface expression of a 72 kD gelatinase that may aid T cells in degrading matrix molecules during interstitial migration.[165] Thus, the α4β1/VCAM-1 pathway can perhaps contribute to lymphocyte migration within extravascular regions where VCAM-1 is expressed.

MAdCAM-1: The integrin counter-receptor for MAdCAM-1 is the α4β7 integrin.[166] This receptor has been detected at low to intermediate levels on most B and T cells in man. High densities can be found on gut-homing lymphoblasts and some memory cell subsets.[167] α4β7 is also found on NK cells, monocytes, mast cells, and basophils. The N-terminal Ig domain (D1) of MAdCAM-1 contains the binding site for α4β7 and is sufficient for adhesion when the integrin is activated, but sequences within D2 were shown to support binding, especially when α4β7 is in a low activation state.[168] In accordance with the structural requirements for integrin-IgSF interactions (see above), mutations in residues Leu_{40}, Asp_{41} or Thr_{42} in the CD loop but not in other residues of D1 were found to abrogate α4β7 binding.[168,169]

In addition to its function as a protein ligand for α4β7, MAdCAM-1 is also a facultative ligand for L-selectin. The mucin domain of murine HEV-derived MAdCAM-1 was shown to be decorated with the peripheral node addressin, a unique O-linked carbohydrate epitope that is recognized by the rat monoclonal antibody MECA-79 and supports L-selectin binding.[170] As shown for its murine counterpart, primate MAdCAM-1 is also a potent ligand for α4β7 and does not interact with α4β1.[171] The mucin-like domain in human and macaque MAdCAM-1 is even larger than that in the mouse, but primate MAdCAM-1 lacks a third IgA-like Ig loop that constitutes the membrane-proximal domain of the murine form.[171] Whether human

MAdCAM-1 can be decorated with L-selectin ligands has not been determined.

Experiments on MAdCAM-1 function under flow have been performed using the murine molecule. Immobilized purified MAdCAM-1 from mouse MLN supported lymphocyte attachment and rolling on a glass surface which could be mediated by binding of α4β7, or L-selectin, or both. In contrast, recombinant MAdCAM-1 which lacked PNAd glycosylation did not support L-selectin adhesion, but supported avid binding via α4β7.[88,170] These *in vitro* observations are consistent with intravital microscopy observations of lymphocytes in murine Peyer's patch HEV.[172] Primary interactions in these vessels were nearly completely blocked by a monoclonal antibody to MAdCAM-1. Initial attachment of resting lymphocytes was predominantly mediated by L-selectin (presumably via interactions with the mucin domain of MAdCAM-1), whereas α4β7 was required for slowing of the rolling cell following contact and as a bridging molecule for LFA-1-dependent arrest. Importantly, when a gut-homing T cell line was used that expressed constitutively activated α4β7, or when α4β7 affinity was increased on peripheral blood lymphocytes by activation or exposure to Mn^{2+}, the cells were able to undergo attachment and sticking that was independent of selectins and β2 integrins, respectively. Thus, MAdCAM-1 is not only a receptor for primary adhesion, but can also support secondary interactions.[88,172]

Endothelial Immunoglobulins in Leukocyte Homeostasis

The average residence time of lymphocytes in the peripheral blood is quite short (~25 min) because most blood-borne lymphocytes migrate (home) rapidly to secondary lymphoid organs or other sites within the body (reviewed in refs. 7 and 173-175). Homing is tissue-specific and many lymphocyte subsets preferentially follow distinct migratory routes.[7,175] Homing to LN and Peyer's patches occurs nearly exclusively via a subset of postcapillary venules in these organs that feature a conspicuous cuboid endothelium.[176] These HEV express site-specific ligands for lymphocyte homing receptors, the so-called vascular addressins.[177] Intravascular interactions between vascular addressins and homing receptors are integral parts of adhesion cascades that involve endothelial IgSF (Fig. 3-1) and are thought to be largely responsible for subset- and organ-specific homing.[3,6,7,178] In addition, IgSF molecules have other important functions in lymphocyte homeostasis such as costimulation during antigen recognition and induction or prevention of cell death.

Unlike lymphocytes, neutrophils are short-lived (~ 6 - 12 hours). As neutrophils prepare to die, they leave the vascular compartment and do not recirculate. This continuous loss of neutrophils is balanced by the production of new cells in the bone marrow. There is ample evidence that adhesion molecules, including IgSF, are involved in neutrophil homeostasis both at the level of the bone marrow and during emigration. An imbalance in these mechanisms is reflected by a change in circulating leukocyte counts. This becomes apparent in humans who suffer from inherited leukocyte adhesion deficiency syndromes[115,179] and in animals that have been rendered genetically deficient in adhesion receptors (Table 3-4).

ICAM-1: Several observations point to a contribution by endothelial ICAMs to lymphocyte homing to lymphoid organs. Firstly, LFA-1, the main lymphocyte counter-receptor for ICAM-1 and ICAM-2, plays a dominant role in the mediation of activation-induced sticking of lymphocytes in HEV of both Peyer's patches and peripheral lymph nodes.[172,180a,180b] Secondly, immunoelectron microscopy of lymph node HEV indicates that ICAM-1 is expressed at very high density (5- to 30-fold higher than in non-HEV) and concentrated on microfolds and branched ridges in these specialized microvessels.[29] Thirdly, patients suffering from rheumatoid arthritis who had been treated with anti-ICAM-1 were found to have elevated levels of circulating T cells throughout the treatment period.[181] Finally, mice with a partial loss of ICAM-1 or CD18 were reported to have reduced T cell populations in their intestinal mucosa.[182] These findings are consistent with the observation that peripheral blood lymphocyte counts in ICAM-1-deficient mice are two to three-fold higher than in wild type litter mates[183] and suggest a potential role for ICAM-1 in lymphocyte trafficking. However, ICAM-1 null animals did not reveal any abnormalities in the relative or absolute frequency of T and B cell subsets in thymus, spleen, Peyer's patches, or lymph nodes.[183,184] Thus, other effects of ICAM-1 deficiency might be responsible for the subtle lymphocytosis in knockout mice.

ICAM-1 deficiency also affects circulating neutrophil counts, which were found to be three-fold higher than in wild-type litter mates.[183,184] This has been interpreted as a requirement for ICAM-1 in neutrophil adhesion to and diapedesis across EC. In addition, interaction with Mac-1 on neutrophils may constitute another mechanism through which ICAM-1 may regulate neutrophil homeostasis. Recent studies have shown that ligation of Mac-1 triggers an oxidative burst in neutrophils which, in turn, induces a cascade of events leading to apoptosis.[185] Finally, ICAM-1-mediated crosslinking of β2 integrins is also important during antigen-induced lymphocyte activation. Binding of ICAM-1 on APC to LFA-1 on T or B cells provides a strong accessory signal during antigen recognition.[186-188] This is supported by studies on the role of ICAM-1 in MLR which indicate that ICAM-1 expression is essential on stimulator cells but is not required on responder cells.[184]

Table 3-4. Phenotype of IgSF mutant mice

Mouse Strain	Phenotype	Reference
Mutants expressing residual ICAM-1 variants in some organs (targeted in exon 5)	Granulocytosis, reduced neutrophil accumulation in peritonitis, blunted DTH response, ICAM-1-deficient lymphocytes fail to stimulate allogeneic cells in MLR	184
	Decreased or increased mortality in bacterial meningitis in infant mice following *H. influenzae* type B or S. pneumoniae, respectively	270
	Partial (60 - 70%) reduction in neutrophil migration in peritonitis in either ICAM-1 or P-selectin-deficient mice; complete absence of neutrophils in peritonitis in ICAM-1+P-selectin double mutants, normal neutrophil accumulation in pneumonia	271
	Two functional ICAM-1 variants in thymus and lung	217
	Mutant mice not protected from rejection of cardiac allografts	272
	Increased bacteremia upon i.p. injection of *P. aeruginosa*; enhanced skin lesions upon i.d. injection of *S. aureus*; no difference in mortality upon i.p. injection of *P. aeruginosa, S. aureas* or *E. coli*	273

Table 3-4 (continued). Phenotype of IgSF mutant mice

Mouse Strain	Phenotype	Reference
Mutants expressing residual ICAM-1 variants in some organs (targeted in exon 5)	No difference in trauma-induced leukocyte rolling in cremaster muscle venules, but rolling was absent in a strain deficient in ICAM-1 + P-selectin, whereas P-selectin KO mice had low level rolling	116
	LPS-induced pneumonia: no protection in mutants, but WT mice are protected by anti-ICAM-1 MAb	261
	Impaired defense against disseminated candidiasis	274
	Collagen-induced arthritis: decreased incidence, but affected mutants have equivalent severity and time course of disease	275
	Cobra venom factor-induced pneumonia: no protection in ICAM-1 or ICAM-1+P-selectin mutants, but WT mice are protected by MAb	262
	P. aeruginosa-induced pneumonia: no protection in ICAM-1mutants, but WT mice are protected by MAb	276

Table 3-4 (continued). Phenotype of IgSF mutant mice

Mouse Strain	Phenotype	Reference
ICAM-1-deficient (targeted in exon 4)	Granulocytosis and lymphocytosis; resistance to septic shock induced by high dose LPS in non-sensitized mice or by SEB in D-galactosamine sensitized mice; blunted DTH response; lymphocytes fail to stimulate allogeneic cells in MLR	183
	Decreased mortality, infarct size and CBF reduction after 45 min MCA occlusion	277
	SEB superantigen challenge: normal anergy and apoptosis, but decreased proliferation of lymphocytes	278
	Decreased infarct size after 3 hours MCA occlusion	279
	Staphylococcal bacteremia: increased mortality, but decreased frequency and severity of arthritis	280
	Decreased renal injury after kidney ischemia/reperfusion	281
	Ovalbumin-induced allergic pulmonary inflammation: abrogation of eosinophilia, reduced lymphocyte migration, normal macrophage accumulation in the lung	250

Table 3-4. Phenotype of IgSF mutant mice

Partial loss of ICAM-1	Impaired mucosal T cell compartment	182
Congenic C57BL/6 mutants (mivit/mivit), selective defect in epidermal ICAM-1	Muted contact hypersensitivity response	282
VCAM-1-deficient	Embryonic lethal	241,242
VCAM-1 hypomorph (D4 deleted)	95% reduced mutant VCAM-1 mRNA levels	243
	Normal myelopoiesis and lymphopoiesis	196
	Ovalbumin-induced allergic pulmonary inflammation: abrogation of eosinophilia, reduced lymphocyte migration, normal macrophage accumulation in the lung	250

ICAM-2: Constitutive ICAM-2 levels on resting EC are higher than those of ICAM-1, but ICAM-2 is not upregulated during inflammation. Thus, it has been proposed that a major role of ICAM-2 may be in leukocyte migration in the absence of inflammation (e.g., during physiologic homing) and within the first hours following an inflammatory insult, whereas cytokine-responsive EC adhesion molecules (especially ICAM-1 and VCAM-1) may dominate thereafter.[6]

Analogous to ICAM-1, binding of ICAM-2 (and its non-endothelial homologue ICAM-3) to LFA-1 can also provide a costimulatory signal to CD4[+] lymphocytes exposed to anti-CD3 antibodies.[111,189] Thus, ICAM-2 may function during the initiation of T cell activation and antigen recognition, especially in environments where little or no ICAM-1 is expressed. In addition, the ICAM-2/LFA-1 pathway was recently reported to have a selective role in the activation-induced death of T cells responding to antigen rechallenge.[190]

PECAM-1: In contrast to its function in neutrophil and monocyte migration, the role of PECAM-1 in lymphocyte homing is probably minor. Peripheral blood lymphocytes that express little or no PECAM-1 were found to have a higher capacity for transendothelial migration than cells with high surface expression of PECAM-1, indicating that PECAM-1 on the lymphocyte is not necessary for diapedesis.[191] However, both PECAM-1[+] and PECAM-1[-] lymphocytes transmigrated through monolayers of transfected fibroblasts that expressed PECAM-1, but not ICAM-1, suggesting that endothelial PECAM-1 can play a role.[192] Moreover, PECAM-1 inhibition with a monoclonal antibody to D1 was reported to reduce the spontaneous transendothelial migration of CD56[+] NK cells *in vitro*[45] and also blocked the migration of mononuclear cells via afferent lymphatics to MLN in a peritonitis model.[139] Thus, PECAM-1 may have a role in the immune surveillance of peripheral tissues.

PECAM-1 is also involved in the regulation of T cell activation and proliferation. Lymphocyte responses to MLR were blocked by a monoclonal antibody to a membrane-proximal epitope associated with D6 of PECAM-1 and by a synthetic peptide mimicking this region.[130] Similarly, a PECAM-1/Ig fusion protein was shown to downregulate the response of anti-CD3 treated CD31[-] T helper clones, presumably by binding to a 120 kD ligand on the lymphocyte.[133] In addition to functioning in cell-cell communication via binding to a heterotypic ligand on the responding cell, PECAM-1 can also transmit signals to its host cell. Crosslinking of PECAM-1 was found to modulate integrin function on myeloid, lymphoid and progenitor cells, and several studies have shown that anti-PECAM-1 monoclonal antibodies trigger signals and augment colony formation in CD34[+] cord blood progenitor cultures.[43,45,142,143,193,194] Thus, a number of PECAM-1-dependent mechanisms are conceivable that may influence leukocyte homeostasis.

VCAM-1: EC in bone marrow sinusoids constitutively express VCAM-1 on their lumenal surface, and VCAM-1 has been implicated in the traffic of cells between blood and bone marrow.[56,87] Consistent with this idea, it has been shown that homing of hematogenic progenitor cells to the bone marrow of lethally irradiated mice is significantly reduced by neutralizing monoclonal antibodies to VCAM-1 or α4 integrin. Monoclonal antibody treatment was accompanied by increased numbers of progenitors in the spleen and peripheral blood of recipients, suggesting that VCAM-1 is a site-specific vascular addressin for cell migration to the bone marrow.[57] In addition, VCAM-1 may prevent the premature exit of newly formed cells from the bone marrow, since it is expressed on bone marrow stroma cells and binds to progenitor cells, megakaryocytes, and immature myeloid and erythroid cells. The α4β1/VCAM-1 pathway also supports bone marrow stroma cell interactions with lymphocytes, especially B cell precursors, and promotes B cell maturation *in vitro.*[59] In addition, VCAM-1 may contribute to extramyeloid hematopoiesis; monoclonal antibodies to VCAM-1 were shown to disrupt erythroblastic islands that form in spleens of phlebotomized mice.[195] It should be noted, however, that VCAM-1 is probably not essential for hematopoiesis, since mice with a hypomorphic mutation in VCAM-1 had normal hematopoietic activity and showed no abnormalities in lymphoid or myeloid differentiation.[196]

Another constitutive function of VCAM-1 may be the regulation of leukocyte activation, maturation, and proliferation within lymphoid tissues. It has been shown that follicular dendritic cells in germinal centers express high levels of VCAM-1, which supports B cell binding and proliferation[58] and inhibits B cell apoptosis.[197] In addition, VCAM-1 functions as a T cell receptor-dependent costimulator of resting CD4+ lymphocytes.[198-200] A recent study has shown that simultaneous crosslinking of CD3 and α4β1 on Jurkat or CD4+ T cells induces tyrosine phosphorylation of pp125FAK, a kinase that is thought to link integrin-mediated signaling to the Ras pathway.[201] VCAM-1-mediated α4β1 ligation on B cells was also shown to induce tyrosine phosphorylation of a 110 kD protein.[202] Similarly, eosinophil incubation on VCAM-1 triggered tyrosine kinase activation and resulted in spontaneous superoxide generation and enhanced formyl peptide-induced oxidative burst.[203]

VCAM-1 has also been implicated in thymic T cell development. Most immature (pre-selection) CD4+CD8+ thymocytes were found to bind avidly to VCAM-1, whereas non-adherent thymocytes displayed a phenotype that was typical for more mature cells that had already undergone positive selection.[204] Moreover, the migration of a subset of CD4+CD45RC- (recently activated/memory) T cells to the thymus was inhibited by anti-VCAM-1 in rats.[205] Thus, as in the bone marrow, thymic VCAM-1 may support homing and prevent premature release of distinct cell subsets. In contrast, monoclonal antibodies to VCAM-1 have no effect on lymphocyte homing to peripheral lymph nodes or Peyer's patches,[206] and inhibition of α4β1 does not alter the intravascular behavior of lymphocytes in lymph node HEV.[180b] VCAM-1 is sparse or undetectable in HEV of non-inflamed secondary lymphoid organs in humans, adult mice, or sheep.[207-209] However, low levels were observed in HEV of murine fetal lymph nodes,[210] and HEV in rat peripheral lymph nodes reportedly express VCAM-1 under basal conditions and upregulate expression upon cytokine treatment.[211] VCAM-1 expression could also be induced in sheep LN HEV upon antigenic challenge[207] and was observed in reactive LN and in Hodgkin's disease in humans.[212] These studies suggest that in most species VCAM-1 plays a relatively minor role in physiologic homing to lymphoid organs (except perhaps the thymus), but it is very likely to contribute to lymphocyte trafficking in some pathologic settings.

MAdCAM-1: MAdCAM-1 is essential for lymphocyte migration to non-pulmonary mucosa-associated lymphoid tissues. The isolated molecule was shown to support selective binding of mucosa-specific, but not of peripheral lymph node-specific, lymphocytes.[213] Accordingly, lymphocyte homing to Peyer's patches and to intestinal lamina propria is markedly reduced in mice treated with monoclonal antibodies to MAdCAM-1 or $\alpha 4\beta 7$,[81] but not with anti-VCAM-1 or a peptide that blocks $\alpha 4$ integrin binding to fibronectin.[206] These observations are consistent with the phenotype of $\beta 7$ integrin-deficient mice which selectively lack lymphocytes in the gut.[214] MAdCAM-1 and $\alpha 4\beta 7$ are also partially required for lymphocyte homing to MLN, but are not required for homing to peripheral lymph nodes or the spleen in adult animals.[81,83] In contrast, fetal and newborn mice express MAdCAM-1 not only in mucosal sites but also in peripheral lymph nodes. PNAd, the vascular addressin in peripheral lymph nodes of adults, is absent in newborn animals. PNAd becomes detectable only several days after birth, at which time MAdCAM-1 begins to disappear from peripheral lymph node HEV. MAdCAM-1 in fetal lymph nodes mediates recruitment of a unique CD4⁺CD3⁻ $\alpha 4\beta 7^{hi}$ lymphocyte subset. Migration of this subset precedes postnatal homing of naive lymphocytes that utilize the L-selectin/PNAd pathway.[210] This differential regulation of MAdCAM-1 in fetal and adult lymph nodes suggests a broader role for MAdCAM-1 in the ontogeny of the peripheral immune system. The early dominance of the mucosal homing pathway may also reflect lymphoid phylogeny in higher animals, since organized lymphoid tissues are thought to have developed first in the alimentary tract, whereas other lymphoid organs most likely arose later.

Little is known so far about the role of MAdCAM-1 in cell-cell communication. However, crosslinking of $\alpha 4\beta 7$ by immobilized monoclonal antibody was shown to costimulate human peripheral blood T cells exposed to submitogenic levels of anti-CD3 antibodies.[215] Thus, as shown for ICAM-1 and VCAM-1, it is possible that

MAdCAM-1 has an immunomodulatory function in addition to its role in adhesion.

Endothelial Immunoglobulins in Inflammation

Most *in vivo* studies on the role of endothelial IgSF members in inflammation have employed inhibitors of molecular interactions such as monoclonal antibodies or soluble receptors. These investigations have identified a large array of pathologic conditions in which IgSF play a role. However, some studies have reported results that are in apparent conflict with others that have addressed similar questions using different experimental or analytic approaches. Moreover, the majority of experimental protocols have addressed the function of IgSF during relatively short periods of time or in acute inflammation. Many questions on the role of IgSF and the effects of long-term inhibition of IgSF in chronic inflammatory settings remain to be answered. The usefulness of protein-based IgSF inhibitors may be limited for such investigations because they can be recognized by the recipient's immune system and may lead to allergic reactions or rapid inactivation by anti-idiotype antibodies.

Future studies using inhibitors of endothelial IgSF are likely to employ small molecules which are currently a subject of intense research by a number of pharmaceutical and biotechnology companies (reviewed in Chapters 8 and 9). In addition, mice that have a targeted disruption or mutation in one or more endothelial IgSF genes are likely to yield significant new insights in the role of these molecules, especially in chronic settings. Animals that are completely or partially deficient in ICAM-1 or VCAM-1 have already been studied extensively. Table 3-4 provides a partial summary of the phenotypes of these IgSF mutant mice.

ICAM-1: ICAM-1 upregulation on EC has been found in a host of pathologic conditions including allergy, atherosclerosis, and during reperfusion following tissue ischemia and endotoxemia.[97,216] Neutralizing antibodies to ICAM-1 were observed to exert beneficial effects in many experimental models of these and other conditions such as allograft rejection and autoimmune disease. These studies have been reviewed in depth elsewhere.[5,97,181] Two strains of gene-targeted mice have been generated. Initial studies in one strain with a targeted mutation in exon 5 found that uninflamed and LPS-treated tissues did not show immunoreactivity to an anti-ICAM-1 antibody.[184] However, the lung and thymus in these mice were later found to express low levels of two alternatively spliced variants of ICAM-1 that can interact with LFA-1 and are upregulated by cytokines.[217] The second strain which was targeted in exon 4 [183] has not been found to express residual isoforms of ICAM-1 protein (J.-C. Gutierrez-Ramos, personal communication). Studies with these and other mutant mouse strains confirm the role of ICAM-1 in a host of pathologic conditions (Table 3-4).

ICAM-2: ICAM-2 on non-inflamed EC supports most of the β2 integrin-dependent adhesion of leukocytes and participates in chemoattractant-induced lymphocyte diapedesis.[37,91] However, ICAM-2 is a minor contributor to adhesion to or migration across cytokine-stimulated EC. Thus, it has been proposed that ICAM-2 may play a role during leukocyte sticking in early acute inflammatory settings, but may become less important as inflammation progresses. However, this notion has not been tested *in vivo*.

Recent studies have suggested a role for ICAM-2 as a target for killing by IL-2 activated killer cells. Cells that were susceptible to killing were found to redistribute ICAM-2 to uropod-like structures through a mechanism involving the cytoskeletal membrane linker molecule ezrin. In contrast, cells that were resistant to killing displayed equivalent levels of ICAM-2, but in a uniform surface distribution. Expression of human ezrin in these resistant cells induced the formation of uropods with highly concentrated ICAM-2. This sensitized the transfectants to IL-2-activated killing that was blocked by a monoclonal antibody to ICAM-2. Based on these observations, it has been speculated that ICAM-2 redistribution to uropods or microvilli may occur in virus-infected or mutated cells as a defense mechanism to enhance recognition and lysis by killer cells.[218] It is interesting to note in this context that a synthetic peptide resembling residues 21-42 of D1 in ICAM-2 was found to enhance NK cell cytotoxicity and migration via allosteric activation of Mac-1, [119,219] providing a possible mechanism for ICAM-2-dependent NK cell-mediated killing.

PECAM-1: In both normal and pathologic mammalian tissues, PECAM-1 is routinely detected on EC, platelets, and some leukocytes. No dramatic changes in endothelial PECAM-1 have been reported in a variety of inflammatory conditions. These observations suggest that PECAM-1 does not initiate leukocyte recruitment by itself, but they do not exclude an active role during a later step such as transendothelial migration. This idea is supported by the finding that polyclonal antibodies to PECAM-1 block neutrophil recruitment to the inflamed lung and peritoneal cavity in rats and to human skin grafts transplanted onto immunodeficient mice.[137] A similar treatment in mice also reduced leukocyte accumulation in the inflamed peritoneal cavity and migration of mononuclear cells via afferent lymphatics to the draining MLN.[139] More recently, the effect of PECAM-1 inhibition was studied by intravital and electron microscopy of mesenteric venules. Leukocyte rolling and intravascular adhesion in IL-1β-treated venules was not affected. However, leukocyte emigration in response to IL-1β was blocked; cells became trapped within the vessel wall and did not pass through the basement membrane. Interestingly, anti-PECAM-1 did not block formyl-peptide-induced leukocyte extravasation in the same model, suggesting that PECAM-1 plays a role in some, but not all, inflammatory settings.[140] Intravital microscopy studies also indicate that PECAM-1 can participate in events that may lead to vessel occlusion by platelet aggregates on damaged EC. Laser-induced platelet adhesion and aggregation in murine pia mater arterioles was delayed by treatment with a neutralizing monoclonal antibody to PECAM-1.[220] Postischemic tissue damage that occurs after restoration of blood flow (ischemia/reperfusion injury) also appears to involve PECAM-1. A recent study in a feline model of myocardial ischemia/reperfusion injury has shown that anti-PECAM-1 reduced the area of myocardial necrosis by more than half. This was attributed to the antibody-mediated inhibition of transendothelial neutrophil migration.[221]

VCAM-1: Endothelial upregulation of VCAM-1 has been reported in a host of pathologic conditions in humans and animals. Most of these settings involve local or systemic actions of cytokines that induce VCAM-1 gene transcription (see Chapter 5). While induction of VCAM-1 within an inflamed tissue is mostly (but not always) paralleled or followed by an influx of mononuclear cells, it often remains unclear whether and to what extent VCAM-1 is causally involved in the inflammatory process. Other endothelial adhesion molecules (e.g., ICAM-1 and E-selectin) are frequently coexpressed with VCAM-1 and may be similarly important. Nevertheless, VCAM-1 has received particular attention in a wide range of inflammatory conditions including asthma/allergy;[222] DTH;[223] inflammation of the CNS;[224,225] liver, and skin; autoimmune diabetes;[226] GvHD;[227] and allograft rejection.[228,229]

VCAM-1 is also deemed important in the pathogenesis of atherosclerosis. Metabolic or nutritional factors such as hyperlipidemia, a cholesterol-rich diet,[230] oxidized or glycated LDL,[231] diabetic metabolism,[232] advanced glycation end products,[233] and lysophosphatidylcholine, an atheroma component,[20] can promote VCAM-1 upregulation in arteries of rabbits. VCAM-1 was also found in atherosclerotic plaques of human coronary arteries, but mostly within areas of neovascularization and inflammation at the base of the plaque and less frequently in the arterial lumen.[234] Several studies also suggest a role for VCAM-1 in rheumatoid diseases and autoimmune vasculitis. Synovial fibroblasts express increased levels of VCAM-1 in rheumatoid arthritis or osteoarthritis. T cells isolated from synovial membranes of patients with rheumatoid arthritis were found to bind avidly to VCAM-1 and exhibit a strong proliferative response when exposed to submitogenic levels of anti-CD3 combined with VCAM-1.[235-237] Endothelial VCAM-1 has also been detected in affected organs of patients suffering from systemic lupus erythematosus, Wegener's granulomatosis, and other vasculitic conditions.[238,239] The plasma of such patients contains anti-neutrophil cytoplasmic antibodies (ANCA) including antibodies to proteinase 3. These antibodies were recently found to induce VCAM-1 expression on HUVEC and subsequent T cell adhesion, suggesting that ANCA may play a direct role in VCAM-1-dependent cell recruitment in some forms of vasculitis.[240]

Because a homozygous deficiency in VCAM-1 is developmentally lethal due to defective placentation and cardiovascular development, there are currently no published data on inflammatory models in VCAM-1 knockout mice.[241,242] However, a small fraction (~0.1%) of homozygous-deficient embryos was found to survive to adulthood. Furthermore, a strain of hypomorph animals with a targeted deletion in D4 of VCAM-1 has been generated recently.[243] These mutants were found to be viable and express less than 5% of normal VCAM-1 mRNA levels. These animals are likely to be immensely useful for future studies on the role of VCAM-1 in inflammation.

A large number of studies have made use of neutralizing antibodies to block VCAM-1 in vivo. One major focus has been the role of VCAM-1 in transplantation immunity. VCAM-1 inhibition was found to prolong the survival of cardiac allografts in mice.[229,244] Similarly, anti-VCAM-1 was reported to suppress the rejection of allogeneic pancreatic islets in non-sensitized mice. However, the monoclonal antibody had no effect when recipients were previously sensitized to donor tissue.[245] In contrast, in a mouse model of skin transplantation, anti-VCAM-1 induced prolonged graft survival in immunized, but not in naive, animals.[246] Transplanted organs in anti-VCAM-1-treated animals were frequently found to contain significant numbers of lymphocytes, suggesting that the therapeutic effect may be due in part to inhibition of VCAM-1-dependent lymphocyte stimulation. Consistent with this concept, monoclonal antibodies to VCAM-1 were found to inhibit mixed lymphocyte reactions across major and minor histocompatibility barriers and to reduce the incidence, severity, and mortality of graft-versus-host disease in mice transplanted with mismatched bone marrow.[227]

VCAM-1 may also play a role in the pathogenesis of autoimmune conditions such as juvenile diabetes. Chronic treatment of non-obese diabetic (NOD) mice with monoclonal antibodies to VCAM-1 was reported to prevent insulitis and diabetes, and anti-VCAM-1 as well as soluble VCAM-1/Ig fusion proteins blocked adoptive transfer diabetes in SCID mice.[226,247] In another adoptive transfer model of experimental allergic encephalitis (EAE), anti-α4 effectively prevented the accumulation of myelin basic protein-specific T cells in the brain and the development of EAE.[248] However, anti-α4 monoclonal antibodies had no effect in viral encephalitis although endothelial VCAM-1 expression was found to be increased in this setting.[249] Thus, the role of the α4β1/VCAM-1 pathway in lymphocyte migration to the inflamed CNS depends on the underlying disease condition. VCAM-1 inhibition has also been tested in several allergic settings: anti-VCAM-1 reduced leukocyte influx in a cutaneous DTH reaction in rhesus monkeys;[223] and blocking monoclonal antibodies were found to inhibit antigen-induced influx of eosinophils and T cells into the trachea of mice;[222] more recent studies in an asthma model have shown that eosinophil migration to allergen-challenged lungs does not occur in VCAM-1 hypomorph mice.[250]

MAdCAM-1: MAdCAM-1 expression can be induced by chronic inflammatory stimuli in nonlymphoid tissues.[177] Exposure of murine brain-derived bEnd.3 endothelioma cells to TNF-α, IL-1 or LPS stimulated surface expression of functional (i. e., α4β7-binding) MAdCAM-1, reaching a peak after 18 hours.[251] The TNF effect depended on binding of NF-κB (predominantly p65 homodimer) to two regulatory sites within the MAdCAM-1 promoter.[252] Treatment of bEnd.3 cells with IFN-γ did not induce MAdCAM-1 and suppressed its induction when TNF-α, IL-1, or LPS were added subsequently.[251] In contrast to these *in vitro* observations, *in vivo* expression of IFN-γ in pancreatic beta cells of transgenic mice resulted in insulitis and concomitant upregulation of endothelial MAdCAM-1 and ICAM-1. This effect was also observed in IFN-γ transgenic mice that lacked mature lymphocytes, suggesting that MAdCAM-1 induction *in vivo* can be a direct consequence of IFN-γ expression.[253] Thus, IFN-γ may be able to induce directly or suppress indirectly (via inhibition of other cytokine effects) MAdCAM-1 expression in EC.
    Venules in chronically inflamed tissues assume an HEV-like phenotype and are often (but not always) found to express MAdCAM-1. For instance, MAdCAM-1 was detected in vessels adjacent to and within inflamed islets, but not in inflamed salivary glands of NOD mice.[254,255] During late stages of *Borrelia burgdorferi*-induced chronic arthritis, MAdCAM-1 was observed on newly generated venules in synovial lesions of SCID mice.[256] AKR mice with thymic hyperplasia were found to develop HEV within the hyperplastic thymic medulla that expressed both MAdCAM-1 and PNAd. Adhesion assays on frozen thymic sections of these animals showed that both addressins may be involved in lymphocyte traffic to the hyperplastic thymus.[257] Low MAdCAM-1 levels have also been detected in spinal cord venules of Biozzi AB/H mice with relapsing episodes of chronic EAE,[258] but others have reported that T cells from brains of mice with EAE bind poorly to MAdCAM-1.[259] Functional MAdCAM-1 was also found on choroid plexus epithelial cells in two mouse strains with EAE *in vivo* and in response to cytokines *in vitro*, whereas choroid plexus EC did not express MAdCAM-1 in these studies.[68]
    It has also been shown that monoclonal antibodies to both MAdCAM-1 and β7 can block lymphocyte migration and colonic inflammation in SCID mice that were reconstituted with CD4+ CD45RB[hi] T cells.[260a] *In vivo* studies in Cotton-top Tamarin monkeys (*Saguinus oedipus*) suggest that MAdCAM-1 may also play a role in inflammatory bowel disease in primates. These animals spontaneously develop a chronic colitis that is similar to ulcerative colitis in humans. Their condition improved histologically and clinically after treatment with blocking monoclonal antibodies to either chain of α4β7.[260b]

FUTURE APPLICATIONS AND ANTICIPATED DEVELOPMENT

Basic research has accumulated a wealth of data (of which only a small fraction has been presented here) that point to endothelial IgSF molecules as promising therapeutic targets for leukocyte-mediated diseases. Clinical studies on the potential for therapeutic interventions using antibodies that block IgSF or their counter-receptors have recently begun. For a full understanding of what can reasonably be expected from such treatments both therapeutically and in terms of potential side effects, further evaluation of the physiologic role of these molecules is needed. In particular, it will be necessary to dissect the role of IgSF in the normal immune surveillance of an organism. Which molecules contribute to the migration and function of which leukocyte subset in what organ? How and when are endothelial IgSF involved in inter- and intracellular interactions that go beyond the mere mechanical stabilization of cell-cell contacts? What are the bottom line effects of these multi-faceted functions on an organism? How do these functions change during disease, both qualitatively and quantitatively? Undoubtedly, such insights will arise increasingly from the use of mice that are genetically deficient or mutated in one or more of these adhesion molecules. Recent studies suggest that this approach is not just a mere substitute for antibody experiments in normal animals, but can yield results that demand careful reassessment of previously held beliefs.[261,262]

A precise knowledge of the transcriptional regulation of IgSF may be useful to explore and implement novel therapeutic approaches. Studies have already been undertaken in transgenic mice that were induced to express human CD59 under control of the human ICAM-2 promoter. Transgene expression in all organs in these animals was uniformly and selectively detected on vascular EC, but not on other cell types.[263] Future efforts may be directed toward exploiting the selective transcriptional activation of IgSF members in diseased tissues to achieve site-specific expression of transgenes in gene therapy. In addition to the use of mutant animals and inhibitory antibodies, other avenues of interference with intravascular adhesion events are being explored as well. For instance, inhibiting the transcriptional induction of IgSF in addition to or instead of interference with their function offers a therapeutic alternative that may be especially useful in attenuating, rather than abolishing, critical adhesive events in chronic inflammatory settings.

ACKNOWLEDGMENTS

I wish to thank Drs. M. Briskin, E. C. Butcher, T. D. Diacovo, W. A. Muller and T. A. Springer for comments and constructive criticism on this manuscript. Special thanks to Dr. S. E. Erlandsen, S. H. Hasslen for providing electron micrographs of immunogold labeled lymphocytes. This work was supported by grants from the NIH.

REFERENCES

1.   Fung YC. Biomechanics: Mechanical Properties of Living Tissues. New York: Springer-Verlag, 1981: 1-433.
2.   Springer TA. Adhesion receptors of the immune system. Nature 1990; 346:425-433.
3.   Butcher EC. Leukocyte-endothelial cell recognition: Three (or more) steps to specificity and diversity. Cell 1991; 67:1033-1036.
4.   Hynes RO. Integrins: Versatility, modulation, and signaling in cell adhesion. Cell 1992; 69:11-25.
5.   Carlos TM, Harlan JM. Leukocyte-endothelial adhesion molecules. Blood 1994; 84:2068-2101.
6.   Springer TA. Traffic signals for lymphocyte recirculation and leukocyte emigration: The multi-step paradigm. Cell 1994; 76:301-314.
7.   Butcher EC, Picker LJ. Lymphocyte homing and homeostasis. Science 1996; 272:60-66.
8.   Arfors K-E,Lundberg C, Lindbom L, et al. A monoclonal antibody to the membrane glycoprotein complex CD18 inhibits polymorphonuclear leukocyte accumulation and plasma leakage *in vivo*. Blood 1987; 69:338-340.
9.   von Andrian UH, Chambers JD, McEvoy LM, et al. Two-step model of leukocyte-endothelial cell interaction in inflammation: Distinct roles for LECAM-1 and the leukocyte β_2 integrins *in vivo*. Proc Natl Acad Sci USA 1991; 88:7538-7542.
10.   Lawrence MB, Springer TA. Leukocytes roll on a selectin at physiologic flow rates: distinction from and prerequisite for adhesion through integrins. Cell 1991; 65:859-873.
11.   Hammer DA, Tirrell M. Biological adhesion at interfaces. Annu Rev Mater Sci 1996; 26:651-691.
12.   Alon R, Hammer DA, Springer TA. Lifetime of the P-selectin: carbohydrate bond and its response to tensile force in hydrodynamic flow. Nature 1995; 374:539-542.
13.   von Andrian UH, Hasslen SR, Nelson RD, et al. A central role for microvillous receptor presentation in leukocyte adhesion under flow. Cell 1995; 82:989-999.
14.   Williams AF, Barclay AN. The immunoglobulin superfamily: Domains for cell surface recognition. Annu Rev Immunol 1988; 6:381-405.
15.   Harrelson AL, Goodman CS. Growth cone guidance in insects: Fasciculin II is a member of the immunoglobulin superfamily. Science 1988; 242:700-708.
16.   Williams AF. A year in the life of the immunoglobulin superfamily. Immunol Today 1987; 8:298-303.

92

17. Dustin ML, Rothlein R, Bhan AK, et al. Induction by IL-1 and interferon, tissue distribution, biochemistry, and function of a natural adherence molecule (ICAM-1). J Immunol 1986; 137:245-254.

18. Takei F. Inhibition of mixed lymphocyte response by a rat monoclonal antibody to a novel murine lymphocyte activation antigen (MALA-2). J Immunol 1985; 134:1403-1407.

19. Rothlein R, Czajkowski M, O'Neil MM, et al. Induction of intercellular adhesion molecule 1 on primary and continuous cell lines by pro-inflammatory cytokines. Regulation by pharmacologic agents and neutralizing antibodies. J Immunol 1988; 141:1665-1669.

20. Kume N, Cybulsky MI, Gimbrone MA,Jr. Lysophosphatidylcholine, a component of atherogenic lipoproteins, induces mononuclear leukocyte adhesion molecules in cultured human and rabbit arterial endothelial cells. J Clin Invest 1992; 90:1138-1144.

21. Bouillon M, Tessier P, Boulianne R, et al. Regulation by retinoic acid of ICAM-1 expression on human tumor cell lines. Biochim Biophys Acta 1991; 1097:95-102.

22. Karmann K, Hughes CCW, Schechner J, et al. CD40 on human endothelial cells: Inducibility by cytokines and functional regulation of adhesion molecule expression. Proc Natl Acad Sci USA 1995; 92:4342-4346.

23. Nagel T, Resnick N, Atkinson WJ, et al. Shear stress selectively upregulates intercellular adhesion molecule-1 expression in cultured human vascular endothelial cells. J Clin Invest 1994; 94:885-891.

24. Griffioen AW, Damen CA, Blijham GH, et al. Tumor angiogenesis is accompanied by a decreased inflammatory response of tumor-associated endothelium. Blood 1996; 88:667-673.

25. Weiss JM, Pilarski KA, Weyl A, et al. Prostaglandin E1 inhibits TNF alpha-induced T-cell adhesion to endothelial cells by selective down-modulation of ICAM-1 expression on endothelial cells. Exp Dermatol 1995; 4:302-307.

26. Gerritsen ME, Carley WW, Ranges GE, et al. Flavonoids inhibit cytokine-induced endothelial cell adhesion protein gene expression. Am J Pathol 1995; 147:278-292.

27. Panés J, Gerritsen ME, Anderson DC, et al. Apigenin inhibits tumor necrosis factor-induced intercellular adhesion molecule-1 upregulation in vivo. Microcirc 1996; 3:279-286.

28. Oppenheimer-Marks N, Davis LS, Bogue DT, et al. Differential utilization of ICAM-1 and VCAM-1 during the adhesion and transendothelial migration of human T lymphocytes. J Immunol 1991; 147:2913-2921.

29. Sasaki K, Okouchi Y, Rothkötter H-J, et al. Ultrastructural localization of the intercellular adhesion molecule (ICAM-1) on the cell surface of high endothelial venules in lymph nodes. Anat Rec 1996; 244:105-111.

30. Bradley JR, Thiru S, Pober JS. Hydrogen peroxide-induced endothelial retraction is accompanied by a loss of the normal spatial organization of endothelial cell adhesion molecules. Am J Pathol 1995; 147:627-641.

31. Carpen O, Pallai P, Staunton DE, et al. Association of intercellular adhesion molecule-1 (ICAM-1) with actin-containing cytoskeleton and α-actinin. J Cell Biol 1992; 118:1223-1234.

32. Erlandsen SL, Hasslen SR, Nelson RD. Detection and spatial distribution of the β_2 integrin (Mac-1) and L-selectin (LECAM-1) adherence receptors on human neutrophils by high-resolution field emission SEM. J Histochem Cytochem 1993; 41:327-333.

33. Pavalko FM, Walker DM, Graham L, et al. The cytoplasmic domain of L-selectin interacts with cytoskeletal proteins via α-actinin: Receptor

positioning in microvilli does not require interactions with α-actinin. J Cell Biol 1995; 129:1155-1164.

34.   Reilly PL, Woska JR,Jr., Jeanfavre DD, et al. The native structure of intercellular adhesion molecule-1 (ICAM-1) is a dimer: Correlation with binding to LFA-1. J Immunol 1995; 155:529-532.

35.   Miller J, Knorr R, Ferrone M, et al. Intercellular adhesion molecule-1 dimerization and its consequences for adhesion mediated by lymphocyte function associated-1. J Exp Med 1995; 182:1231-1241.

36.   Staunton DE, Dustin ML, Springer TA. Functional cloning of ICAM-2, a cell adhesion ligand for LFA-1 homologous to ICAM-1. Nature 1989; 339:61-64.

37.   de Fougerolles AR, Stacker SA, Schwarting R, et al. Characterization of ICAM-2 and evidence for a third counter-receptor for LFA-1. J Exp Med 1991; 174:253-267.

38.   Diacovo TG, de Fougerolles AR, Bainton DF, et al. A functional integrin ligand on the surface of platelets: Intercellular adhesion molecule-2. J Clin Invest 1994; 94:1243-1251.

39.   Nortamo P, Li R, Renkonen R, et al. The expression of human intercellular adhesion molecule-2 is refractory to inflammatory cytokines. Eur J Immunol 1991; 21:2629-2632.

40.   Renkonen R, Paavonen T, Nortamo P, et al. Expression of endothelial adhesion molecules *in vivo*: Increased endothelial ICAM-2 expression in lymphoid malignancies. Am J Pathol 1992; 140:763-767.

41.   Newman PJ. The role of PECAM-1 in vascular cell biology. Ann N Y Acad Sci 1994; 714:165-174.

42.   DeLisser HM, Newman PJ, Albelda SM. Molecular and functional aspects of PECAM-1/CD31. Immunol Today 1994; 15:490-495.

43.   Piali L, Albelda SM, Baldwin HS, et al. Murine platelet endothelial cell adhesion molecule (PECAM-1)/CD31 modulates β_2 integrins on lymphokine-activated killer cells. Eur J Immunol 1993; 23:2464-2471.

44.   Stockinger H, Schreiber W, Majdic O, et al. Phenotype of human T cells expressing CD31, a molecule of the immunoglobulin supergene family. Immunol 1992; 75:53-58.

45.   Berman ME, Xie Y, Muller WA. Roles of platelet/endothelial cell adhesion molecule-1 (PECAM-1, CD31) in natural killer cell transendothelial migration and $\beta2$ integrin activation. J Immunol 1996; 156:1515-1524.

46.   Stockinger H, Gadd SJ, Eher R, et al. Molecular characterization and functional analysis of the leukocyte surface protein CD31. J Immunol 1990; 145:3889-3897.

47.   Zehnder JL, Hirai K, Shatsky M, et al. The cell adhesion molecule CD31 is phosphorylated after cell activation. Down-regulation of CD31 in activated T lymphocytes. J Biol Chem 1992; 267:5243-5249.

48.   Demeure CE, Byun DG, Yang LP, et al. CD31 (PECAM-1) is a differentiation antigen lost during human CD4 T-cell maturation into TH1 or TH2 effector cells. Immunol 1996; 88:110-115.

49a.   Muller WA, Rapti CM, McDonnell SL, et al. A human endothelial cell-restricted externally disposed plasmalemmal protein enriched in intercellular junctions. J Exp Med 1989; 170:399-414.

49b.   Albelda SM, Oliver PD, Romer LH, et al. EndoCAM: A novel endothelial cell-cell adhesion molecule. J Cell Biol 1990; 110:1227-1237.

50.   Newman PJ, Berndt MC, Gorski J, et al. PECAM-1 (CD31) cloning and relation to adhesion molecules of the immunoglobulin gene superfamily. Science 1990; 247:1219-1222.

51. Albelda SM, Muller WA, Buck CA, et al. Molecular and cellular properties of PECAM-1 (endoCAM/CD31): A novel vascular cell-cell adhesion molecule. J Cell Biol 1991; 114:1059-1068.

52. Newman PJ, Hillery CA, Albrecht R, et al. Activation-dependent changes in human platelet PECAM-1: phosphorylation, cytoskeletal association, and surface membrane redistribution. J Cell Biol 1992; 119:239-246.

53. Romer LH, McLean NV, Yan H-C, et al. IFN-γ and TNF-α induce redistribution of PECAM-1 (CD31) on human endothelial cells. J Immunol 1995; 154:6582-6592.

54. Rival Y, Del Maschio A, Rabiet M-J, et al. Inhibition of platelet endothelial cell adhesion molecule-1 synthesis and leukocyte transmigration in endothelial cells by the combined action of TNF-α and IFN-γ. J Immunol 1996; 157:1233-1241.

55. Stewart RJ, Kashxur TS, Marsden PA. Vascular endothelial platelet endothelial adhesion molecule-1 (PECAM-1) expression is decreased by TNF-α and IFN-γ. Evidence for cytokine-induced destabilization of messenger ribonucleic acid transcripts in bovine endothelial cells. J Immunol 1996; 156:1221-1228.

56. Jacobsen K, Kravitz J, Kincade PW, et al. Adhesion receptors on bone marrow stromal cells: *in vivo* expression of vascular cell adhesion molecule-1 by reticular cells and sinusoidal endothelium in normal and γ-irradiated mice. Blood 1996; 87:73-82.

57. Papayannopoulou T, Craddock C, Nakamoto B, et al. The VLA4/VCAM-1 adhesion pathway defines contrasting mechanisms of lodgement of transplanted murine hemopoietic progenitors between bone marrow and spleen. Proc Natl Acad Sci USA 1995; 92:9647-9651.

58. Freedman AS, Munro JM, Rice GE, et al. Adhesion of human B cells to germinal centers in vitro involves VLA-4 and INCAM-110. Science 1990; 249:1030-1033.

59. Miyake K, Medina K, Ishihara K, et al. A VCAM-like adhesion molecule on murine bone marrow stromal cells mediates binding of lymphocyte precursors in culture. J Cell Biol 1991; 114:557-565.

60. Edwards JC. The nature and origins of synovium: experimental approaches to the study of synoviocyte differentiation. J Anat 1994; 184:493-501.

61. Rosen GD, Sanes JR, LaChance R, et al. Roles for the integrin VLA-4 and its counter receptor VCAM-1 in myogenesis. Cell 1992; 69:1107-1119.

62. Tanaka Y, Morimoto I, Nakano Y, et al. Osteoblasts are regulated by the cellular adhesion through ICAM-1 and VCAM-1. J Bone Miner Res 1995; 10:1462-1469.

63. Rice GE, Munro JM, Corless C, et al. Vascular and nonvascular expression of INCAM-110: A target for mononuclear leukocyte adhesion in normal and inflamed human tissues. Am J Pathol 1991; 138:385-393.

64. Birdsall HH, Lane C, Ramser MN, et al. Induction of VCAM-1 and ICAM-1 on human neural cells and mechanisms of mononuclear leukocyte adherence. J Immunol 1992; 148:2717-2723.

65. Sebire G, Hery C, Peudenier S, et al. Adhesion proteins on human microglial cells and modulation of their expression by IL1-α and TNF-α. Res Virol 1993; 144:47-52.

66. Jonjic N, Martin-Padura I, Pollicino T, et al. Regulated expression of vascular cell adhesion molecule-1 in human malignant melanoma. Am J Pathol 1992; 141:1323-1330.

67. Brockmeyer C, Ulbrecht M, Schendel DJ, et al. Distribution of cell adhesion molecules (ICAM-1, VCAM-1, ELAM-1) in renal tissue during allograft rejection. Transplantation 1993; 55:610-615.

68. Steffen BJ, Breier G, Butcher EC, et al. ICAM-1, VCAM-1, and MAdCAM-1 are expressed on choroid plexus epithelium but not endothelium and mediate binding of lymphocytes *in vitro*. Am J Pathol 1996; 148:1819-1838.

69. Osborn L, Hession C, Tizard R, et al. Direct cloning of vascular cell adhesion molecule 1 (VCAM-1), a cytokine-induced endothelial protein that binds to lymphocytes. Cell 1989; 59:1203-1211.

70. Li H, Cybusky MI, Gimbrone MA,Jr., et al. Inducible expression of vascular cell adhesion molecule-1 (VCAM-1) by vascular smooth muscle cells *in vitro* and within rabbit atheroma. Am J Pathol 1993; 143:1551-1559.

71. Sironi M, Sciacca FL, Matteucci C, et al. Regulation of endothelial and mesothelial cell function by interleukin-13: Selective induction of vascular cell adhesion molecule-1 and amplification of interleukin-6 production. Blood 1994; 84:1913-1921.

72. Palkama T, Majuri M-L, Mattila P, et al. Regulation of endothelial adhesion molecules by ligands binding to the scavenger receptor. Clin Exp Immunol 1993; 92:353-360.

73. Swerlick RA, Lee KH, Li L-J, et al. Regulation of vascular cell adhesion molecule 1 on human dermal microvascular endothelial cells. J Immunol 1992; 149:698-705.

74. Hauser IA, Johnson DR, Madri JA. Differential induction of VCAM-1 on human iliac venous and arterial endothelial cells and its role in adhesion. J Immunol 1993; 151:5172-5185.

75. Morisaki N, Kanzaki T, Tamura K, et al. Specific inhibition of vascular cell adhesion molecule-1 expression by type IV collagen in endothelial cells. Biochem Biophys Res Commun 1995; 214:1163-1167.

76. Walpola PL, Gotlieb AI, Cybulsky MI, et al. Expression of ICAM-1 and VCAM-1 and monocyte adherence in arteries exposed to altered shear stress. Arterioscler Thromb Vasc Biol 1995; 15:2-10.

77. Ando J, Tsuboi H, Korenaga R, et al. Down-regulation of vascular adhesion molecule-1 by fluid shear stress in cultured mouse endothelial cells. Ann N Y Acad Sci 1995; 748:148-156.

78. Damle NK, Eberhardt C, Van der Vieren M. Direct interaction with primed CD4[+] CD45R0[+] memory T lymphocytes induces expression of endothelial leukocyte adhesion molecule-1 and vascular cell adhesion molecule-1 on the surface of vascular endothelial cells. Eur J Immunol 1991; 21:2915-2923.

79. Bombara MP, Webb DL, Conrad P, et al. Cell contact between T cells and synovial fibroblasts causes induction of adhesion molecules and cytokines. J Leukoc Biol 1993; 54:399-406.

80. Yellin MJ, Winikoff S, Fortune SM, et al. Ligation of CD40 on fibroblasts induces CD54 (ICAM-1) and CD106 (VCAM-1) up-regulation and IL-6 production and proliferation. J Leukoc Biol 1995; 58:209-216.

81. Streeter PR, Lakey-Berg E, Rouse BTN, et al. A tissue-specific endothelial cell molecule involved in lymphocyte homing. Nature 1988; 331:41-46.

82. Streeter PR, Rouse BTN, Butcher EC. Immunohistologic and functional characterization of a vascular addressin involved in lymphocyte homing into peripheral lymph nodes. J Cell Biol 1988; 107:1853-1862.

83. Kraal G, Schornagel K, Streeter PR, et al. Expression of the mucosal vascular addressin, MAdCAM-1, on sinus-lining cells in the spleen. Am J Pathol 1995; 147:763-771.

84. Neumann B, Machleidt T, Lifka A, et al. Crucial role of 55-kilodalton TNF receptor in TNF-induced adhesion molecule expression and leukocyte organ infiltration. J Immunol 1996; 156:1587-1593.

85. Eugster HP, Muller M, Karrer U, et al. Multiple immune abnormalities in tumor necrosis factor and lymphotoxin-α-double-deficient mice. Int Immunol 1996; 8:23-36.

86. Kratz A, Campos-Neto A, Hanson MS, et al. Chronic inflammation caused by lymphotoxin is lymphoid neogenesis. J Exp Med 1996; 183:1461-1472.

87. Mazo IB, Guttierez-Ramos JC, Frenette P, Wagner DD, Hynes RO, von Andrian UH. Hematopoietic progenitor cell rolling in bone marrow microvessels: parallel contributions by endothelial selectins and VCAM-1. J Exp Med 1998; 188:465-474.

88. Berlin C, Bargatze RF, von Andrian UH, et al. α4 integrins mediate lymphocyte attachment and rolling under physiologic flow. Cell 1995; 80:413-422.

89. Alon R, Kassner PD, Carr MW, et al. The integrin VLA-4 supports tethering and rolling in flow on VCAM-1. J Cell Biol 1995; 128:1243-1253.

90. Lawrence MB, Berg EL, Butcher EC, et al. Rolling of lymphocytes and neutrophils on peripheral node addressin and subsequent arrest on ICAM-1 in shear flow. Eur J Immunol 1995; 25:1025-1031.

91. Roth SJ, Carr MW, Rose SS, et al. Characterization of transendothelial chemotaxis of T lymphocytes. J Immunol Methods 1995; 100:97-116.

92. Muller WA, Weigl SA, Deng X, et al. PECAM-1 is required for transendothelial migration of leukocytes. J Exp Med 1993; 178:449-460.

93. Diamond MS, Springer TA. The dynamic regulation of integrin adhesiveness. Curr Biol 1994; 4:506-517.

94. Piali L, Hammel P, Uherek C, et al. CD31/PECAM-1 is a ligand for $\alpha_v\beta_3$ integrin involved in adhesion of leukocytes to endothelium. J Cell Biol 1995; 130:451-460.

95. Huang C, Springer TA. A binding interface on the I domain of lymphocyte function associated antigen-1 (LFA-1) required for specific interaction with intercellular adhesion molecule 1 (ICAM-1). J Biol Chem 1995; 270:19008-19016.

96. D'Souza SE, Haas TA, Piotrowicz RS, et al. Ligand and cation binding are dual functions of a discrete segment of the integrin β_3 subunit: Cation displacement is involved in ligand binding. Cell 1994; 79:659-667.

97. Simmons DL. The role of ICAM expression in immunity and disease. Cancer Surv 1995; 24:141-155.

98. Mould AP. Getting integrins into shape: recent insights into how integrin activity is regulated by conformational changes. J Cell Sci 1996; 109:2613-2618.

99. Rothlein R, Dustin ML, Marlin SD, et al. A human intercellular adhesion molecule (ICAM-1) distinct from LFA-1. J Immunol 1986; 137:1270-1274.

100. Diamond MS, Staunton DE, Marlin SD, et al. Binding of the integrin Mac-1 (CD11b/CD18) to the third Ig-like domain of ICAM-1 (CD54) and its regulation by glycosylation. Cell 1991; 65:961-971.

101. de Fougerolles AR, Diamond MS, Springer TA. Heterogenous glycosylation of ICAM-3 and lack of interaction with Mac-1 and p150,95. Eur J Immunol 1995; 25:1008-1012.

102. Languino LR, Plescia J, Duperray A, et al. Fibrinogen mediates leukocyte adhesion to vascular endothelium through an ICAM-1-dependent pathway. Cell 1993; 73:1423-1434.

103. D'Souza SE, Byers-Ward VJ, Gardiner EE, et al. Identification of an active sequence within the first immunoglobulin domain of intercellular cell adhesion molecule-1 (ICAM-1) that interacts with fibrinogen. J Biol Chem 1996; 271:24270-24277.

104.   Sriramarao P, Languino LR, Altieri DC. Fibrinogen mediates leukocyte-endothelium bridging *in vivo* at low shear forces. Blood 1996; 88:3416-3423.
105.   Ugarova TP, Budzynski AZ, Shattil SJ, et al. Conformational changes in fibrinogen elicited by its interaction with platelet membrane glycoprotein GPIIb-IIIa. J Biol Chem 1993; 268:21080-21087.
106.   Kirchhausen T, Staunton DE, Springer TA. Location of the domains of ICAM-1 by immunolabeling and single-molecule electron microscopy. J Leukocyte Biol 1993; 53:342-346.
107.   Staunton DE, Dustin ML, Erickson HP, et al. The arrangement of the immunoglobulin-like domains of ICAM-1 and the binding sites for LFA-1 and rhinovirus. Cell 1990; 61:243-254.
108.   Ockenhouse CF, Betageri R, Springer TA, et al. Plasmodium falciparum-infected erythrocytes bind ICAM-1 at a site distinct from LFA-1, Mac-1, and human rhinovirus. Cell 1992; 68:63-69.
109.   McCourt PAG, Ek B, Forsberg N, et al. Intercellular adhesion molecule-1 is a cell surface receptor for hyaluronan. J Biol Chem 1996; 269:30081-30084.
110.   Rosenstein Y, Park JK, Hahn WC, et al. CD43, a molecule defective in Wiskott-Aldrich syndrome, binds ICAM-1. Nature 1991; 354:233-235.
111.   de Fougerolles AR, Qin X, Springer TA. Characterization of the function of ICAM-3 and comparison to ICAM-1 and ICAM-2 in immune responses. J Exp Med 1994; 179:619-629.
112a.   Manjunath N, Correa M, Ardman M, et al. Negative regulation of T cell adhesion and activation by CD43. Nature 1995; 377:535-538.
112b.   Stockton BM, Ardman B, Cheng G, Manjunath N, von Andrian UH. Negative regulation of T cell homing by CD43. Immunity 1998; 8:373-381.
113.   Argenbright LW, Letts LG, Rothlein R. Monoclonal antibodies to the leukocyte membrane CD18 glycoprotein complex and to intercellular adhesion molecule-1 inhibit leukocyte-endothelial adhesion in rabbits. J Leukoc Biol 1991; 49:253-257.
114.   Jones DA, McIntire LV, Smith CW, et al. A two-step adhesion cascade for T cell/endothelial cell interactions under flow conditions. J Clin Invest 1994; 94:2443-2450.
115.   Anderson DC, Springer TA. Leukocyte adhesion deficiency: An inherited defect in the Mac-1, LFA-1, and p150,95 glycoproteins. Ann Rev Med 1987; 38:175-194.
116.   Kunkel EJ, Jung U, Bullard DC, et al. Absence of trauma-induced leukocyte rolling in mice deficient in both P-selectin and intercellular adhesion molecule 1. J Exp Med 1996; 183:57-65.
117.   Bargatze RF, Kurk S, Butcher EC, et al. Neutrophils roll on adherent neutrophils bound to cytokine-induced endothelial cells via L-selectin on the rolling cells. J Exp Med 1994; 180:1785-1792.
118.   Casasnovas JM, Springer TA, Liu J-h, et al. The crystal structure of ICAM-2 reveals presentation of a distinctive integrin recognition surface. Nature 1997;387:312-315.
119.   Xie J, Li R, Kotovuori P, et al. Intercellular adhesion molecule-2 (CD102) binds to the leukocyte integrin CD11b/CD18 through the A domain. J Immunol 1995; 155:3619-3628.
120.   Li R, Xie J, Kantor C, et al. A peptide derived from the intercellular adhesion molecule-2 regulates the avidity of the leukocyte integrins CD11b/CD18 and CD11c/CD18. J Cell Biol 1995; 129:1143-1153.
121.   Diacovo TG, Roth SJ, Buccola JM, et al. Neutrophil rolling, arrest, and transmigration across activated, surface-adherent platelets via sequential

98

action of P-selectin and the β_2-integrin CD11b/CD18. Blood 1996; 88:146-157.

122. Xie Y, Muller WA. Molecular cloning and adhesive properties of murine platelet/endothelial cell adhesion molecule 1. Proc Natl Acad Sci USA 1993; 90:5569-5573.

123. Simmons DL, Walker C, Power C, et al. Molecular cloning of CD31, a putative intercellular adhesion molecule closely related to carcinoembryonic antigen. J Exp Med 1990; 171:2147-2152.

124. Goldberger A, Middleton KA, Oliver JA, et al. Biosynthesis and processing of the cell adhesion molecule PECAM-1 includes production of a soluble form. J Biol Chem 1994; 269:17183-17191.

125. Ohto H, Maeda H, Shibata Y, et al. A novel leukocyte differentiation antigen: Two monoclonal antibodies TM2 and TM3 define a 120-kd molecule present on neutrophils, monocytes, platelets,and activated lymphoblasts. Blood 1985; 66:873-881.

126. Muller WA, Berman ME, Newman PJ, et al. A heterophilic adhesion mechanism for platelet/endothelial cell adhesion molecule-1 (CD31). J Exp Med 1992; 175:1401-1404.

127. Liao F, Huynh HK, Eiroa A, et al. Migration of monocytes across endothelium and passage through extracellular matrix involve separate molecular domains of PECAM-1. J Exp Med 1995; 182:1337-1343.

128. Sun J, Williams J, Yan HC, et al. Platelet endothelial cell adhesion molecule-1 (PECAM-1) hemophilic adhesion is mediated by immunoglobulin-like domains 1 and 2 and depends on the cytoplasmic domain and the level of surface expression. J Biol Chem 1996; 271:18561-18570.

129. Sun QH, DeLisser HM, Zukowski MM, et al. Individually distinct Ig homology domains in PECAM-1 regulate hemophilic binding and modulate receptor affinity. J Biol Chem 1996; 271:11090-11098.

130. Zehnder JL, Shatsky M, Leung LL, et al. Involvement of CD31 in lymphocyte-mediated immune responses: importance of the membrane-proximal immunoglobulin domain and identification of an inhibiting CD31 peptide. Blood 1995; 85:1282-1288.

131. DeLisser HM, Yan HC, Newman PJ, et al. Platelet/endothelial cell adhesion molecule-1 (CD31)-mediated cellular aggregation involves cell surface glycosaminoglycans. J Biol Chem 1991; 268:16037-16046.

132. Buckley CD, Doyonnas R, Newton JP, et al. Identification of $\alpha(V)\beta(3)$ as a heterotypic ligand for CD31/PECAM-1. J Cell Sci 1996; 109(Part2):437-445.

133. Prager E, Sunderplassmann R, Hansmann C, et al. Interaction of CD31 with a heterophilic counterreceptor involved in downregulation of human T cell responses. J Exp Med 1996; 184:41-50.

134. Yan HC, Baldwin HS, Sun J, et al. Alternative splicing of a specific cytoplasmic exon alters the binding characteristics of murine platelet/endothelial cell adhesion molecule-1 (PECAM-1). J Biol Chem 1995; 270:23672-23680.

135. Baldwin HS, Shen HM, Yan HC, et al. Platelet endothelial cell adhesion molecule-1 (PECAM-1/CD31): alternatively spliced, functionally distinct isoforms expressed during mammalian cardiovascular development. Devel 1994; 120:2539-2553.

136. DeLisser HM, Chiltowsky J, Yan H-C, et al. Deletions in the cytoplasmic domain of platelet-endothelial cell adhesion molecule-1 (PECAM-1,CD31) result in changes in ligand binding properties. J Cell Biol 1994; 124:195-203.

137. Vaporciyan AA, DeLisser HM, Yan H-C, et al. Involvement of platelet-endothelial cell adhesion molecule-1 in neutrophil recruitment *in vivo*. Science 1993; 262:1580-1582.

138. Muller WA, Weigl SA. Monocyte-selective transendothelial migration: dissection of the binding and transmigration phases by an *in vitro* assay. J Exp Med 1992; 176:819-828.

139. Bogen S, Pak J, Garifallou M, et al. Monoclonal antibody to murine PECAM-1 (CD31) blocks acute inflammation *in vivo*. J Exp Med 1994; 179:1059-1064.

140. Wakelin MW, Sanz M-J, Dewar A, et al. An anti-platelet-endothelial cell adhesion molecule-1 antibody inhibits leukocyte extravasation from mesenteric microvessels *in vivo* by blocking the passage through the basement membrane. J Exp Med 1996; 184:229-239.

141. Ayalon O, Sabanai H, Lampugnani MG, et al. Spatial and temporal relationships between cadherins and PECAM-1 in cell-cell junctions of human endothelial cells. J Cell Biol 1994; 126:247-258.

142. Berman ME, Muller WA. Ligation of platelet/endothelial cell adhesion molecule 1 (PECAM-1/CD31) on monocytes and neutrophils increases binding capacity of leukocyte CR3 (CD11b/CD18). J Immunol 1995; 154:299-307.

143. Leavesley DI, Oliver JM, Swart BW, et al. Signals from platelet/endothelial cell adhesion molecule enhance the adhesive activity of the very late antigen-4 integrin of human CD34+ hemopoietic progenitor cells. J Immunol 1994; 153:4673-4683.

144. Cybulsky MI, Fries JWU, Williams AJ, et al. Alternative splicing of human VCAM-1 in activated vascular endothelium. Am J Pathol 1991; 138:815-820.

145. Hession C, Tizard R, Vassallo C, et al. Cloning of an alternate form of vascular cell adhesion molecule-1 (VCAM1). J Biol Chem 1991; 266:6682-6685.

146. Moy P, Lobb R, Tizard R, et al. Cloning of an inflammation-specific phosphatidyl inositol-linked form of murine vascular cell adhesion molecule-1. J Biol Chem 1993; 268:8835-8841.

147. Osborn L, Vassallo C, Browning BG, et al. Arrangement of domains, and amino acid residues required for binding of vascular cell adhesion molecule-1 to its counter-receptor VLA-4 ($\alpha 4\beta 1$). J Cell Biol 1994; 124:601-608.

148. Elices MJ, Osborn L, Takada Y, et al. VCAM-1 on activated endothelium interacts with the leukocyte integrin VLA-4 at a site distinct from the VLA-4/fibronectin binding site. Cell 1990; 60:577-584.

149. Vonderheide RH, Springer TA. Lymphocyte adhesion through VLA-4: Evidence for a novel binding site in the alternatively spliced domain of VCAM-1 and an additional $\alpha 4$ integrin counter-receptor on stimulated endothelium. J Exp Med 1992; 175:1433-1442.

150. Osborn L, Vassallo C, Benjamin CD. Activated endothelium binds lymphocytes through a novel binding site in the alternately spliced domain of vascular cell adhesion molecule-1. J Exp Med 1992; 176:99-107.

151. Chan BM, Elices MJ, Murphy E, et al. Adhesion to vascular cell adhesion molecule 1 and fibronectin: Comparison of $\alpha 4\beta 1$ (VLA-4) and $\alpha 4$ $\beta 7$ on the human B cell line JY. J Biol Chem 1992; 267:8366-8370.

152. Ruegg C, Postigo AA, Sikorski EE, et al. Role of integrin $\alpha 4\beta 7/\alpha 4\beta$ P in lymphocyte adherence to fibronectin and VCAM-1 and in homotypic cell clustering. J Cell Biol 1992; 117:179-189.

153. Chiu HH, Crowe DT, Renz ME, et al. Similar but nonidentical amino acid residues on vascular cell adhesion molecule-1 are involved in the interaction with $\alpha4\beta1$ and $\alpha4\beta7$ under different activity states. J Immunol 1995; 155:5257-5267.

154. Jones EY, Harlos K, Bottomley MJ, et al. Crystal structure of an integrin-binding fragment of vascular cell adhesion molecule-1 at 1.8 Å resolution. Nature 1995; 373:539-544.

155. Wang J-h, Pepinsky RB, Stehle T, et al. The crystal structure of an N-terminal two-domain fragment of vascular cell adhesion molecule 1 (VCAM-1): A cyclic peptide based based on the domain 1 C-D loop can inhibit VCAM-1-$\alpha4$ integrin interaction . Proc Natl Acad Sci USA 1995; 92:5714-5718.

156. Rice GE, Munro JM, Bevilacqua MP. Inducible cell adhesion molecule 110 (INCAM-110) is an endothelial receptor for lymphocytes: A CD11/CD18-independent adhesion mechanism. J Exp Med 1990; 171:1369-1374.

157. Sriramarao P, von Andrian UH, Butcher EC, et al. L-selectin and very late antigen-4 integrin promote eosinophil rolling at physiological shear rates in vivo. J Immunol 1994; 153:4238-4246.

158. Luscinskas FW, Ding H, Lichtman AH. P-selectin and vascular cell adhesion molecule 1 mediate rolling and arrest, respectively, of CD4$^+$ T lymphocytes on tumor necrosis factor α-activated vascular endothelium under flow. J Exp Med 1995; 181:1179-1186.

159. Luscinskas FW, Ding H, Tan P, et al. L- and P-selectins, but not CD49d (VLA-4) integrins, mediate monocyte initial attachment to TNF-α-activated vascular endothelium under flow in vitro. J Immunol 1996; 156:326-335.

160. Chuluyan EH, Issekutz AC. VLA-4 integrin can mediate CD11/CD18 independent transendothelial migration of human monocytes. J Clin Invest 1993; 92:2768-2777.

161. Chuluyan HE, Osborn L, Lobb R, et al. Domains 1 and 4 of vascular cell adhesion molecule-1 (CD106) both support very late activation antigen-4 (CD49d/CD29)-dependent monocyte transendothelial migration. J Immunol 1995; 155:3135-3144.

162. Sano H, Nakagawa N, Nakajima H, et al. Role of vascular cell adhesion molecule-1 and platelet-activating factor in selective eosinophil migration across vascular endothelial cells. Int Arch Allergy Immunol 1995; 107:533-540.

163. Kavanaugh AF, Lightfoot E, Lipsky PE, et al. Role of CD11/CD18 in adhesion and transendothelial migration of T cells: Analysis utilizing CD18-deficient T cell clones. J Immunol 1991; 146:4149-4156.

164. Chan P-Y, Aruffo A. VLA-4 integrin mediates lymphocyte migration on the inducible endothlial cell ligand VCAM-1 and the extracellular matrix ligand fibronectin. J Biol Chem 1993; 268:24655-24664.

165. Romanic AM, Madri JA. The induction of 72-kD gelatinase in T cells upon adhesion to endothelial cells is VCAM-1 dependent. J Cell Biol 1994; 125:1165-1178.

166. Berlin C, Berg EL, Briskin MJ, et al. $\alpha4$ $\beta7$ integrin mediates lymphocyte binding to the mucosal vascular addressin MAdCAM-1. Cell 1993; 74:185-195.

167. Rott LS, Briskin MJ, Andrew DP, et al. A fundamental subdivision of circulating lymphocytes defined by adhesion to mucosal addressin cell adhesion molecule-1. Comparison with vascular cell adhesion molecule-1 and correlation with $\beta7$ integrins and memory differentiation. J Immunol 1996; 156:3727-3736.

168. Briskin MJ, Rott L, Butcher EC. Structural requirements for mucosal vascular addressin binding to its lymphocyte receptor α4β7. Common themes among integrin-Ig family interactions. J Immunol 1996; 156:719-726.

169. Viney JL, Jones S, Chiu HH, et al. Mucosal addressin cell adhesion molecule-1. A structural and functional analysis demarcates the integrin binding motif. J Immunol 1996; 157:2488-2497.

170. Berg EL, McEvoy LM, Berlin C, et al. L-selectin-mediated lymphocyte rolling on MAdCAM-1. Nature 1993; 366:695-698.

171. Shyjan AM, Bertagnolli M, Kenney CJ, et al. Human mucosal addressin cell adhesion molecule-1 (MAdCAM-1) demonstrates structural and functional similarities to the α4β7-integrin binding domains of murine MAdCAM-1, but extreme divergence of mucin-like sequences. J Immunol 1996; 156:2851-2857.

172. Bargatze RF, Jutila MA, Butcher EC. Distinct roles of L-selectin and integrins α4β7 and LFA-1 in lymphocyte homing to Peyer's patch-HEV in situ: The multistep model confirmed and refined. Immunity 1995; 3:99-108.

173. Gowans JL, Knight EJ. The route of re-circulation of lymphocytes in the rat. Proc R Soc Lond B 1964; 159:257-282.

174. Pabst R. The spleen in lymphocyte migration. Immunol Today 1988; 9:43-45.

175. Mackay C. Homing of naive, memory and effector lymphocytes. Current Opinion in Immunology 1993; 5:423-427.

176. Marchesi VT, Gowans JL. The migration of lymphocytes through the endothelium of venules in lymph nodes: an electron microscope study. Proc R Soc Lond B 1963; 159:283-290.

177. Girard J-P, Springer TA. High endothelial venules (HEVs): Specialized endothelium for lymphocyte migration. Immunol Today 1995; 16:449-457.

178. Springer TA. Traffic signals on endothelium for lymphocyte recirculation and leukocyte emigration. Annu Rev Physiol 1995; 57:827-872.

179. Etzioni A, Frydman M, Pollack S, et al. Recurrent severe infections caused by a novel leukocyte adhesion deficiency. N Engl J Med 1992; 327:1789-1792.

180a. Hamann A, Westrich DJ, Duijevstijn A, et al. Evidence for an accessory role of LFA-1 in lymphocyte-high endothelium interaction during homing. J Immunol 1988; 140:693-699.

180b. Warnock RA, Askari S, Butcher EC, von Andrian UH. Molecular mechanisms of lymphocyte homing to peripheral lymph nodes. J Exp Med 1998; 187:205-216.

181. Rothlein R, Jaeger JR. Treatment of inflammatory diseases with a monoclonal antibody to intercellular adhesion molecule 1. Ciba Found Symp 1995; 189:200-211.

182. Huleatt JW, Lefrancois L. β2 integrins and ICAM-1 are involved in establishment of the intestinal mucosal T cell compartment. Immunity 1996; 5:263-273.

183. Xu H, Gonzalo JA, St.Pierre Y, et al. Leukocytosis and resistance to septic shock in intercellular adhesion molecule 1-deficient mice. J Exp Med 1994; 180:95-109.

184. Sligh Jr. JE, Ballantyne CM, Rich S, et al. Inflammatory and immune responses are impaired in mice deficient in intercellular adhesion molecule 1. Proc Natl Acad Sci USA 1993; 90:8529-8533.

185. Coxon A, Barkalow FJ, Askari S, et al. A novel role for the β2 integrin CD11b/CD18 in neutrophil apoptosis: A homeostatic mechanism in inflammation. Immunity 1996; 5:653-666.

102

186. Van Seventer GA, Shimizu Y, Horgan KJ, et al. The LFA-1 ligand ICAM-1 provides an important costimulatory signal for T cell receptor-mediated activation of resting T cells. J Immunol 1990; 144:4579-4586.
187. Tohma S, Hirohata S, Lipsky PE. The role of CD11a/CD18-CD54 interactions in human T cell-dependent B cell activation. J Immunol 1991; 146:492-499.
188. Kuhlman P, Moy VT, Lollo BA, et al. The accessory function of murine intercellular adhesion molecule-1 in T lymphocyte activation. J Immunol 1991; 146:1773-1782.
189. Damle NK, Klussman K, Aruffo A. Intercellular adhesion molecule-2, a second counter-receptor for CD11a/CD18 (leukocyte function-associated antigen-1), provides a costimulatory signal for T-cell receptor-initiated activation of human T cells. J Immunol 1992; 148:665-671.
190. Tchilian EZ, Anderson G, Moore NC, et al. Involvement of LFA-1/ICAM-2 adhesive interactions and PKC in activation-induced cell death following SEB rechallenge. Immunology 1996; 87:566-572.
191. Bird IN, Spragg JH, Ager A, et al. Studies of lymphocyte transendothelial migration: Analysis of migrated cell phenotypes with regard to CD31 (PECAM-1), CD45RA and CD45RO. Immunol 1993; 80:553-560.
192. Zocchi MR, Ferrero E, Leone BE, et al. CD31/PECAM-1-driven chemokine-independent transmigration of human T lymphocytes. Eur J Immunol 1996; 26:759-767.
193. Tanaka Y, Albelda SM, Horgan KJ, et al. CD31 expressed on distinctive T cell subsets is a preferential amplifier of β1 integrin-mediated adhesion. J Exp Med 1992; 176:245-253.
194. Elmarsafy S, Carosella ED, Agrawal SG, et al. Functional role of PECAM-1/CD31 molecule expressed on human cord blood progenitors. Leukemia 1996; 10:1340-1346.
195. Sadahira Y, Yoshino T, Monobe Y. Very late activation antigen 4-vascular cell adhesion molecule 1 interaction is involved in the formation of erythroblastic islands. J Exp Med 1995; 181:411-415.
196. Friedrich C, Cybulsky MI, Gutierrez-Ramos J-C. Vascular cell adhesion molecule-1 expression by hematopoiesis-supporting stromal cells is not essential for lymphoid or myeloid differentiation *in vivo* or *in vitro*. Eur J Immunol 1996; 26:2773-2780.
197. Koopman G, Keehnen RMJ, Lindhout E, et al. Adhesion through the LFA-1 (CD11a/CD18)-ICAM-1 (CD54) and the VLA-4 (CD49d)-VCAM-1 (CD106) pathways prevents apoptosis of germinal center B cells. J Immunol 1994; 152:3760-3767.
198. Damle NK, Aruffo A. Vascular cell adhesion molecule 1 induces T-cell antigen receptor-dependent activation of CD4+ T lymphocytes. Proc Natl Acad Sci USA 1991; 88:6403-6407.
199. Van Seventer GA, Newman W, Shimizu Y, et al. Analysis of T cell stimulation by superantigen plus major histocompatibility complex class II molecules or by CD3 monoclonal antibody: Costimulation by purified adhesion ligands VCAM-1, ICAM-1, but not ELAM-1. J Exp Med 1991; 174:901-913.
200. Damle NK, Klussman K, Linsley PS, et al. Differential costimulatory effects of adhesion molecules B7, ICAM-1, LFA-3, and VCAM-1 on resting and antigen-primed CD4+ T lymphocytes. J Immunol 1992; 148:1985-1992.
201. Maguire JE, Danahey KM, Burkly LC, et al. T cell receptor- and β 1 integrin-mediated signals synergize to induce tyrosine phosphorylation of focal adhesion kinase (pp125FAK) in human T cells. J Exp Med 1995; 182:2079-2090.

202. Freedman AS, Rhynhart K, Nojima Y, et al. Stimulation of protein tyrosine phosphorylation in human B cells after ligation of the β 1 integrin VLA-4. J Immunol 1993; 150:1645-1652.

203. Nagata M, Sedgwick JB, Bates ME, et al. Eosinophil adhesion to vascular cell adhesion molecule-1 activates superoxide anion generation. J Immunol 1995; 155:2194-2202.

204. Salomon DR, Mojcik CF, Chang AC, et al. Constitutive activation of integrin α4β1 defines a unique stage of human thymocyte development. J Exp Med 1994; 179:1573-1584.

205. Bell EB, Sparshott SM, Ager A. Migration pathways of CD4 T cell subsets *in vivo*: the CD45RC- subset enters the thymus via α 4 integrin-VCAM-1 interaction. Int Immunol 1995; 7:1861-1871.

206. Hamann A, Andrew DP, Jablonski-Westrich D, et al. Role of α4-integrins in lymphocyte homing to mucosal tissues *in vivo*. J Immunol 1994; 152:3282-3293.

207. Mackay CR, Marston W, Dudler L. Altered patterns of T cell migration through lymph nodes and skin following antigen challenge. Eur J Immunol 1992; 22:2205-2210.

208. Tanaka H, Saito S, Sasaki H, et al. Morphological aspects of LFA-1/ICAM-1 and VLA4/VCAM-1 adhesion pathways in human lymph nodes. Pathol Int 1994; 44:268-279.

209. Kraal G, Schornagel K, Savelkoul H, et al. Activation of high endothelial venules in peripheral lymph nodes. The involvement of interferon-gamma. Int Immunol 1994; 6:1195-1201.

210. Mebius RE, Streeter PR, Michie S, et al. A developmental switch in lymphocyte homing receptor and endothelial vascular addressin expression regulates lymphocyte homing and permits CD4+CD3- cells to colonize lymph nodes. Proc Natl Acad Sci USA 1996; 93:11019-11024.

211. May MJ, Entwistle G, Humphries MJ, et al. VCAM-1 is a CS1 peptide-inhibitable adhesion molecule expressed by lymph node high endothelium. J Cell Sci 1993; 106:109-119.

212. Ruco LP, Pomponi D, Pigott R, et al. Expression and cell distribution of the intercellular adhesion molecule, vascular cell adhesion molecule, endothelial leukocyte adhesion molecule and endothelial cell adhesion molecule (CD31) in reactive human lymph nodes and in Hodgkin's disease. Am J Pathol 1992; 140:1337-1344.

213. Nakache M, Berg EL, Streeter PR, et al. The mucosal vascular addressin is a tissue-specific endothelial cell adhesion molecule for circulating lymphocytes. Nature 1989; 337:179-181.

214. Wagner N, Lohler J, Kunkel EJ, et al. Critical role for β7 integrins in formation of the gut-associated lymphoid tissue. Nature 1996; 382:366-370.

215. Teague TK, Lazarovits AI, McIntyre BW. Integrin α 4 β 7 co-stimulation of human peripheral blood T cell proliferation. Cell Adhes Commun 1994; 2:539-547.

216. Panes J, Perry MA, Anderson DC, et al. Regional differences in constitutive and induced ICAM-1 expression *in vivo*. Am J Physiol 1995; 269(6 pt 2):H1955-H1964.

217. King PD, Sandberg ET, Selvakumar A, et al. Novel isoforms of murine intercellular adhesion molecule-1 generated by alternative RNA splicing. J Immunol 1995; 154:6080-6093.

218. Helander TS, Carpen O, Turunen O, et al. ICAM-2 redistributed by ezrin as a target for killer cells. Nature 1996; 382:265-268.

219.   Li R, Nortamo P, Valmu L, et al. A peptide from ICAM-2 binds to the leukocyte integrin CD11a/CD18 and inhibits endothelial cell adhesion. J Biol Chem 1993; 268:17513-17518.

220.   Rosenblum WI, Murata S, Nelson GH, et al. Anti-CD31 delays platelet adhesion/aggregation at sites of endothelial injury in mouse cerebral arterioles. Am J Pathol 1994; 145:33-36.

221.   Murohara T, Delyani JA, Albelda SM, et al. Blockade of platelet endothelial cell adhesion molecule-1 protects against myocardial ischemia and reperfusion in cats. J Immunol 1996; 156:3550-3557.

222.   Nakajima H, Sano H, Nishimura T, et al. Role of vascular cell adhesion molecule 1/very late activation antigen 4 and intercellular adhesion molecule 1/lymphoycte function -associated antigen 1 interactions in antigen-induced eosinophil and T cell recruitment into the tissue. J Exp Med 1994; 179:1145-1154.

223.   Silber A, Newman W, Sasseville VG, et al. Recruitment of lymphocytes during cutaneous delayed hypersensitivity in nonhuman primates is dependent on E-selectin and VCAM-1. J Clin Invest 1994; 93:1554-1563.

224.   Baron JL, Madri JA, Ruddle NH, et al. Surface expression of α4 integrin by CD4 T cells is required for their entry into brain parenchyma. J Exp Med 1993; 177:57-68.

225.   Steffen BJ, Butcher EC, Engelhardt B. Evidence for involvement of ICAM-1 and VCAM-1 in lymphocyte interaction with endothelium in experimental autoimmune encephalomyelitis in the central nervous system in the SJL/J mouse. Am J Pathol 1994; 145:189-201.

226.   Jakubowski A, Rosa MD, Bixler S, et al. Vascular cell adhesion molecule (VCAM)-Ig fusion protein defines distinct affinity states of the very late antigen-4 (VLA-4) receptor. Cell Adhes Commun 1995; 3:131-142.

227.   Schlegel PG, Vaysburd M, Chen Y, et al. Inhibition of T cell costimulation by VCAM-1 prevents murine graft-versus-host disease across minor histocompatibility barriers. J Immunol 1995; 155:3856-3865.

228.   Ferran C, Peuchmaur M, Desruennes M, et al. Implications of de novo ELAM-1 and VCAM-1 expression in human cardiac allograft rejection. Transplantation 1993; 55:605-609.

229.   Orosz CG, Van Buskirk A, Sedmak DD, et al. Role of the endothelial adhesion molecule VCAM in murine cardiac allograft rejection. Immunol Lett 1992; 32:7-12.

230.   Cybulsky MI, Gimbrone MA,Jr.. Endothelial expression of a mononuclear leukocyte adhesion molecule during atherogenesis. Science 1991; 251:788-791.

231.   Khan BV, Parthasarathy SS, Alexander RW, et al. Modified low density lipoprotein and its constituents augment cytokine-activated vascular cell adhesion molecule-1 gene expression in human vascular endothelial cells. J Clin Invest 1995; 95:1262-1270.

232.   La Riviere G, Gebbinck JWTMK, Driessens MHE, et al. Pertussis toxin inhibition of T-cell hybridoma invasion is reversed by manganese-induced activation of LFA-1. journal of Cell Science 1994; 107:551-559.

233.   Vlassara H, Fuh H, Donnelly T, et al. Advanced glycation endproducts promote adhesion molecule (VCAM-1, ICAM-1) expression and atheroma formation in normal rabbits. Molec Med 1995; 1:447-456.

234.   O'Brien KD, Allen MD, McDonald TO, et al. Vascular cell adhesion molecule-1 is expressed in human coronary atherosclerotic plaques. Implications for the mode of progression of advanced coronary atherosclerosis. J Clin Invest 1993; 92:945-951.

235. van Dinther-Janssen ACHM, Horst E, Koopman G, et al. The VLA-4/VCAM-1 pathway is involved in lymphocyte adhesion to endothelium in rheumatoid synovium. J Immunol 1991; 147:4207-4210.

236. Postigo AA, Garcia-Vicuña R, Diaz-Gonzalez F, et al. Increased binding of synovial T Lymphocytes from Rheumatoid Arthritis to Endothelial-Leukocyte Adhesion Molecule-1 (ELAM-1) and Vascular Cell Adhesion Molecule-1 (VCAM-1). J Clin Invest 1992; 89:1445-1452.

237. Morales-Ducret J, Wayner E, Elices MJ, et al. Alpha 4/beta 1 integrin (VLA-4) ligands in arthritis. Vascular cell adhesion molecule-1 expression in synovium and on fibroblast-like synoviocytes. J Immunol 1992; 149:1424-1431.

238. Belmont HM, Buyon J, Giorno R, et al. Up-regulation of endothelial cell adhesion molecules characterizes disease activity in systemic lupus erythematosus. The Shwartzman phenomenon revisited. Arthritis Rheum 1994; 37:376-383.

239. Rastaldi MP, Ferrario F, Tunesi S, et al. Intraglomerular and interstitial leukocyte infiltration, adhesion molecules, and interleukin-1 α expression in 15 cases of antineutrophil cytoplasmic autoantibody-associated renal vasculitis. Am J Kidney Dis 1996; 27:48-57.

240. Mayet WJ, Schwarting A, Orth T, et al. Antibodies to proteinase 3 mediate expression of vascular cell adhesion molecule-1 (VCAM). Clin Exp Immunol 1996; 103:259-267.

241. Kwee L, Baldwin HS, Shen HM, et al. Defective development of the embryonic and extraembryonic circulatory systems in vascular cell adhesion molecule (VCAM-1) deficient mice. Devel 1995; 121:489-503.

242. Gurtner GC, Davis V, Li H, et al. Targeted disruption of the murine VCAM1 gene: essential role of VCAM-1 in chorioallantoic fusion and placentation. Genes Dev 1995; 9:1-14.

243. Li H, Iiyama K, DiChiara M, et al. A hypomorphic VCAM-1 mutation rescues the embryonic lethal null phenotype and reveals a requirement for VCAM-1 in inflammation. FASEB J 1996; 10:A1281.

244. Isobe M, Suzuki J, Yagita H, et al. Immunosuppression to cardiac allografts and soluble antigens by anti-vascular cellular adhesion molecule-1 and anti-very late antigen-4 monoclonal antibodies. J Immunol 1994; 153:5810-5818.

245. Stegall MD, Ostrowska A, Haynes J, et al. Prolongation of islet allograft survival with an antibody to vascular cell adhesion molecule 1. Surgery 1995; 118:369-370.

246. Gorcyznski RM, Chung S, Fu XM, et al. Manipulation of skin graft rejection in alloimmune mice by anti-VCAM-1:VLA-4 but not anti-ICAM-1:LFA-1 monoclonal antibodies. Transpl Immunol 1995; 3:55-61.

247. Tsukamoto K, Yokono K, Amano K, et al. Administration of monoclonal antibodies against vascular cell adhesion molecule-1/very late antigen-4 abrogates predisposing autoimmune diabetes in NOD mice. Cell Immunol 1995; 165:193-201.

248. Yednock TA, Cannon C, Fritz LC, et al. Prevention of experimental autoimmune encephalomyelitis by antibodies against $\alpha 4 \beta 1$ integrin. Nature 1992; 356:63-66.

249. Irani DN, Griffin DE. Regulation of lymphocyte homing into the brain during viral encephalitis at various stages of infection. J Immunol 1996; 156:3850-3857.

250. Gonzalo J-A, Lloyd CM, Kremer L, et al. Eosinophil recruitment to the lung in a murine model of allergic inflammation. The role of T cells, chemokines, and adhesion receptors. J Clin Invest 1996; 98:2332-2345.

106

251. Sikorski EE, Hallmann R, Berg EL, et al. The Peyer's Patch high endothelial receptor for lymphocytes, the mucosal vascular addressin, is induced on a murine endothelial cell line by tumor necrosis factor-α and IL-1. J Immunol 1993; 151:5239-5250.

252. Takeuchi M, Baichwal VR. Induction of the gene encoding mucosal vascular addressin cell adhesion molecule 1 by tumor necrosis factor α is mediated by NF-κ B proteins. Proc Natl Acad Sci USA 1995; 92:3561-3565.

253. Lee M-S, Sarvetnick N. Induction of vascular addressins and adhesion molecules in the pancreas of IFN-γ transgenic mice. J Immunol 1994; 152:4597-4603.

254. Hänninen A, Taylor C, Streeter PR, et al. Vascular addressins are induced on islet vessels during insulitis in nonobese diabetic mice and are involved in lymphoid cell binding to islet endothelium. J Clin Invest 1993; 92:2509-2515.

255. Faveeuw C, Gagnerault M-C, Lepault F. Expression of homing and adhesion molecules in infiltrated islets of langerhans and salivary glands of nonobese diabetic mice. J Immunol 1994; 152:5969-5978.

256. Schaible UE, Vestweber D, Butcher EG, et al. Expression of endothelial cell adhesion molecules in joints and heart during Borrelia burgdorferi infection in mice. Cell Adhes Commun 1994; 2:465-479.

257. Michie SA, Streeter PR, Butcher EC, et al. L-selectin and $\alpha 4\beta 7$ integrin homing receptor pathways mediate peripheral lymphocyte traffic to AKR mouse hyperplastic thymus. Am J Pathol 1995; 147:412-421.

258. O'Neill JK, Butter C, Baker D, et al. Expression of vascular addressins and ICAM-1 by endothelial cells in the spinal cord during chronic relapsing experimental allergic encephalomyelitis in the Biozzi AB/H mouse. Immunology 1991; 72:520-525.

259. Engelhardt B, Conley FK, Kilshaw PJ, et al. Lymphocytes infiltrating the CNS during inflammation display a distinctive phenotype and bind to VCAM-1 but not to MAdCAM-1. Int Immunol 1995; 7:481-491.

260a. Picarella D, Hurlbut P, Rottman J, et al. Monoclonal antibodies specific for ß7 and MAdCAM-1 reduce inflammation in the colon of scid mice reconstituted with CD45RB^{hi} CD4$^+$ T cells. J Immunol 1997; 158:2099-2106.

260b. Podolsky DK, Lobb R, King N, et al. Attenuation of colitis in the cotton top tamarin by anti-$\alpha 4$ integrin monoclonal antibody. J Clin Invest 1993; 92:372-380.

261. Kumasaka T, Quinlan WM, Doyle NA, et al. Role of the intercellular adhesion molecule-1(ICAM-1) in endotoxin-induced pneumonia evaluated using ICAM-1 antisense oligonucleotides, anti-ICAM-1 monoclonal antibodies, and ICAM-1 mutant mice. J Clin Invest 1996; 97:2362-2369.

262. Doerschuk CM, Quinlan WM, Doyle NA, et al. The role of P-selectin and ICAM-1 in acute lung injury as determined using blocking antibodies and mutant mice. J Immunol 1996; 157:4609-4614.

263. Cowan PJ, Shinkel TA, Witort EJ, et al. Targeting gene expression to endothelial cells in transgenic mice using the human intercellular adhesion molecule 2 promoter. Transplantation 1996; 62:155-160.

264. Staunton DE, Marlin SD, Stratowa C, et al. Primary structure of intercellular adhesion molecule 1 (ICAM-1) demonstrates interaction between members of the immunoglobulin and integrin supergene families. Cell 1988; 52:925-933.

265. Simmons D, Makgoba MW, Seed B. ICAM, an adhesion ligand of LFA-1, is homologous to the neural cell adhesion molecule NCAM. Nature 1988; 331:624-627.

266. Horley KJ, Carpenito C, Baker B, et al. Molecular cloning of murine intercellular adhesion molecule (ICAM-1). EMBO J 1989; 8:2889-2896.

267. Xu H, Tong IL, de Fougerolles AR, et al. Isolation, characterization, and expression of mouse ICAM-2 complementary and genomic DNA. J Immunol 1992; 149:2650-2655.

268. Briskin MJ, McEvoy LM, Butcher EC. MAdCAM-1 has homology to immunoglobulin and mucin-like adhesion receptors and to IgA1. Nature 1993; 363:461-464.

269. Hession C, Moy P, Tizard R, et al. Cloning of murine and rat vascular cell adhesion molecule-1. Biochem Biophys Res Commun 1992; 183:163-169.

270. Tan TQ, Smith CW, Hawkins EP, et al. Hematogenous bacterial meningitis in an intercellular adhesion molecule-1-deficient infant mouse model. J Infect Dis 1995; 171:342-349.

271. Bullard DC, Qin L, Lorenzo I, et al. P-selectin/ICAM-1 double mutant mice: acute emigration of neutrophils into the peritoneum is completely absent but is normal into pulmonary alveoli. J Clin Invest 1995; 95:1782-1788.

272. Schowengerdt KO, Zhu JY, Stepkowski SM, et al. Cardiac allograft survival in mice deficient in intercellular adhesion molecule-1. Circulation 1995; 92:82-87.

273. Sarman G, Shappell SB, Mason EOJr, et al. Susceptibility to local and systemic bacterial infections in intercellular adhesion molecule 1-deficient transgenic mice. J Infect Dis 1995; 172:1001-1006.

274. Davis SL, Hawkins EP, Mason EOJr, et al. Host defenses against disseminated candidiasis are impaired in intercellular adhesion molecule 1-deficient mice. J Infect Dis 1996; 174:435-439.

275. Bullard DC, Hurley LA, Lorenzo I, et al. Reduced susceptibility to collagen-induced arthritis in mice deficient in intercellular adhesion molecule-1. J Immunol 1996; 157:3153-3158.

276. Qin L, Quinlan WM, Doyle NA, et al. The roles of CD11/CD18 and ICAM-1 in acute pseudomonas aeruginosa-induced pneumonia in mice. J Immunol 1996; 157:5016-5021.

277. Connolly ES,Jr., Winfree CJ, Springer TA, et al. Cerebral protection in homozygous null ICAM-1 mice after middle cerebral artery occlusion: Role of neutrophil adhesion in the pathogenesis of stroke. J Clin Invest 1996; 97:209-216.

278. Gonzalo JA, Martinez-A C, Springer TA, et al. ICAM-1 is required for T cell proliferation but not for anergy or apoptosis induced by *Staphylococcus aureus* enterotoxin B *in vivo*. Int Immunol 1995; 7:1691-1698.

279. Soriano SG, Wang YF, Lipton SA, et al. Intercellular adhesion molecule (ICAM-1) deficient mice are resistant to cerebral ischemia-reperfusion injury. Ann Neurol 1996; 39:618-624.

280. Verdrengh M, Springer TA, Gutierrez-Ramos J-C, et al. Role of intercellular adhesion molecule 1 in pathogenesis of staphylococcal arthritis and in host defense against staphylococcal bacteremia. Infect Immun 1996; 64:2804-2807.

281. Kelly KJ, Williams WW,Jr., Colvin RB, et al. Intercellular adhesion molecule-1 deficient mice are protected against ischemic renal injury. J Clin Invest 1996; 97:1056-1063.

282. Nordlund JJ, Csato M, Babcock G, et al. Low ICAM-1 expression in the epidermis of depigmenting C57BL/6J-mivit/mivit mice: a possible cause of muted contact sensitization. Exp Dermatol 1995; 4:20-29.

Chapter 4

REGULATION OF INTERCELLULAR ADHESION MOLECULE (ICAM) GENE EXPRESSION

Thomas P. Parks[*] and Mary E. Gerritsen[§]

[*]Cellegy Pharmaceuticals, Inc.
349 Oyster Point Boulevard, Suite 200
South San Francisco, CA 94080

[§]Department of Cardiovascular Research
Genentech
1 DNA Way
South San Francisco, CA 94080

INTRODUCTION

The human intercellular adhesion molecule (ICAM) family is composed of five members: ICAM-1, ICAM-2, ICAM-3, ICAM-4 (LW), and telencephalon (Fig. 4-1). All are type 1 transmembrane glycoproteins belonging to a subset of the immunoglobulin (Ig) supergene family. Another unifying feature of this family of molecules is their ability to bind the integrin LFA-1 (lymphocyte function-associated antigen type 1, $\alpha_L\beta_2$), a counter-receptor expressed on all leukocytes. Functional adhesive interactions between the ICAMs and LFA-1 portend important roles for these molecules in cell-cell communication. Unlike LFA-1, the ICAMs are regulated primarily by expression levels, rather than by changes in binding affinity. The ICAMs are expressed constitutively at different levels on a wide range of cell types, indicating distinct patterns of gene regulation. ICAM-1 is the only member of the ICAM family whose expression, and thus function, is inducibly regulated. This review will discuss the current state of knowledge with respect to expression regulation of ICAM family members, with a primary focus on the inducible expression of ICAM-1.

110

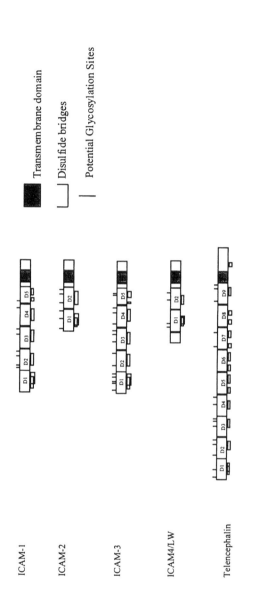

Fig. 4-1: The ICAM Gene Family (see also Table 4-1): D represents extracellular Ig-like domain.

ICAM STRUCTURE, FUNCTION, AND DISTRIBUTION

ICAM-1

ICAM-1 was first identified with a novel monoclonal antibody (mAb, RR1/1) that inhibited LFA-1-dependent homotypic lymphoblast aggregation.[1] At about the same time as the identification of ICAM-1 by Rothlein et al,[1] several other ICAM-1-specific mAbs were described which recognized membrane glycoproteins associated with cell adhesion,[2] leukocyte activation,[3] and melanoma tumor progression.[4] A panel of these and other ICAM-1-specific mAbs were used to classify ICAM-1 as CD54 at the Leukocyte Typing IV Workshop.[5] In contrast to the leukocyte specific expression of LFA-1, ICAM-1 is widely expressed on both hematopoietic cells, such as tissue macrophages, mitogen-stimulated T lymphocyte blasts, germinal center dendritic cells, and B lymphocytes, and on non-hematopoietic cells, including endothelial cells, fibroblasts, keratinocytes, certain epithelial cells, and other mesenchymal cells or cell lines.[6] Although the basal expression of ICAM-1 on non-hematopoietic cells is generally low, surface expression is markedly upregulated by a variety of mediators, including the inflammatory triad of interleukin-1α (IL-1α), tumor necrosis factor-α (TNF-α), and lipopolysaccharide (LPS), as well as interferon γ (IFNγ). The cloning of ICAM-1 cDNAs by two groups in 1988 revealed that ICAM-1 is a member of the immunoglobulin (Ig) gene superfamily, possessing 5 Ig-like domains, and closely related to two other adhesion molecules of the Ig superfamily, vascular cell adhesion molecule-1 (VCAM-1) and neural cell adhesion molecule (NCAM).[7,8] Human ICAM-1 is a type I transmembrane glycoprotein containing 505 amino acids and a predicted molecular mass of 55 kDa. The actual molecular mass ranges from 76 to 114 kDa depending on tissue-specific glycosylation at eight potential N-glycosylation sites. In endothelial cells, ICAM-1 exhibits an apparent molecular mass of 90-100 kDa on SDS-polyacrylamide gels.[9] The extracellular domain consists of 453 predominantly hydrophobic amino acids, followed by a hydrophobic transmembrane domain of 24 residues, and a relatively short (28 residues) and charged cytoplasmic domain.[9]

ICAM-1 was the first ligand or counter-receptor identified for the leukocyte integrin LFA-1(CD11a/CD18), as demonstrated by specific LFA-1 binding to purified ICAM-1[10,11] or to "null" cells expressing an ICAM-1 cDNA.[8] Subsequently, ICAM-1 was shown to be a ligand for two other α_2 integrins, Mac-1 (CD11b/CD18)[12] and p150,95 (CD11c/CD18),[13,14] that are expressed predominantly on monocytes, macrophages, granulocytes, and large granular lymphocytes. ICAM-1 has also been subverted by several pathogens to infect host cells and evade immune surveillance. In this respect, ICAM-1 serves as the major group receptor for rhinoviruses[15] and *Plasmodium falciparum* infected erythrocytes.[16] ICAM-1 has also been reported to bind to soluble fibrinogen,[17] CD43,[18] and hyaluranon.[19]

ICAM-1 has been implicated in a variety of immune and inflammatory responses, including antigen presentation, cell-mediated cytolysis, lymphocyte homotypic aggregation, leukocyte-endothelial interactions, and leukocyte trafficking to sites of inflammation.[20,21] As a ligand for Mac-1, ICAM-1 participates in neutrophil-endothelial interactions, trans-endothelial migration, and adhesion dependent respiratory burst. Targeted disruption of the ICAM-1 gene in mice has revealed many abnormalities in inflammatory responses, such as impaired neutrophil extravasation, defective T cell proliferation in mixed allogeneic lymphocyte cultures, resistance to septic shock, reduced suseptibility to cerebral ischemia-reperfusion injury, and protection against radiation induced pulmonary inflammation and experimentally induced colitis.[22-26] ICAM-1 thus plays important roles in host defense, bridging both innate and adaptive immune responses. It follows that ICAM-1 also plays pathophysiological roles when these immune responses are excessive, making ICAM-1 an attractive target for therapeutic intervention in diseases associated with e.g., transplantation, autoimmunity, ischemia-reperfusion injury, and rhinovirus infection.

High constitutive ICAM-1 expression is often observed on malignant melanomas and renal cell carcinomas, but is found less frequently on gastric carcinomas and breast carcinomas, and only rarely on colorectal carcinomas.[27-30] ICAM-1 showed an expression pattern on primary melanomas correlating with vertical tumor thickness, the most predictive parameter for the development of metastasis in melanoma, and expression of ICAM-1 has been shown to be a marker of poor prognosis in stage I tumors. Interfering with ICAM-1 expression inhibited experimental metastasis of melanomas in nude mice.[31] The role of ICAM-1 in melanoma tumor progression and metastisis is unclear. The metastatic cascade is similar in some respects to the extravasation of leukocytes at sites of inflammation. On the other hand, tumor expression of ICAM-1 would be expected to increase tumor cells vulnerability to immune attack. Tumor cells may also shed ICAM-1, which may play a role in immune evasion, a function that remains to be determined.

A soluble, circulating form of ICAM-1 (cICAM-1) has been described in normal human serum, and in the media of cultured JY cells and peripheral blood monocytes, but not human umbilical vein endothelial cells (HUVECs).[32,33] Soluble forms of the adhesion molecules ICAM-1, VCAM-1, and E-selectin were detected by Pigott et al[34] in supernatants from cytokine activated endothelial cells. The molecular weights of the released molecules were consistent with the generation of soluble forms by cleavage at a site close to the point of membrane insertion. Recently, a distinct monensin-sensitive ICAM-1 secretory pathway was identified in human synovial endothelial cells (HSECs) that was largely absent in HUVECs.[35] Curiously, circulating ICAM-1 was present in the serum of ICAM-1 knockout mice which lacked detectable membrane-bound ICAM-1.[36] From spleen RNA, these investigators identified 3 alternatively spliced isoforms of ICAM-1, lacking 2 or 3 extracellular domains, but including the transmembrane domain. It is still not clear how soluble ICAM-1 is generated, but it may be derived from the processing of a membranous precursor during secretion. The plasma levels of cICAM-1 are generally elevated during infection, and it may be a marker for inflammatory processes. Elevated levels of cICAM-1 have been observed in certain diseases, including leukocyte adhesion deficiency, allograft rejection, malignant melanoma, rheumatoid arthritis, and atopic dermatitis.[32,33,37-40] It has been proposed that cICAM-1 may interfere with normal leukocyte adhesion, but this role remains unclear.

ICAM-2

The existence of an additional LFA-1 ligand was inferred by the inability of ICAM-1-specific mAbs to completely block LFA-1-dependent lymphocyte adhesion to endothelial monolayers.[9] ICAM-2 (CD102) was subsequently identified using a functional cloning approach in which a cDNA library derived from LPS-stimulated HUVECs was expressed in COS cells and screened for clones binding to LFA-1 substrates in the presence of ICAM-1 mAb.[41] The resultant ICAM-2 cDNA predicted an integral membrane protein of 28.4 kDa with two amino terminal extracellular domains sharing 35% sequence identity with the first two Ig-like domains of ICAM-1. Human ICAM-2 runs as a 55-60 kDa band on SDS-polyacrylamide gels.[42,43] Like ICAM-1, ICAM-2 interacts with LFA-1 and Mac-1.[41,43-45] ICAM-2 is expressed constitutively at low levels on resting lymphocytes, monocytes, platelets, megakaryocytes, and leukocytic cell lines,[43,46-48] and it is the only ICAM expressed on platelets. ICAM-2 is expressed constitutively at high levels on vascular endothelium.[43] Unlike ICAM-1, however, ICAM-2 is not upregulated by inflammatory stimuli.[42,43] This observation and the ability of endothelial ICAM-2 to interact with lymphocytes, led investigators to speculate that the molecule might function in normal lymphocyte recirculation in the body. Recently, McLaughlin et al[49,50] reported that TNF-α and IL-1β were capable of down-regulating ICAM-2 expression on human endothelial cells. ICAM-2 was also found to be highly expressed on endothelial cells in malignant lymphoma, suggesting that it may be upregulated under some circumstances.[51] *In vitro* studies have suggested a role for ICAM-2 in the transendothelial migration of lymphocytes,[52] monocytes,[53] neutrophils,[54] and eosinophils.[55] Targeted deletion of the ICAM-2 gene in mice demonstrated that ICAM-2 plays a crucial role in eosinophil trafficking and localization in the lung during allergic inflammation.[55] Eosinophil accumulation in the airway lumen was significantly delayed in ICAM-2 deficient mice, accompanied by a prolonged accumulation of eosinophils in the lung interstitium and heightened airway hyperresponsiveness during the development of allergic lung inflammation. In this study, ICAM-2 was not essential for lymphocyte homing under basal conditions, or for the development of leukocytes, with the exception of megakaryocyte progenitors, which were significantly reduced.[55]

ICAM-3

The existence of ICAM-3 was predicated on the basis that both ICAM-1 and ICAM-2 mAbs were unable to completely block LFA-1-dependent homotypic T cell aggregation.[43] Subsequently, three groups independently identified ICAM-3 (or ICAM-R, CD50) as the third LFA-1 ligand.[46,56,57] ICAM-3, like ICAM-1, was found to possess five extracellular Ig-like domains. ICAM-3 is heavily glycosylated, with 15 potential N-glycosylation sites. The mature cell-surface form of the ICAM-3 has a molecular mass which varies from 116 to 140 kDa in a cell type-specific fashion.[56] ICAM-3 binds LFA-1 but apparently not Mac-1. In addition, ICAM-3 has been identified as the preferred ligand for $\alpha_d\beta_2$, a recently discovered β_2 integrin highly expressed on macrophages and on subsets of blood leukocytes.[58] The expression of ICAM-3 is restricted to monocytes, lymphocytes, and granulocytes, but not endothelial cells.[46] ICAM-3 is the predominant ICAM expressed on resting T lymphocytes and is the only ICAM molecule highly expressed on neutrophils.[46,59-61] ICAM-3 may play a role in initiating immune responses, perhaps by contributing to transient lymphocyte interactions and serving as a co-stimulatory molecule.[62]

ICAM-4 (LW), Telencephalon, and ICAM-1

The LW (Landsteiner-Weiner) blood group system is defined by the LW^a and LW^b antigens that are recognized by specific antisera, and reside on a 42 kDa glycoprotein restricted to erythrocytes and erythroid precursors. The LW locus is composed of a single gene with two alleles (i.e., LW^a and LW^b). In 1994, Bailly et al cloned the gene encoding the LW-protein and discovered that the deduced amino acid sequence contained two extracellular Ig-like domains, resembling ICAM-2, with about 30% homology to the first two domains of other ICAM family members.[63] The LW protein was shown to be a functional ICAM molecule by binding to leukocyte LFA-1 and Mac-1 in a static adhesion assay.[64] Conversely, purified leukocyte integrins bound to erythrocytes, and the interaction was blocked with anti-LW or anti-integrin antibodies. As the fourth member of the ICAM family to be identified, the LW glycoprotein has been renamed ICAM-4. The role of ICAM-4 is unclear, but may play a role in erythrocyte turnover in the spleen.[64]

Telencephalon (TLN) is a neural glycoprotein first identified with a monoclonal antibody directed against a synaptosome preparation from rabbit olfactory bulb.[65] TLN is expressed exclusively in the telencephalon of the mammalian brain, on neuronal cell bodies and dendrites, but not axons.[66] In addition, TLN is developmentally regulated, first appearing around birth when dendritic outgrowth, branching, and synapse formation occur in the telencephalon.[65] Rabbit, mouse, and human TLN cDNAs have been cloned,[67,68] revealing TLN to be a novel member of the ICAM family, with nine Ig-like domains and a molecular mass of about 130 kDa. Domains 2-5 of telencephalon are most similar to the corresponding domains of ICAM-1 and ICAM-3, whereas domains 6-8 resemble ICAM domain 5, and domain 9 is divergent from the ICAMs, more closely resembling the C2-type Ig-like domains found in a number of neural adhesion proteins. Even though the amino acid identity of TLN domain 1 is somewhat less than the corresponding domain of other ICAMs (25-38%), it contains the four cysteine residues capable of forming two intradomain disulfide bridges found in other family members, and has retained a conserved motif thought to be required for ICAM/β_2 integrin interactions.[67,68] Telencephalon has been shown to bind LFA-1, suggesting that this molecule may participate in integrin-mediated cell-cell interactions in the brain.[67,69] In the central nervous system, LFA-1 is constitutively expressed on microglia, a brain-type macrophage of hematopoietic origin. Thus TLN and LFA-1 may mediate neuron/microglia interactions in the telencephalon of the normal brain, or TLN may be a neuronal target molecule for LFA-1-expressing activated microglia or infiltrating leukocytes in some brain diseases.

In a search for genes downstream of the rat growth hormone (GH) locus, Ono et al[70] recently identified a gene specifically expressed in the testis that encodes a type I transmembrane protein with five extracelular Ig-like domains, resembling ICAM-1 and ICAM-3. The gene, named testicular adhesion molecule 1 (TCAM-1), has the same genomic organization as the ICAM-1 gene, and is located between the GH and BAF60b (a component of mammalian SWI/SNF complexes) genes. Currently, little is known about the cell-surface expression of TCAM-1, and consequently its potential biological roles.

The structural features of the ICAM family members are summarized in Figure 4-1 and Table 4-1. As discussed above, the ICAMs are members of a subset of the Ig superfamily of cell surface glycoproteins, characterized by the presence of extracellular Ig domains. Ig domains are protein modules of approximately 80-100 amino acids with a characteristic β-sheet structure stabilized by a highly conserved disulfide bond. The interactions between ICAM Ig-like domains and β_2-integrins have been intensively studied by a number of groups (see 62). Crystal structures of the two amino-terminal Ig-like domains of ICAM-2[71] and ICAM-1[72,73] has led to a detailed description of the ICAM surfaces that interact with LFA-1, via the inserted domain (I-domain) of the LFA-1 α-chain (CD11a), and rhinoviruses. ICAM-1 contains 5 Ig domains of the C2 set with a short hinge region separating the third and fourth Ig domains.[7,8] ICAM-3 also contains 5 Ig domains which are structurally similar to the corresponding domains of ICAM-1 with 52% sequence identity.[46,56,57] Domain 1 of ICAM-1 and domain 2 of ICAM-3 exhibit 77% sequence identity. ICAM-2 contains 2 Ig domains which are most closely related to the first 2 domains of ICAM-1, with 36% and 37% sequence identity, respectively.[74] ICAM-4 also possesses 2 Ig domains, the first one sharing 30% sequence identity with those of the other ICAMs. Telencephalon has nine Ig domains; the first eight have homology to the Ig domains of ICAM subfamily members, whereas domain 9 is divergent. TLN domains 1-5 exhibit 50% and 55% amino acid identity with ICAM-1 and ICAM-3, respectively. Members of the

ICAM subgroup all possess a characteristic arrangement of four cysteine residues in domain 1 capable of forming two intradomain disulfide bonds, a feature shared by adhesion molecules that bind integrins, including VCAM-1 and MAdCAM-1.[75]

ICAM-1, 2 and 3 were all defined functionally as ligands or counter-receptors for LFA-1. LFA-1 binds to the most distal Ig domain, i.e., domain 1, of ICAM-1[76] ICAM-2,[47,77] ICAM-3[78,79] and ICAM-4.[80] Rhinoviruses[74,76] and malaria-infected erythrocyte[81,82] also bind to domain 1 of ICAM-1, but to regions distinct from the LFA-1 binding site. The LFA-1 recognition site in TLN resides within domains 1-5.[69] Mac-1 binds to domain 3 of ICAM-1[12] and domain 1 of ICAM-2,[45] but does not appear to bind to ICAM-3[78,83] or TLN.[69] The Mac-1 binding site in ICAM-4 spans domains 1 and 2 and is distinct from the LFA-1 site.[80]

In contrast to the extracellular Ig-like domains, the cytoplasmic regions of the different ICAM molecules exhibit little homology. However, the cytoplasmic regions of individual ICAMs retain considerable amino acid sequence conservation across species, with important functional implications. ICAM-1, ICAM-2, and ICAM-3 appear to form transient associations with the cytoskeleton and localize to uropods during cellular interactions or upon cellular activation. The cytoplasmic regions of these molecules bind to ERM (ezrin, radixin, and moesin) proteins, which function to cross-link integral membrane proteins with cortical actin filaments and are involved in the formation of e.g., cell adhesion sites.[84] The ERM proteins are also involved in signal transduction pathways such as Rho signaling. In addition to their adhesive and co-stimulatory roles, via integrin activation, engagement of ICAM-1 and ICAM-3, in particular, can also transmit signals into cells, a process termed outside-in signaling (reviewed in 62).

THE *ICAM* GENES

The human *ICAM-1*, *ICAM-3*, *LW* (ICAM-4), and *Tlcn* (telencephalon) genes are clustered in chromosome region 19p13.2-p13.3.[67,85-89] Kilgannon et al[90] have recently mapped the *Tlcn* (or *ICAM-5*) gene to an 80 kb region of 19p13.2 that also contains *ICAM-1* and *ICAM-3*. The *ICAM-2* gene has been mapped to chromosome region 17q23-q25.[91] Ono et al[70] stated that they found a human homologue of the rat TCAM-1 gene downstream of the human growth hormone locus at 17q22-24, i.e., in the vicinity of *ICAM-2*. The murine *ICAM-1* and *Tlcn* genes are also closely linked on proximal mouse chromosome 9, which shares homology with human chromosmes 19p and 11q.[92,93] In light of the chromosomal clustering, and structural and functional homology of the genes, the ICAM family likely arose from a common ancestral gene by gene duplication and, in the case of *ICAM-2*, chromosomal translocation. Gene duplication is a basic mechanism for generating protein families, including the Ig superfamily.[94] Evolution of the ICAM subgroup has produced a variety of molecular forms with similar functional capabilities, but with extremely diverse expression patterns.

The human and murine *ICAM-1* genes are highly conserved. Both genes span over 13 kb and consist of 7 exons and 6 introns.[95,96] Organization of the human ICAM-1 gene is depicted in Figure 4-2. Exon 1 contains 5' untranslated sequence and the signal polypeptide, exons 2 to 6 contain extracellular Ig-like domains 1 to 5, and exon 7 encodes the membrane spanning region, the cytoplasmic domain, and 3' untranslated sequence. Each Ig-like domain is encoded by a single exon, a characteristic found in most members of the Ig superfamily.[94] The splice junctions in the human and murine genes also exhibit a high degree of sequence conservation. Alternative RNA spicing has been observed in mice suggesting the possible existence of additional biologically relevant ICAM-1 isoforms.[97] Evidence for similar alternative splicing of human ICAM-1 RNA has not been reported. However, Wakatsuki et al[98] recently identified a human mRNA which specifically encoded a soluble ICAM-1. These investigators suggested that alternative splicing may have been responsible for a 19 base deletion and frameshift mutation, just upstream of the transmembrane region of the coding sequence, resulting in loss of the transmembrane and cytoplasmic domains.

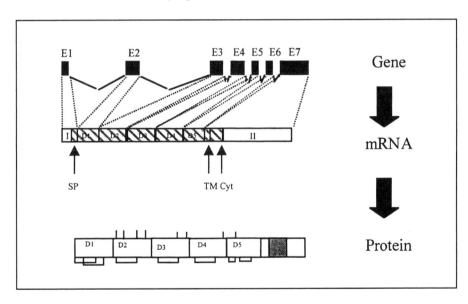

Figure 4-2. Structural organization of the ICAM-1 gene. Abbreviations: SP (signal peptide), TM (transmembrane domain), Cyt (cytoplasmic domain), E1-7 (exons 1-7), D1-5 (Ig-like domains 1-5), I (5' untranslated region), II (3' untranslated region)

The *ICAM-2* gene is present in a single copy in the mouse genome and contains four exons spanning about 5.0 kb of DNA[99] The exon/intron architecture correlates to the structural domains of the protein and resembles that of the *ICAM-1* gene, with one exon encoding the signal peptide, one exon for each of the Ig-like domains, and one containing the transmembrane and cytoplasmic domains.

Table 4-1.
Structural and Functional Characteristics of the ICAM Family

Characteristic	ICAM-1	ICAM-2	ICAM-2	ICAM-4	Telencephalin
Synonyms	CD54	CD102	CD50	LW Antigen	ICAM-5
Ig Domains	5	2	5	2	9
Native Mr (kDa)	76-114	55-60	110-140	42	130(rabbit, human)
Core Mr	55	28.4	57	26.5	93 (rabbit)
Amino Acids	505	254	518	241	883 (rabbit) 897 (human)
Potential N-glycosylation sites	8	6	15	4	14 (rabbit) 15 (human)
Ligands	LFA-1 (D1) MAC-1 (D3) P150,95 CD43 Fibrinogen (D1) Hyaluronan Rhinovirus (D1) Malaria-infected erythrocytes (D1)	LFA1(D1) MAC1 (D1)	LFA1](D1) αdβ2	LFA1 (D1) Mac-1 (D1&D2)	LFA1 (D1-5))

Table 4-1. (Continued)

Characteristic	ICAM-1	ICAM-2	ICAM-2	ICAM-4	Telencephalin
Distribution	Monocytes Endothelium Lymphocytes Epithelial Cells Keratinocytes Fibroblasts	Endothelium Monocytes Platelets Lymphocytes Dendritic Cells	Lymphocytes Monocytes Neutrophils Dendritic Cells	Erythrocytes	Neurons (Itelencephalon)
Expression/ Regulation	Constitutive and Strongly and Widely Inducible	Constitutive	Constitutive	Constitutive	Constitutive
Chromosomal Location	19p13.2	17q23-q25	19p13.2-p13.3	19p13.3	19p13.2

The murine and human *Tlcn* *(ICAM-5)* genes have been characterized and exhibit an overall genomic organization similar to the other *ICAM* genes.[90,93] The mouse and human genes contain 11 exons. Exon 1 contains the translation initiation codon (ATG) and signal peptide sequence, followed by 9 exons, each encoding a single extracellular Ig-like domain, and an exon containing both the transmembrane and cytoplasmic domains.[93] The human *Tlcn/ICAM-5* gene is located only 5 kb downstream of the *ICAM-1* gene, juxtaposing the two ICAM family members with the most highly restricted as well as the most promiscuous tissue expression patterns. In addition, the genes are oriented such that the 5' regulatory region of the *Tlcn/ICAM-5* gene lies within the 5 kb region that separates the two genes. Since chromatin remodeling accompanies *ICAM-1* gene expression, Kilgannon et al[90] suggested that negative transcriptional elements must exist to prevent inappropriate expression of the adjacent *Tlcn/ICAM-5* gene. The human *ICAM-3* gene is located about 30-40 kb downstream of *Tlcn/ICAM-5*. The mouse does not appear to have an *ICAM-3* gene.[90]

ICAM EXPRESSION AND ITS REGULATION

A characteristic of ICAM-1 is its widespread tissue distribution, particularly following induction by a variety of stimuli. Immunohistochemical staining reveals low level constitutive ICAM-1 expression on endothelial and epithelial cells, macrophages, fibroblast-like cells, and dendritic cells in human thymus, lymph nodes, intestine, skin, kidney, and liver.[6] On hematopoietic cells, ICAM-1 is expressed on bone marrow progenitor cells, on cells of the lymphoid lineage and myeloid lineage, up to the myelocyte and erythroblast stage, while ICAM-1 remains expressed in the monocytic and lymphoid lineages.[20] Many cells, including those which consititutively express, or do not express ICAM-1, can markedly upregulate the expression of this adhesion molecule upon stimulation with a variety of stimuli. Foremost among the inducing agents are inflammatory cytokines, such as TNF-α, IL-1β, and IFNγ, as well as LPS (Table 4-2). Inflamed tissues generally stain strongly positive for ICAM-1 *in vivo*, particularly the vascular endothelium (see 13). The pro-inflammatory phorbol esters, such as phorbol myristate acetate (PMA), are also potent inducers of ICAM-1 expression on a variety of cells.[100-103] ICAM-1 expression is upregulated by agents and conditions as divergent as hormones, neurotransmitters, autocoids, metal ions, oxidants, hypoxia-reoxygenation, ionizing radiation, and a wide range of infectious agents (Table 4-1). In this regard, ICAM-1 induction often represents a primary response to stress, occurring in parallel with the induction of cytokine genes which can further amplify ICAM-1 expression.

The sensitivity of cells and tissues to different stimuli varies widely and depends on many factors, including the presence of appropriate receptors, signaling pathways, transcriptional and post-transcriptional mechanisms. Isolated endothelial cells in culture can respond quite

Table 4-2. Stimuli and Conditions Known to Affect ICAM-1 Expression

Category	Factors	Responsive Cell Types
	I. Inducing Factors	
Cytokines	TNF-α	Numerous
	IFNγ	Numerous
	IL-1β	Numerous
	IL-3	Eosinophils
	IL-4	Lymphocytes
		Fibroblasts
		Monocytes
		Mast Cells
	IL-5	Eosinophils
	IL-6	Hepatocytes
Autocoids	Thrombin	Endothelial Cells
	Endothelin 1-3	Endothelial Cells
	Substance P	Endothelial Cells
	Histamine	Epithelial Cells
		Keratinocytes
Hormones and Growth Factors	Retinoic Acid	Fibroblasts
		Glial Cells
		Various Tumor Cells

Table 4-2. (continued)

Hormones and Growth Factors (continued)	Estradiol	Endothelial Cells
	PDGF	Smooth Muscle Cells
	GM-CSF	Eosinophils
Myeloid Leukemia Blasts		
Radiation	X-Rays	Endothelial Cells
		Epithelial Cells
		Keratinocytes
293 Cells	UV-A,B	Endothelial Cells
Oxidants	H2O2	Endothelial Cells
	Metal Ions (Ni++, Co++, Zn++)	Numerous Cell Types
	Oxidized LDL	Endothelial Cells
Mechanical Forces	Fluid Shear Stress	Endothelial Cells
Infectious Agents	LPS	Endothelial Cells
		Epithelial Cells
		Monocytes
	Cytomegalus Virus	Endothelial Cells
	Parainfluenza Virus	Epithelial Cells
	P.falciparum	Endothelial Cells
	Rhinovirus	Epithelial Cells
	RSV	Epithelial Cells

Table 4-2. (continued)

Infectious Agents (continued)	Rhinovirus	Epitheilial Cells
	RSV	Epithelial Cells
	B. bergdorferi	Endothelial Cells
	Invasive enteric bacteria	Epithelial Cells
	HIV (gp120)	Glial Cells
	HTLV (Tax1)	T-lymphocytes
		Lymphocytes
		Epithelial Cells
	EBV	Hematopoietic Cell Lines
Other	Phorbol Esters	Most Cell Types
	IgM Cross Linking	B-Cells
	CD40 Ligand	Endothelial Cells
		B-Cells
	PDTC	Endothelial Cells
	Poly (I:C)	Endothelial Cells
	II. Inhibiting Factors	
Cytokines	Il-4	Endothelial Cells
	IL-10	Monocytes
		Endothelial Cells

Table 4-2. (continued)

Growth Factors	TGFb	Fibroblasts Glial Cells
Steroids	Glucocorticoids	Adenocarcinoma Cells Epithelial Cells Monocytes Eosinophils Macrophages Endothelial Cells
Other	Nicotinamide	Endothelial Cells T-Cells
	Retinoic Acid	Monocytes
	UV-B	Monocytes

differently depending on their source within the vasculature. TNF-α and IL-1β are typically more powerful stimulants of ICAM-1 expression on human umbilical vein endothelial cells or bovine aortic endothelial cells than is IFNγ.[6,104,105] Human synovial endothelial cells are equally responsive to IFNγ and TNF-α, and the combination of the two cytokines leads to a synergistic upregulation of ICAM-1.[106] Among glial cells in the CNS, astrocytes respond to TNF-α > IL-1β > IFNγ,[107] while microglial cells (of hematopoietic origin) respond only to IFNγ.[108] IFNγ can stimulate the expression of ICAM-1 and MHC class II molecules on non-professional antigen presenting cells (APCs), such as epithelial and endothelial cells.[109] Engagement of CD40 (a member of the TNF receptor superfamily) by CD40 ligand (present on activated T-cells) or cross-linking surface IgM by antigen upregulates ICAM-1 expression on B lymphocytes.[110,111]

The induction of ICAM-1 expression can be inhibited by several agents, including glucocorticoids (e.g., dexamethasone), transforming growth factor β (TGFβ), interleukin-10 (IL-10). and nitric oxide (NO). The cell specific effects of these agents are discussed in detail below.

ICAM-1 expression is complex in that it is both constitutive and inducible; and both modes of expression vary widely across different cell types. ICAM-1 expression is also controlled at multiple levels. Early studies established that inducible ICAM-1 expression required *de novo* RNA and protein synthesis, and that ICAM-1 mRNA levels preceded and paralleled the appearance of ICAM-1 protein on the cell surface.[6] These observations suggested that ICAM-1 gene expression was controlled primarily at the transcriptional level. TNF-α was subsequently shown to directly increase the rate of ICAM-1 gene transcription in HUVECs.[101] PMA-stimulated ICAM-1 expression, however, was found to be more complex. PMA can act post-trancriptionally via a PKC-dependent pathway to decrease the degradation rate of ICAM-1 mRNA, thus allowing accumulation of the labile message. Wertheimer et al[101] suggested that multiple copies of the putative destabilizing motif, AUUUA, in the 3' untranslated region (UTR) of ICAM-1 mRNA might contribute to message stability. Ohh et al[112] observed that PMA and IFNγ were both capable of stabilizing ICAM-1 mRNA in the murine monocytic cell line P388D1. Using chimeric ICAM-1 genes stably expressed in mouse L cells, these investigators showed that the AUUUA motifs in the 3' UTR of the ICAM-1 message played a role in PMA-regulated ICAM-1 mRNA turnover. However, the stabilization of ICAM-1 message by IFNγ occurred through a different mechanism. The IFNγ-responsive sequences mediating ICAM-1 mRNA stabilization appeared to reside, at least in part, within the coding region, rather than the 3'-UTR.[112,113] IFNγ did not affect ICAM-1 message stability in astrocytes or microglia cells of the central nervous system (CNS).[108] It is likely, therefore, that the role of message stability in ICAM-1 gene expression will vary depending on the stimulus and cell type examined. In addition to regulation at the transcriptional and mRNA levels, it is also possible that ICAM-1 expression can be influenced at the translational and/or post-translational levels. The kinetics of ICAM-1 surface expression on endothelial cells elicited by TNF-α and PMA are very similar even though the kinetics of ICAM-1 mRNA induction are quite different.[101] PMA-induced message typically lags 2 hours behind its TNF-α-induced counterpart, suggesting that one or more steps between mRNA and surface expression may be accelerated. These steps might include translation, glycosylation, trafficking to the plasma membrane, or other steps. Although ICAM-1 cell surface expression induced by cytokines persists for at least 72 hr (even in the absence of detectable ICAM-1 mRNA), phorbol ester-induced ICAM-1 expression is transient, lasting only one day, suggesting that the turnover of ICAM-1 in the plasma membrane might be differentially regulated.

Transcriptional Regulation of the ICAM Gene

The 5' flanking region of the human ICAM-1 gene was cloned and sequenced independently by several groups in the early 1990s.[95,114-116] The corresponding upstream sequences of the murine ICAM-1 gene have also been determined (96, 117, and John Monroe, University of Pennsylvania, personal communication), and are provided for comparison (Fig. 4-3). Noteworthy is the high degree of sequence conservation across species, particularly in the proximal region containing numerous putative regulatory elements. Based on sequence analysis, Wawryk et al[115] described the 5' upstream region and first exon of the ICAM-1 gene as a CpG island, in that it was (G+C)-rich and possessed a high frequency of the CpG dinucleotide. In addition, the ICAM-1 promoter region appeared to be hypomethylated irrespective of ICAM-1 expression levels in the tissues examined. Thus it appeared unlikely that ICAM-1 expression was regulated by gene methylation. These characteristics of the ICAM-1 promoter are similar in many respects to the promoters of 'housekeeping genes'. Wawryk et al[115] described a constitutive DNase I-hypersensitive site 1.5 kb upstream of the transcription start site in the Burkitt lymphoma cell line Raji and the human colon carcinoma cell line 1215. However, no mention was made of additional DNase I hypersensitive sites following cellular activation. Dnase hypersensitive sites in chromatin are often used to identify regulatory regions of genes. In keeping with the strongly regulated nature of ICAM-1 expression, numerous potential cis-acting regulatory elements corresponding to consensus sequence motifs of known transcription factor binding sites have also been described (Fig. 4-3 and Table 4-2, see below). These early studies also established that the 5' flanking sequences of the ICAM-1 gene possessed both constitutive and inducible promoter activity when subcloned into heterologous reporter gene constructs.[95,115,116]

Transcription Start Sites: The presence of two apparently functional TATA boxes and three reported transcription start sites provided early indications of the complexity of ICAM-1 transcriptional regulation. Voraberger et al[95] identified two transcription initiation sites 319 and 41 bp upstream of the AUG translation initiation codon in HS913T fibrosarcoma cells and A549 adenocarcinoma cells (Fig. 4-3). Consensus TATA boxes were found about 25 bp upstream of each start site. These investigators observed that the distal start site was utilized following PMA induction, whereas constitutive and TNF-α-stimulated transcription occurred at the proximal start site. Degitz et al[116] localized the ICAM-1 transcription start site to 39-40 bp upstream of the translation initiation site in IFNγ-stimulated primary human keratinocytes and the A431 epidermoid squamous cell carcinoma line. Wawryk et al[115] reported a start site about 124 bp upstream of the translation initiation site in Raji cells using primer extension analysis. Although the sequence AAATAA is present 27 bp upstream, Wawryk et al[115] did not determine whether it functioned as a TATA element. In TNF-α- and PMA-stimulated human umbilical vein endothelial cells, ICAM-1 transcription begins 40-41 bp upstream of the translation initiation site. Although the proximal TATA box appears to direct both constitutive and cytokine-inducible transcription in many cases, the roles of additional upstream start sites remain enigmatic[142]. The ICAM-1 cDNA isolated by Staunton et al[7] from an HL60 cell library, for example, ends 17 bp upstream of the proximal transcription start site.

ICAM-1 exhibits multiple mRNA species in most tissues, the most abundant transcript being 3.0-3.3 kb, but 2.4 kb and 1.9 kb forms have also been reported.[7,8,101] The molecular basis for the multiple forms of mRNA is not clear. Until recently, there was no evidence for alternative spicing of human ICAM-1 RNA, nor could multiple transcription start sites explain the mRNA species observed RNA processing, particularly at the 3'-UTR, might contribute to the multiple mRNA species observed.[8]

Inducible Regulation

ICAM-1 can be upregulated by a variety of stimuli (summarized in Table 4-2). Recent publications have established that these stimuli can act through various signal transduction pathways leading to the activation of different transcription factors which can act individually or in concert to activate ICAM-1 gene transcription. The transcriptional activation by various classes of stimuli are summarized in Table 4-3 and discussed in detail below.

Cytokines: The cytokines, TNF-α, IL-1β, and IFNγ, are probably some of the most intensively studied regulators of inducible ICAM-1 expression. These agents have all been shown to regulate ICAM-1 gene expression at the transcriptional level. The IL-1β receptor (IL1-R) and the two TNF-α receptors (75 kDa TNF-RI and 55 kDa TNF-RII) both employ NF-κB in nuclear signaling to the same responsive elements on the promoter. The IFNγ receptor, however, utilizes the JAnus Kinase (JAK)/Signal Transducer and Activator of Transcription (STAT) signaling pathway to activate the promoter through a distinct response element.

```
                                                     NF-κB
HUMAN GAATTCAGAACTCCTCAGCCCCCCAAGAAAAAAATATCCCCG|GGAAATTCG|TGGGAAT  -1335 (-1295)

MOUSE -------------------------------------------------------

                                           AP-1
HUMAN GACCGAGGCGGGGGAAATATGCGTCTCTGGATGGCCAG|TGACTCG|CAGCCCCCTTCCCCG  -1275 (-1235)
      :::    :::    :      :: :   ::: ::
MOUSE -----AGGCAGG-------------------ATCGGGAGT---TCAAGGTCATCCTCAACT  -1058

HUMAN ATAGGAAGGGCCTGCGCGTCCGGGGACCCTTCGCTTCCCCTTCTGCTGCGCGGACCTCCCT  -1215 (-1175)
      :  :::            :  :     :: :: :  ::::: ::::
MOUSE ACAGG---------GCGTTTGAGG----------TCACCCTGAGCTACTTGAC--ACC     -1020

HUMAN GGCCCCTCGGAGAT-CTCCATGGCGAC-GCCGCGCGCGCCCCACAACAGGAAAGCCTTAG   -1157 (-1117)
      : :  :   :  :: : :   :: :   :  :    :  :   :: ::: : ::
MOUSE GACCTCAAGAAGAGACAGCAGGGGGAGAGATGAGAGGGGG----AAGGGGAGAAACTA--   -966

HUMAN GCGGCGCGGCTTGGTGCTCGGAGACTTAAGAGTACCCAGCCTCGACGTGGTGGATGTCGA   -1097 (-1057)
      :: :::   :GG----:::::   :: :::          ::: :::: :: :::
MOUSE ---GCACGG---GG----CGGGGCC---AGAAGA-------------TGGAGGAAGTGGG   -932

HUMAN GTCTTGGGGTCACACGCACAGGCGGTGGCCAAGCAAACACCCGCTCATATTTAGTGCATG   -1037 (-997)
      :                 :     :  :   :                         :  :
MOUSE GGAAAGGAGTGAGAAGGAGAGG--------------AAACCCCCCCATA--------AGG   -894

HUMAN AGC-CTGGGTTCGAGTTGCCGGAGCCTCGCGGCGTAGGGCAGGGGTTCGAGCGCCCCTTCT   -978 (-938)
      :::::::  :::: :  ::: :::: :  ::::: :   :::::
MOUSE CGCGCTGGG---GAGTGG--GGGATCTCT-GCTTTCTG-AGAGGT---------------   -856

                    AP-2        AP-1/ETS
HUMAN CCCTGCCTCGCCTCTGC|GCCTGGGG|CTGC|TGCCTCAGTTTCCC|AGCGACAGGCAGGGAT   -918 (-878)
      ::::: :::::  : :::    : : :    :: ::::::: :
MOUSE ----GCCTCTCCTCTTC-----GGAG-------TTCTGTTTCTCATCCA-----------   -823

                                               AP-1/ETS
HUMAN TTCGAGCGTCCCCCTCCCCTCCCTCGTCAAGATCCAAGCTAGC|TGCCTCAGTTTCCC|CGC   -858 (-818)
      :  ::           ::::    :::  :  :  :
MOUSE -------GTCT-----------------AAGA---AAGGGAGGTG---CAGGGTTCC---   -796

HUMAN GGAGCCTGGGACGCCAGCGGAGGGGCTCGGCGCGTAGGGATCACGCAGCTTCCTTCCTTT   -798 (-758)
      :::::: :::             :: :: ::::::::::::
MOUSE ----------ACGCCACAGGAC-------------------CAGACATCTTCCTTCCTTT   -765

HUMAN TTCTGGGAGCTGTAAAGACGCCTCCGCGGCCAAGGCCGAAA-GGGGAAGCGAGGAGGCC|G|  -739 (-699)
      :       :::::::: :::::::: :  : ::: :  ::::  :::::: ::::    :::|
MOUSE TGG-GGGAGCTGTCAAGACGCCCCAGAAGTCAAAGCAGAAAAGGGGGAACCGAG---GCCG  -709

           EGR-1
HUMAN |CCGGGGTG|AGTGCCCTCGGGTGTAG-AGAGAGGACGCC--GATTTCCCCGGACGTGGT|GA|  -682 (-642)
      |: :::::  |       :: : :::::: ::::::             ::: ::::|
MOUSE |CGGGGGCG|GGAGCAGTCGTGAGTGGGAGAAAGGGAAGTGAGGACTGCCTC--------TCA  -657

      SSRE       CRE
HUMAN |GACC|CGCGC|TTCGTCA|CTCCCACGGTTAGCGG-TC-GCCGGGAGGTGCCTGGCTCTGCTCT   -624 (-584)
      : :   ::: ::::          ::::::::  :::::::: :::: :::
MOUSE GGCAGGACTTTCTCAC---------AGCGGATCTGTCCGGGCCTGCTTGGATC-GCT--   -610

HUMAN GGCCGCTTCTCGAGAAATGCCCGTGTCAGCTAGGTGTGGACGTGACCTAGGGGGAGGGGC   -564 (-524)
      :: ::::  :: :       :: ::  :::::  :    :::  :::: :::::::::::
MOUSE -GCTTCATCTCTAG--------TGCCAAGTGGGTGGAGCATGGTCCTGGGGAGAGGGGC   -560

                               NF-κB
HUMAN ATCCCTCAGTGGAGGGAGCCCGG|GGAGGATTCCT|GGGCCCCCA---CCCAGGCAGGGGGC   -507 (-467)
      : : : :::: : :::     : ::::::: :::  :::::::
MOUSE ACCTCTGAATGGATGGCGTCCCCTAACGATTCCCAGGGAGCCCTTATCCCCGGCAAGGGCA   -500

HUMAN TCATCCACTCGATTAAAGAGGCCTGCGTAAGCTGGAGAGGG-AGGACTTGAGTTCGGACC   -448 (-408)
      : ::  ::::::     ::::::            :::::::: ::::::  ::::::: :
MOUSE TG-TCTGGTGGGTTAAAGAGGCTTGCAGTAGTTGGGGAAATCAGGACTTGATTTCGGATC   -441

                                          AP-3
HUMAN CCCTCGCAGCCTGGAGTCTCAGTTTACCGCTT|TGTGAAATGGA|CACAATAACAGTCTCCA   -388 (-348)
      :: :::                  ::::::  : :: ::::::
MOUSE C--TCG-------------AGGATCCC----|TGCGAAATGCC|GAGCCTCAGTTTATCC-   -402
```

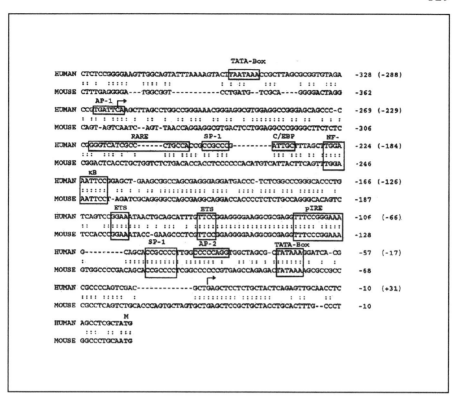

Figure 4-3. The ICAM-1 promoter. Upstream regulatory sequences of the human and mouse gene are shown. The human sequence is a consensus of published sequences by several groups.[38,69,70,71] The mouse sequence combines published[39] and unpublished data (JS Maltzman and JC Monroe, personal communication). Putative regulatory elements are indicated by the boxes, and the arrows indicate the transcriptional start sites of the human ICAM-1 gene. The numbering system is relative to the translation start site (ATG). The numbers in brackets refer to the numbering system relative to the proximal transcription start site.

TNF-α: Direct evidence for TNF-α-regulated ICAM-1 transcription in A549 cells and primary cultures of endothelial cells was provided by nuclear run-on assays,[101,118] a technique widely used to measure the rate of gene transcription initiation. In HUVECs, TNF-α-induced transcription is transient with the maximal rates achieved at 1 hour and returning to near basal levels by 2 hours despite the continued presence of cytokine.[101] This 2 hours transcriptional window is essential for cell surface ICAM-1 expression. If RNA synthesis is blocked with actinomycin D at different time points following TNF-α addition, surface ICAM-1 expression is refractory to inhibition by actinomycin D after 2 hour.[101] The induction of ICAM-1 mRNA by TNF-α is also transient, peaking at 2 hours and declining to near control levels by 24 hours.

Table 4-3.
Functionally Important Regions and Elements of the Human ICAM-1 Promoter

Stimulus	Responsive Region	Transcription Factor
TNF-α	NF-κB/Rel (-187/-170) TGGAAATTCC	RelA RelA/p50 RelA/c-Rel
	C/EBP (-199/-195) ATTGC	C/EBPα C/EBPα/ C/EBPβ C/EBPβ
IL-1β	NF-κB /Rel (-187/-170)	ND
IFNγ	pIRE (-76/-66) TTTCCGGGAAA	STAT1α
IL-6	pIRE	STAT1α

Table 4-3. (Continued)

Retinoic Acid	RARE	RAR RXR (inferred)
	(-226/-210) GGGTCATCGCCCTGCCA	
LPS	NF-κB /Rel (-187/-178)	RelA RelA/c-rel
Tax1	CRE-like (-630/-624) TTCGTCA	Unidentified complexes
	NF-κB (-187-178)	ND
Tax2	NF-κB (-187/-178)	ND
	SP1 (-59/-54) CCGCCC	ND

Table 4-3. (Continued)

	Proximal 282 bp	
Lyso-PC		ND
B-Cell Receptor Cross Linking	egr-1 (-709/-701)	ND ND
H2O2	AP-1/Ets-like (-908/-894) (-834/-821) TGCCTCAGTTCCC	ND
PDTC	(-353/-136)	AP-1 (inferred) (AP-1 at (-284/-278))
UV-A	(-105/-38)	AP-2 (inferred) (AP-2 like at (-48/-41))
Shear Stress	(-716/+2)	ND (SSRE like at (-644/-639))

The induction of ICAM-1 message did not require protein synthesis, indicating that the ICAM-1 gene is an immediate response gene. The protein synthesis inhibitor, cycloheximide, induces ICAM-1 message in the absence of any stimulus, and superinduces message levels in TNF-α-treated HUVECs.[101] Cycloheximide also induces ICAM-1 mRNA either alone or in combination with cytokines in rat astrocytes.[107] The effects of cycloheximide may be a result of message stabilization, direct activation of a nuclear signaling pathway, or removal of a labile inhibitor or transcriptional repressor. Protein synthesis inhibitors can stabilize short-lived mRNA species.[119] The results of nuclear run-on assays also suggested that cycloheximide might induce ICAM-1 transcription (S.J. Wertheimer and T.P. Parks, unpublished observations), as was shown for the E-selectin gene.[120] Certain protein synthesis inhibitors, e.g., anisomycin, are capable of activating the stress kinase limbs of the mitogen-activated protein kinase (MAP kinase) cascade, i.e.,those mediated by p38 and c-Jun N-terminal kinase (JNK)/stress-activated protein kinase(SAPK).[121-124] Transcription of the E-selectin gene is stimulated by several protein synthesis inhibitors at concentrations below those which block protein synthesis.[120] IκBα is a labile inhibitor of the NF-κB/Rel family of transcription factors that sequesters Rel proteins in the cytosol. Blocking the synthesis of IκBα would result in decreased IκBα levels and enhance the translocation of NF-κB /Rel proteins into the nucleus, and/or delay their removal from the nucleus, since IκBα and NF-κB interact in an autoregulatory loop.[125] The transient transcription of the E-selectin gene, for example, has been linked to postinduction transcription repression mediated by IκBα.[126]

The effects of TNF-α are mediated by two ubiquitously expressed receptors of about 55 kDa (TNF-RI) and 75 kDa (TNF-RII).[127] The ability of TNF-α receptors to activate NF-κB appears to be mediated in part by an adaptor protein TNF Receptor Associated Factor 2 (TRAF2).[128] TRAF2 associates directly with the cytoplasmic domain of TNF-RII, whereas it interacts indirectly with TNF-RI via binding to TNF-RI Associated Death Domain protein (TRADD). Overexpression of TRAF2 activates NF-κB, whereas overexpression of a dominant negative TRAF2, lacking the N-terminal RING finger motif, blocks TNF-α-induced NF-κB activation. TRADD can also bind the death domain protein FADD (Fas Receptor associated Death Domain protein) which appears to participate in TNF-RI (as well as Fas) mediated apoptosis. Another death domain-containing molecule, RIP (Receptor Interacting Protein), binds to TRADD and reportedly participates in both NF-κB activation and induction of apoptosis.[129,130] Studies using cells from TRAF2 deficient mice and dominant negative transgenic mice revealed that TRAF2 is required for TNF-α-induced activation of the JNK/SAPK pathway, but not for the NF-κB pathway.[131,132] Cells from RIP deficient mice, on the other hand, were highly sensitive to TNF-α-induced apoptosis that was accompanied by a failure to activate NF-κB.[133] The connection between RIP and NF-κB activation remains unclear. Several groups have identified a protein, TANK (TNF family member associated NF-κB activator) or I-TRAF (TRAF Interacting protein), which associates with the C-domain (or TRAF domain) of several TRAFs, including TRAF2.[134,135] TANK has been shown to synergistically activate NF-κB when co-expressed with TRAF2 in 293 cells, or to inhibit TRAF2-mediated NF-κB activation.[134, 135] These disparate effects probably reflect differences in the stoichiometry of the molecules expressed during transient transfection experiments. The C-terminus of TANK, which does not interact with TRAF2, contains an inhibitory domain which can inhibit activation of NF-κB by TRAF2 or TNF-α.[134] TRAF2 binds NIK (NF-κB inducing kinase) and NIK, in turn, appears to activate the IκB Kinase (IKK) complex by phosphorylating IKK-2. Activated IKK specifically phosphorylates IκBα at Ser 32 and Ser 36,[136,137] tagging the molecule for ubiquitination and destruction via the ubiquitin-proteasome pathway. The signaling events are also discussed in Chapters 7 and 10.

In the first systematic analysis of the ICAM-1 promoter, Voraberger et al[95] identified a TNF-α responsive region in A549 cells between 176 to 329 bp upstream of the ICAM translation start site. Although the transcription factor NF-κB was suspected of playing a role in TNF-α induced gene expression, sequence analysis of the ICAM-1 promoter did not reveal any consensus NF-κB binding motif within the TNF-α-responsive region.[95] Subsequently, van de Stolpe et al[138] identified a region between -227 and -175 (relative to the transcription start site) which responded to both TNF-α and PMA in human 293 embryonal kidney cells. A variant κB site in this region bound TNF-α- (and PMA-) inducible protein complexes from U937 cell nuclear extracts with the appropriate specificity, i.e., they could be competed by unlabeled oligonucleotides containing consensus NF-κB or ICAM-1 NF-κB sites but not irrelevant sequences. Further support for the role of this element was provided by the demonstration that a multimerized (3x) ICAM-1 κB element coupled to a heterologous thymidine kinase promoter could drive transcription of a luciferase gene in 293 cells in response to TNF-α or PMA treatment. Transactivation from this element induced by TNF-α or PMA was suppressed by ectopic expression of the glucocorticoid receptor and addition of dexamethasone.[139] Subsequent studies on Mel JuSo melanoma cells[140] confirmed that the -277/-174 region of the ICAM-1 promoter is essential for TNF-α responsiveness, and antibody supershift studies suggest that the variant κB element binds two complexes composed of RelA/p50 (lower complex) and RelA/c-Rel (upper complex).[140] Independently, two other laboratories also converged on the variant κB element as the TNF-α responsive element.[141,142]

The studies of Hou et al[141] employed a linker scanning mutagenesis strategy to identify two adjacent sites which confer TNF-α responsiveness to the ICAM-1 promoter in HepG2 hepatoma cells. The variant κB element was essential for transactivation, whereas mutation of the adjacent C/EBP (CAAT Enhancer Binding Protein) site reduced ICAM-1 promoter activity by two-fold, leading to the suggestion that the C/EBP and κB elements formed a composite binding site. Antibody supershift studies using a radiolabelled oligonucleotide probe containing the C/EBP site suggested that this site bound C/EBPα, and β, but not γ.

Analogous experiments using an ICAM-1 κB probe identified two complexes containing RelA/p50 (lower complex) and RelA homodimers (upper complex), but an interaction between C/EBP and NF-κB family members was not described.

Ledebur and Parks[142] analyzed the TNF-α-responsive region of the ICAM-1 promoter by deletion analysis, competitive electrophoretic mobility shift assays (EMSA), and site-directed mutagenesis. The TNF-α-responsive region was narrowed down to the variant κB element, and mutation of this element was shown to completely eliminate the functional responses to not only TNF-α, but also to IL-1β, and LPS. In contrast, the response to PMA was incompletely suppressed. The ICAM-1 κB element was shown to bind RelA homodimers (upper complex) and RelA/p50 heterodimers (lower complex) by methylation interference footprinting, antibody supershift, immunoprecipitation and photoaffinity labelling experiments. Ectopic expression of RelA was shown to drive transcription of a luciferase reporter gene under control of the 1.3 kb ICAM-1 promoter or an ICAM-1 κB element coupled to a minimal SV40 promoter. Coexpression of p50 or IκBα inhibited RelA transactivation, indicating that p50 behaved as a transdominant inhibitor rather than a synergistic activator, as observed for the E-selectin and VCAM-1 promoters.[125,143,144] These results suggest that RelA may be primarily responsible for driving ICAM-1 transcription in response to NF-κB-activating stimuli.

Catron et al[145] provided biochemical and functional evidence that the adjacent C/EBP and NF-κB motifs on the ICAM-1 promoter formed a composite element. A TNF-α-induced DNA binding complex was observed in gel shift experiments with epithelial cell nuclear extracts that contained C/EBPβ and RelA. The complex required intact C/EBP and NF-κB binding sites and could be reconstituted using recombinant proteins. Binding was cooperative in that RelA recruited C/EBPβ to the complex. Functional studies revealed that both binding sites were required for maximal activation of the ICAM-1 promoter by TNF-α, and for synergistic activation by ectopically expressed RelA and C/EBPβ in epithelial cells.

The variant κB element in the ICAM-1 promoter preferentially recruits RelA homodimers and RelA/c-Rel heterodimers to the promoter, depending on the cell type.[146,147] Conversion of the ICAM κB element to a HIV-LTR κB element, a known high affinity binding site for RelA/p50 heterodimers, leads to a selective increase in heterodimer binding, suggesting that the variant κB element selected against heterodimer binding.(148, 149, and T.P. Parks, unpublished observations). Interestingly, the switch to a HIV κB element also results in decreased functional activity. This latter effect may be due to the inferior ability of the RelA/p50 heterodimer to transactivate as compared to RelA homodimers.

IL-1β: Although IL-1β is capable of modest ICAM-1 mRNA stabilization, the induction of ICAM-1 message and cell surface expression appears to occur primarily at the transcriptional level[107] Like TNF-α, induction of ICAM-1 mRNA by IL-1β does not require ongoing protein synthesis.[107] IL-1β, like TNF-α, is a potent stimulus for the activation and nuclear translocation of NF-κB. However, the cytoplasmic domains of the IL-1β receptor (IL1-R) and the TBF-α receptors do not share sequence homology. Recently, it was shown that the cytoplasmic domain of the activated IL1-R can recruit and bind a novel protein, IRAK (IL1-R associated kinase), which possesses a kinase-like domain.[150] IRAK, in turn, binds TRAF6, a new member of the growing TRAF family.[151] TRAF6 and TRAF2 both activate NIK, which represents a potential convergence point for TNF-αand IL-1β signaling to NF-κB. Recently, TGFβ -activated kinase (TAK1), a member of the MAP kinase kinase kinase family,[152] has been implicated in IL-1β signaling. In addition to TGFβ, TAK1 can be activated by a number of stressful stimuli, including IL-1β and LPS, and appears to be a signaling intermediate in both the stress kinase (JNK/SAPK and p38) and NF-κB pathways. IL-1β can induce the recruitment of TAK1 to TRAF6, and activated TAK1 can phosphorylate NIK.[153] As noted earlier, NIK was recently shown to interact with IKK1[136,137] a helix-loop-helix and leucine zipper protein originally identified as CHUK (conserved helix-loop-helix ubiquitous kinase)[154] and IKK2. Activation of the IKK complex depends on phosphorylation of its IKK2 subunit.[155]

Several early studies, based primarily on studies with inhibitors, suggested that IL-1β induction of ICAM-1 was dependent on protein kinase C (PKC) activation.[102,156,157] Later studies, however, suggest that neither PKC inhibitors, nor down-regulation of PKC by prolonged exposure to PMA had much effect on ICAM-1 induction by IL-1β or TNF-α in HUVECs.[103] In astrocytes, IL-1β and TNF-α were shown to increase PKC activity[158] and the inhibitors H7, H8, and calphostin C attenuated ICAM-1 expression.[107] In astrocytes, PKC down-regulation also inhibited cytokine responses, but in a biphasic manner. IL-1β-induced ICAM-1 expression, for example, was sensitive to PKC down-regulation 48 following treatment with PMA (87% inhibition), but relatively refractory to PKC depletion 72 post-PMA treatment (18% inhibition). In HUVECs, ICAM-1 expression was still inhibited by H7 following PKC depletion after 72 of PMA-treatment,[103] possibly suggesting the involvement of a phorbol ester insensitive PKC isoform, e.g., PKCζ or λ,[159] or an entirely different kinase, e.g., NIK or IKK. It is important to note that studies relying solely on pharmacological manipulation of PKC should be interpreted with caution since these compounds are not absolutely specific.

IL-1β is a well-known activator of NF-κB in a variety of cell types. In transiently transfected HUVECs, mutation of the ICAM-1 κB site in the context of the 1.3 kb ICAM-1 promoter resulted in the complete loss of IL-1β responsiveness.[142] Activation of the endogenous ICAM-1 gene by IL-1β was suppressed completely in A549 cells stably expressing a dominant active inhibitor of NF-κB, i.e., IκBα with Ser 32 and 36 mutated to Ala (Y. Li and T.P. Parks, unpublished observations). These observations strongly suggest that NF-κB plays a central role in ICAM-1 transcriptional upregulation by IL-1β.

IFNγ: IFNγ markedly upregulates ICAM-1 expression on several cells types, particularly epithelial cells and cells of the monocytic lineage, including human cervical carcinoma (HeLa)[115] lung adenocarcinoma (A549)[95] and squamous carcinoma cell lines (A431).[160] Look et al[161] demonstrated that IFNγ stimulated ICAM-1 gene transcription in human primary tracheal epithelial cells and localized the IFNγ responsive region to 130 to 94 bp upstream of the AUG (-90 to -54 from the transcriptional start site). This region also responded to IFNγ stimulation when coupled to a heterologous promoter. A palindromic interferon response element, pIRE, was identified at -76 to -66 and was is capable of binding nuclear proteins. This site shares greatest sequence homology with IFNγ responsive elements in the human FcγRI and IRF-1 genes[162] and the murine ICSBP gene.[163] The pIRE is functional in HepG2 human hepatoma cells and in the human melanoma cell line Mel JuSo, and binds STAT1α (previously known as p91).[140,141] ICAM-1 appears to function as a primary IFNγ response gene insofar as ICAM-1 message induction by IFNγ did not require protein synthesis.[164] In addition, ICAM-1 induction by IFNγ was completely blocked in STAT1α knockout mice, and in cells expressing a dominant negative Stat1.[165] These observations are consistent with a primary role of STAT1α in mediating the effects of IFNγ.[166,167] STAT1α is activated by JAK1 or JAK2, which phosphorylates STAT1α on tyrosine residues resulting in dimerization, and translocation to the nucleus. Phosphorylation of STAT1α on serine and threonine residues by another signaling pathway may be also be involved in transactivation.[168]

ICAM-1 expression on Mel JuSo melanoma cells is upregulated by IFNα.[140] Treatment of Mel JuSo cells with IFNα induces a nuclear binding complex indistinguishable from that stimulated by IFNγ and which is recognized by antibodies directed against STAT1α. This observation is surprising in light of the finding that class I interferons (IFNα/IFNβ) activate ISGF3 complexes, which consist of STAT1α, STAT2, and p48, where p48 is the DNA binding subunit.[169]

It is possible that factors other than STAT1α may bind to the interferon-responsive element of the ICAM-1 gene and regulate its expression. IFNγ, for example, can upregulate expression of IFN-regulatory factor 1 (IRF-1).[170,171] Although initially described as a factor which binds to a cis-acting regulatory element in the IFNγ promoter,[172-174] IRF-1 is now known to bind GAS (Gamma Activating Sequence)-like sequences in a number of genes including the FcγRI gene, which (as noted above) is similar to the ICAM-1 pIRE. However, it is not known whether IRF-1 binds to the ICAM-1 pIRE, or whether it plays a role in ICAM-1 expression. An intriguing aspect of IRF-1 is that it can be induced by several hormones and cytokines in addition to IFNγ, including IL-6, TNF-α, IL-1β, IFNγ, leukemia inhibitory factor, and prolactin.[174-176]   Recently,   a   novel   STAT-like   factor,   LIL (Lipopolysaccharide and IL-1 induced factor), was shown to be induced by LPS, IL-6, and IL-1β and to bind to a GAS-like sequence in the human prointerleukin-1β gene.[177] Unlike IRF-1, this factor was found to be activated immediately and did not depend on protein synthesis. It will be interesting to determine whether this factor is involved in ICAM-1 upregulation.

Although IFNγ and TNF-α are known to synergize in the expression of several genes during inflammation, including ICAM-1[104,106] it is still not clear to what extent this phenomenon is mediated at the transcriptional level. Hou et al[141] reported that TNF-α and IFNγ activated the ICAM-1 promoter in an additive fashion, whereas Jahnke and Johnson[140] observed synergism. Ohmori et al[178] observed that TNF-α and IFNγ synergistically induced expression of ICAM-1 mRNA in normal, but not STAT1α-deficient mouse fibroblasts. Intact NF-κB and STAT1α binding elements are required for synergistic activation of the IRF-1 promoter by TNF-α and IFNγ. The interactive effects of IFNγ and TNF-α on ICAM-1 expression are probably dependent upon cell-specific characteristics, including nuclear signaling, transcriptional, and post-transcriptional regulatory events. Additional affects of IFNγ which may contribute to its additive/synergistic effects on cytokine induction include stabilization of ICAM-1 message and upregulation of TNF-α receptors.[113]

IL-6:   Interleukin 6 (IL-6) can upregulate ICAM-1 expression on select populations of cells. The proximal region of the ICAM-1 promoter (-76 to -66) confered responsiveness to both IL-6 and IFNγ in HepG2 cells,[179] an effect which was localized to the palindromic I IFNγ response element. The ICAM-1 pIRE also confered both IL-6 and IFNγ responsiveness when linked to heterologous promoters.[179,180] The ICAM-1 pIRE binds both STAT1α and a factor with the characteristics of STAT3 (also known as the acute phase response factor). IFNα activated STAT1α in HepG2 cells and undifferentiated U937 cells, whereas IL-6 activated a STAT3-like factor as well as STAT1α.[179,180] Differentiation of U937 cells with PMA abolished the responses to IL-6, but not to IFNγ,[179] suggesting that IL-6 and IFNγ activated different signaling pathways which converged on a common response element in the ICAM-1 promoter. The induction of ICAM-1 mRNA by IL-6 in mouse macrophages did not require STAT1α. The ability of IL-6 to induce ICAM-1 expression suggested that ICAM-1 may play a role as an acute phase response gene.[181] An interesting feature to IL-6 is that many cells express gp130, a subunit common to many cytokine receptors, but not the IL-6 specific α chain. However, cells that do not express the IL-6 specific α chain (e.g., endothelial cells) can still respond to IL-6 in the presence of the soluble IL-6 receptor α chain.[182]

Phorbol Esters: The phorbol ester PMA is the prototypical activator of PKC and is a potent stimulantor of ICAM-1 expression. Antisense experiments suggested that the PKCα isoform was responsible for PMA upregulation of ICAM-1 in A549 cells.[100] Initial promoter studies by Voraberger et al[95] suggested that the 393 bp region upstream of the ATG was responsive to PMA (≥ 6-fold induction) inA549 cells. The upstream 176 bp also responded to PMA, albeit with half the activity of the 393 bp region. At the time this study was conducted, an AP-1 site at -324 and an AP-2 site at -88 were considered the likely elements mediating PMA responsiveness. Subsequently, van der Stolpe et al[138] provided evidence that PMA responsiveness in 293 cells was largely mediated via the variant κB element -187/-178 bp relative to the transcription start site, an observation confirmed in HUVECs[142] and A431 cells.[183] A residual response to PMA was still observed in promoter constructs harboring mutated κB elements or truncations proximal to the κB element, suggesting that there are probably multiple PMA-responsive elements within the ICAM-1 promoter.[95,138,142]   Several different transcription factors are known to mediate PMA-induced gene transcription, including AP-1, AP-2, AP-3, as well as NF-κB.[184-186] Using promoter deletion constructs linked to a luciferase reporter gene in 293 human embryonal kidney cells, van de Stolpe et al[138] identified 4 regions of the ICAM-1 promoter which were PMA-responsive and contained known PMA responsive enhancer elements: an AP-3-like site at -375/-365; TRE-like (AP-1) sequence at -284/-279; an NF-κB element at -187/-178; and an AP2-like site at -48/-41 relative to the transcription start site. Transactivation from the TRE at -284/-279 appears to be directed by the distal TATA box (-313/-307) which is absent from the corresponding murine promoter (Fig. 4-2). Previously it had been shown that PMA-induced ICAM-1 mRNA is directed by the distal TATA box in A549 cells.[95]

In B-lymphocytes, ICAM-1 is upregulated in response to activation of the B cell antigen receptor (BCR) or by stimulation with PMA, effects which appear to be mediated, at least in part, by early growth response gene -1 (Egr-1).[117] ICAM-1 transcription induced by BCR crosslinking or PMA was absent in B cell lines defective in Egr-1 expression. Egr-1 is encoded by an immediate early gene and is induced transiently in response to BCR cross-linking or PMA. The murine ICAM-1 promoter contains a functional Egr-1 site at -709/-701 (relative to the murine ICAM AUG) whereas the function of the corresponding Egr-1 site in the human promoter has not been established. Mutation of the Egr-1 site within the context of a 1.1 kb murine ICAM-1 promoter construct reduced promoter activity in response to PMA or BCR cross-linking in primary murine B cells. Ectopic expression of Egr-1 transactivated the wild type murine ICAM-1 promoter, but not a promoter harboring a mutated Egr-1 binding site.[187] The potential role of Egr-1 in regulating the ICAM-1 promoter in endothelial cells has not been explored, but may be of potential significance in view of recent observations that shear stress induces both ICAM-1 upregulation[188] and Egr-1 expression.[189]

PMA is a well-known activator of NF-κB[186] and NF-κB activation has been proposed to mediate PMA induced ICAM-1 expression[183] although the mechanism whereby PKC activates NF-κB is unclear. Antioxidants can inhibit PMA-induced NF-κB activation[190] suggesting that PKC-induced reactive oxygen intermediates (ROI) might activate NF-κB. Since PMA promotes the phosphorylation of IκBα, PMA-induced oxidants could stimulate the activity of IKK, or inhibit a phosphatase.[191] PKCα can directly phosphorylate IκBα *in vitro*[192] although the biological significance of this observation is questionable. Recently, a novel IKK-related kinase, named NF-κB-activating kinase (NAK), was identified that may mediate IKK and NF-κB activation in response to PMA and growth factors that stimulate PKCε activity.[193] The same molecule was independently discovered by Pomerantz and Baltimore[194] as a TANK-binding kinase (TBK1) in a yeast two-hybrid screen. TBK1 can form a terniary complex with TANK and TRAF2 when overexpressed in cells. The role of IKK-related kinases in cell signaling to NF-κB is an important topic for future investigation.

CD40: CD40 is a member of the TNFR superfamily, and is activated by CD40 ligand (or gp39), an inducible membrane-bound protein related to TNF-α and is expressed on activated T cells.[195] CD40 is predominantly expressed on B cells, but is also present on epithelial and endothelial cells.[196-198] Engagement of CD40 by membrane-bound or soluble CD40 ligand upregulates ICAM-1 expression on several cell types, including B cells[199] and vascular endothelial cells.[196,199,200] CD40 ligand is a weaker inducer of ICAM-1 expression than TNF-α or IL-1β, but each of these stimuli exhibited homologous desensitization.[200] TRAF2, TRAF3, TRAF5, and TRAF6 can associate with the cytoplasmic domain of CD40 and, with the exception of TRAF3, can activate NF-κB when ectopically expressed. It has also been reported that the cytoplasmic domain of CD40 binds JAK3 and upon CD40 engagement, JAK3 and STAT3 are activated.[199] Subsequently, *in vitro* binding experiments indicated that TRAF2 and TRAF3 bound to the CD40 cytoplasmic domain with considerably higher affinity than TRAF5 or TRAF6.[201] Using CD40 mutant constructs that were incapable of interacting with either TRAF2, TRAF3, or Jak3, it was demonstrated that an intact TRAF2 binding site was essential for CD40-mediated NF-κB and JNK/SAPK activation. In addition, it was shown that CD40-mediated upregulation of ICAM-1 was dependent on TRAF-2 and NF-κB.

Oxidants and Oxidative Stress: Reactive oxygen intermediates are formed during reperfusion of ischemic tissues, by inflammatory cells at sites of injury and inflammation, by oxidation of phospholipids at atherosclerotic lesions, and by a variety of stimuli which enhance tissue production of reactive oxygen species and thus promote oxidative stress. These latter stimuli include ionizing radiation and exposure of cells to cytokines or drugs.

ICAM-1 contributes to reperfusion injury of tissues following hypoxia/ischemia.[202] Reactive oxygen intermediates, such as H_2O_2, probably contribute to ICAM-1 upregulation, PMN accumulation, and tissue damage associated with reperfusion injury. Hydrogen peroxide (H_2O_2) upregulates ICAM-1 expression on human endothelial cells[203-205] and keratinocytes.[206] In HUVECs, addition of H_2O_2 increases ICAM-1 mRNA levels within 30 min, with maximal levels achieved at 1 hour.[204] ICAM-1 gene expression is transcriptionally regulated by H_2O_2, and an oxidant responsive region has been localized to sequences between 981 and 769 bp upstream of the translation initiation codon.[207] Within this region are two 16 bp repeats containing putative binding sites for AP-1 and Ets. A similar composite AP-1/Ets element is found in the macrophage scavenger receptor gene and confers H_2O_2 responsiveness to a heterologous minimal promoter.[208] Mutation of either the AP-1 or Ets binding sites within the 16 bp repeats abolishes oxidant-induced DNA binding activity and deletion of the region containing these elements from the ICAM-1 promoter prevented H_2O_2-inducible transcriptional activity. The nature of the factors bound to the AP-1/Ets element has not been reported. Although H_2O_2 activates NF-κB in a murine osteoblastic cell line[209] and in HeLa and Jurkat cells,[210] it did not activate NF-κB in HUVECs.[203,207] In addition, an inhibitor of NF-κB activation, pyrrolidine dithiocarbamate (PDTC), did not inhibit H_2O_2-induced ICAM-1 mRNA.[207] Interestingly, both TNF-α and H_2O_2 can induce DNA binding activity to the AP-1/Ets composite element as well as to an AP-1 binding site. Despite these observations, activation of the ICAM-1 promoter by H_2O_2 and TNF-α appeared to be mediated by different cis-elements. As discussed above, TNF-α induction is absolutely dependent on the variant κB element at -187/-178, although it is likely that upstream sequences may contribute to induction. Thus H_2O_2 may act through a novel response element independent of NF-κB. PDTC has been widely used as an inhibitor of NF-κB activation, an attribute widely ascribed to its anti-oxidant properties.[190] Although PDTC invariably inhibits cytokine-induced VCAM-1 and E-selectin expression on endothelial cells, it may be less effective on ICAM-1 expression,[211,212] although there are conflicting reports in endothelial cells[213] and fibroblasts. Conversely, H_2O_2 upregulates ICAM-1 but not VCAM or E-selectin expression on endothelial cells.[214]

PDTC alone can induce ICAM-1 expression at the message and protein level in HUVECs,[211] an action dependent upon a region at -353/-136 (relative to the transcription start site). PDTC induces binding of c-Fos and c-Jun to the ICAM-1 AP-1 site at -284, but inhibited TNF-α-induced NF-κB binding activity to the κB element at -187. Thus, it would appear that oxidants (e.g., H_2O_2) and anti-oxidants (e.g., PDTC) are capable of inducing ICAM-1 expression possibly through activation of AP-1 (and possibly Ets and other factors), and independently of NF-κB. A possible resolution of this apparent paradox is that PDTC may behave as a pro-oxidant. An anti-oxidant response element (ARE)[215] resembling an AP-1 element has been described in the promoters of the oxy-protective enzyme gene, glutathione-S-transferase Ya subunit (GST Ya) and NAD(P)H: quinone oxidoreductase. The anti-oxidants butylated hydoxyanisole (BHA) and PDTC induce AP-1 binding activity and GST Ya gene expression in hepatoma HepG2 cells, an effect mediated by pro-oxidant activity.[216] Exogenous catalase inhibited hydroxyl radical formation and GST Ya gene expression induced by the BHA metabolite *tert*-butylhydroquinone (TBHQ), indicating the intermediate formation of H_2O_2 during the auto-oxidation of these compounds.[216] Similarly, the induction of AP-1 binding activity and GST Ya gene expression by PDTC, BHA, or TBHQ is inhibited by the thiols *N*-acetyl cysteine (NAC) and reduced glutathione (GSH). Thus, BHA, TBHQ, and PDTC induction of AP-1 and GST Ya expression is probably mediated by oxidative stress, rather than through the anti-oxidant properties of these reagents. In hepatoma cells, BHA and TBHQ also induce NF-κB binding activity, an effect blocked by NAC and GSH. Pinkus et al[216] suggested that the ability of PDTC to inhibit NF-κB activation may be related to its metal chelator properties rather than an antioxidant effect of its thiol group. The metal-chelating ability of dithiocarbamates has been linked to inhibition of copper/zinc superoxide dismutase activity which leads to the potentiation of oxygen toxicity in animal tissues and decreased glutathione peroxidase activity and thiol.[217] A variety of metals are known to upregulated ICAM-1 expression (Table 4-2). In this regard, it is interesting to note that NAC inhibited ICAM-1 expression induced by H_2O_2 and PDTC,[203,207,211] Although PDTC reduces the intracellular oxidative state of endothelial cells, as measured by FACS analysis of dichlorofluorescein-stained cells,[218] this compound can also promote the oxidation of GSH to GSSG in thymocytes.[219] The disparate effects of antioxidants on AP-1 and NF-κB activation may reflect differences in the cell types and concentrations used.

Ionizing radiation (X-rays) produces an acute inflammatory response, characterized by edema and leukocyte infiltration. This response is mediated, at least in part, by cytokine release and adhesion molecule upregulation. For example, ICAM-1 expression is increased in the brains of irradiated mice[220] and on keratinocytes and endothelial cells of irradiated human split-skin cultures.[221] Ionizing radiation increases ICAM-1 mRNA and protein expression on several different human cell lines, including HL60, HeLa, and HaCaT (nontumorigenic human keratinocyte cell line).[221] Hallahan and co-workers demonstrated that thoracic irradiation caused a dose-dependent increase in ICAM-1 expression in the pulmonary microvascular endothelium.[222] These investigators also showed that ICAM-1 was required for inflammatory cell infiltration into the irradiated lung, using both anti-ICAM-1 monoclonal antibodies and ICAM-1 knock-out mice.[222]

ICAM-1 gene expression by ionizing radiation does not require *de novo* protein synthesis or the release of preformed cellular activators. Ionizing radiation induces ICAM-1 and E-selectin expression on human endothelial cells independently of cytokine production.[223,224] Transient transfection experiments demonstrated that the 1.2 kb 5' flanking sequence of the ICAM-1 gene was activated by ionizing radiation.[224]

Although the mechanism of ICAM-1 transcription induction by ionizing radiation remains unclear, it rapidly induced NF-κB and AP-1 DNA binding activity.[225-227] E-selectin upregulation by X-irradiation has been linked to activation of NF-κB.[224]

Ultraviolet (UV) radiation upregulated ICAM-1 expression in a number of different human cell lines and skin cultures.[221] Both UVA (320-340 nm) and UVB (280-320 nm) directly induced ICAM-1 expression on human dermal microvessel endothelial cells (HDMEC), although conflicting results have been reported for UVB-induced ICAM-1 expression on HDMEC or skin biopsies (Table 4-2). In addition, UVB inhibited ICAM-1 expression at the transcriptional level in human monocytes (Table 4-2).

UVA radiation induced ICAM-1 mRNA and protein expression in human keratinocytes and activated the ICAM-1 promoter in transient transfection experiments with reporter gene constructs (Table 4-2). The effects of UVA radiation were inhibited by singlet oxygen quenchers and mimicked by a singlet oxygen-generating system. UVA radiation and singlet oxygen both induce activation of an AP-2 like nuclear factor and activation of the ICAM-1 promoter by these stimuli required the proximal region of the promoter (-105/-38, relative to the transcription start site) containing a putative AP-2 element. Although UVB also induces ICAM-1 cell surface expression and promoter activation it did not induce AP-2 or require the AP-2 binding site for activity (Table 4-2).

Fluid Shear Stress: ICAM-1 message and cell surface expression are upregulated on vascular endothelial cells exposed to physiologic levels of laminar shear stress.[188] Studies of the platelet-derived growth factor (PDGF) B chain promoter identified a 12 bp shear-stress responsive element (SSRE) which binds nuclear factors induced by shear stress.[228] Subsequently, it was shown that NF-κB RelA/p50 heterodimers will bind to the SSRE, and that fluid shear stress induces NF-κB.[229] The SSRE sequence is present in the promoters of several other genes known to be up regulated by shear stress, including ICAM-1. However, the elements in the ICAM-1 promoter which respond to biomechanical forces thus remain undefined, although the κB element at -187/-178 is a likely candidate.

Retinoids: Retinoic acid (RA) strongly upregulates ICAM-1 expression on several human tumor cell lines.[230,231] ICAM-1 mRNA is maximally induced by RA after 24 in Colo 38 melanoma cells[232] and SK-N-SH human neuroblastoma cells,[233] consistent with ICAM-1 behaving as a late RA response gene (e.g., like the laminin gene).[234] ICAM-1 protein biosynthesis following RA treatment continues to increase for several days after maximal message levels are reached, suggesting additional post-transcriptional regulation.[232] Aoudjit et al[233] mapped a RA responsive region in the promoter 393 to 176 bp upstream of the translation initiation codon. Within this region is a potential RA responsive element (RARE) at -266/-250 (Fig. 4-3), harboring a consensus RARE half-site, PuGGTCA, and exhibiting homology to the RARE in the alcohol dehydrogenase gene ADH3.[235] Mutation of the RARE half-site abrogates RA responsiveness of the ICAM-1 promoter,[232] and a fragment of the ICAM-1 promoter (-270/-178) containing the RARE confers RA responsiveness to a minimal (i.e., enhancerless) heterologous promoter.[233] Electrophoretic mobility shift assays demonstrate that the ICAM-1 and ADH-3 RAREs bind similar nuclear protein complexes from RA-treated cells, and that the two sites mutually compete for binding.[233] The composition of the endogenous complexes has not been rigorously determined. Recombinant RAR-α, RAR-β, and RXR-α appear to bind the ICAM-1 RARE in vitro.[232,233] RA-induced ICAM-1 expression in melanoma cells was dependent on RAR-β gene expression.[231] RAR-β was induced by RA and preceded ICAM-1 expression in SK-N-SH cells.[233] Ectopic expression of RAR-β or RXR-α can mediate RA-induced ICAM-1 promoter activation in Cos-1 cells.[233] In contrast to the human tumor cell lines discussed above, RA has little effect on ICAM-1 expression in primary cultures of keratinocytes or endothelial cells.[236] RA inhibited TNF-α-induced VCAM-1 expression at the transcriptional level in human dermal endothelial cells, but did not affect ICAM-1 expression.[237] On the other hand, RA inhibited IFNγ-stimulated ICAM-1 expression in isolated blood monocytes and U937 cells.[238] In U937 cells, RA inhibited the induction of ICAM-1 mRNA by IFNγ with little effect on message stability, suggesting transcriptional repression. The inhibition of ICAM-1 expression in monocytes may provide a mechanism for some of the anti-inflammatory effects of RA.

Infectious Agents and Their Components: ICAM-1 plays an important role in host defense against microorganisms, and a variety of infectious agents are known to upregulate ICAM-1. Lipopolysaccharide (LPS), or endotoxin, from the outer surface of gram-negative bacteria, stimulated ICAM-1 mRNA accumulation and cell surface expression on HUVECs.[239] In transiently transfected HUVECs, LPS activated a 1.3 kB ICAM-1 promoter fragment linked to luciferase.[142] Mutation of the ICAM-1 κB site at -187/-178 completely abrogated LPS responsiveness, implicating NF-κB in LPS-induced nuclear signaling. Recently, evidence has been presented that Toll-like receptor-4 (TLR-4) mediates LPS-induced NF-κB activation in cultured human dermal endothelial cells (HDECs).[240] Poltorak et al[243] first identified a missense mutation in the cytoplasmic domain of TLR-4 that was responsible for LPS hyporesponsiveness in C3H/HeJ mice. As noted earlier, ICAM-1 deficient mice are resistant to the lethal effects of high doses of LPS.[23] LPS signaling in HDECs and THP-1 monocytic cells appeared to employ components of the IL-1β signal transduction cascade; dominant negative mutants of MyD88, IRAK, IRAK2, and TRAF6 inhibited LPS-induced NF-κB activity, whereas a dominant negative mutant of TRAF2, involved in TNF-α signaling, did not interfere with LPS signaling.[241] Expression of dominant negative NIK blocked NF-κB activation by LPS, IL-1, and

TNF-α in endothelial cells. Takeuchi et al[242] recently reported that LPS-stimulated ICAM-1 mRNA expression in bovine aortic endothelial cells was inhibited by treatment with *Clostridium botulinum* C3 transferase, which inactivates Rho by ADP-ribosylation, and by overexpression of dominant negative RhoA. It will be interesting to determine whether Rho kinase contributes to LPS-induced NF-κB activation. *Borrellia burgdorferi*, the causative agent of Lyme Disease, upregulates several adhesion molecules on endothelial cells, including ICAM-1.[244-246] The lipid moiety of the outer surface lipoprotein A (OspA) activates NF-κB and mediates chemokine and adhesion molecule upregulation in human endothelial cells and fibroblasts.[246,247] Unlike OspA, LPS did not activate chemokine or ICAM-1 expression in dermal fibroblasts.[247] It has been suggested that persistent infections with *B. burgdorferi*, and other pathogenic bacteria which lack LPS but produce lipoproteins with OspA-like activities, may be associated with localized inflammation triggered by these lipoproteins.[246] Inflammatory signaling by *B. burgdorferi* lipoproteins is thought to be mediated by TLR-2.[248,249]

At least eight mammalian Toll-like receptor family members have been described to date. The extracellular domains of TLR family members contain repeated leucine-rich motifs similar to other pathogen-associated pattern recognition proteins of the innate immune system.[250] TLR family members likely recognize different pathogenic microorganisms. The cytoplasmic domains of TLRs are homologous to the signaling domain of the IL-1 receptor. Engagement of the TLRs leads to the activation of NF-κB and the induction of cytokines and costimulatory molecules, as well as adhesion molecules, required for the development of an adaptive immune response.

Heliobacter pylori, implicated in the pathogenesis of chronic gastritis, peptic ulcer disease, and gastric carcinoma, has been shown to stimulate ICAM-1 mRNA and cell surface expression in gastric epithelial cells.[251] This response was observed only with cytotoxin-associated gene pathogenicity island-positive (*cag* PAI$^+$) *H. pylori* strains and required contact with epithelial cells. *H. pylori* activated NF-κB, and stimulated ICAM-1 promoter activity in an NF-κB-dependent manner. The mechanism whereby *cag* PAI$^+$ *H. pylori* strains activate NF-κB has not yet been elucidated.

Human intestinal epithelial cells respond to invasive enteric bacterial infection by secreting cytokines and chemokines and by upregulating cell surface ICAM-1 expression.[252] The upregulation of ICAM-1 on epithelial cells following infection appears to result from a direct interaction with the invading bacterium since ICAM-1 co-localizes with the invaded cells. This suggests that the release of soluble factors may play a minor role in mediating ICAM-1 upregulation. Although the invasive bacteria *Y. entercolitica* and *S. dublin*, and the enteroinvasive *E. coli* serotype (O29:NM) upregulate ICAM-1 expression on HT29 and T84 cells, infection with the noninvasive gram negative bacteria *E. coli* DH5, gram positive bacteria *S. bovis*, or stimulation with bacterial LPS, do not result in ICAM-1 upregulation. The upregulation of ICAM-1 surface expression requires viable organisms. Infection of epithelial cells with invasive bacteria induces a rapid but transient increase in ICAM-1 mRNA levels.

A number of viruses can induce ICAM-1, including cytomegalovirus (CMV), parainfluenza virus, respiratory syncytial viral (RSV), and human T-cell leukemia virus type-1 (HTLV-1). As in bacterial infections, virus-induced ICAM-1 expression is exerted through multiple mechanisms, and is often complicated by concurrent cytokine gene expression. RSV infection is one of the most common causes of airway inflammation and injury in young children. The pathophysiology of RSV-induced airway injury involves the recruitment and activation of inflammatory cells. ICAM-1 upregulation can be elicited as a direct response to viral infection or as a secondary response to soluble mediators. UV-inactivated RSV did not induce ICAM-1, suggesting that viral entry and possibly viral gene expression may be required. RSV activates NF-κB in lung epithelial cells[253-255] although it is not known how RSV activates NF-κB. RSV is an RNA virus, and ds RNA (or mimics such as poly (I-C)) can activate NF-κB and upregulate ICAM-1 expression.[256,257]

Recent studies have suggested a role for ICAM-1/LFA-1 interactions in human immunodeficiency virus (HIV)-dependent cell fusion, cytopathicity, and transmission in T lymphocytes and monocytes.[258-261] The HIV major envelope glycoprotein, gp120, induces ICAM-1 gene expression in primary rat astrocytes and microglial cells, primary human astrocytes, and in the human astroglioma cell line CRT.[262] The receptor on glial cells which binds gp120 has not been identified. HIV gp120 stimulates tyrosine phosphorylation of STAT1α and JAK2 in astrocytes, but the functional roles of STAT1α and the ICAM-1 promoter pIRE have not been reported.[262]

ICAM-1 and LFA-3 are strongly upregulated on T lymphocytes infected with human T-cell leukemia virus type-1 (HTLV-1).[263,264] ICAM-1 and LFA-3 are thought to facilitate HTLV-1 transmission by promoting adhesion to and proliferation of target T cells. The pX region of HTLV-1 encodes Tax1, a potent inducer of HTLV-1 and a number of cellular genes.[265] Tax proteins do not directly bind DNA, but interact with several cellular transcription factors, including CREB, NF-κB, and the serum response factor.[266,267] The inducible expression of Tax1 in stably transfected Jurkat T-cells upregulated ICAM-1 mRNA and cell surface expression.[267,268] Tax1 expression activated a 4.4 kb ICAM-1 promoter/ CAT reporter gene construct in cotransfected Jurkat and HeLa cells Tanaka et al[269] identified regions within the ICAM-1 promoter which respond to Tax1 and the closely related Tax2 (from HTLV-2). The Tax1 responsive elements mapped to a CRE-like site at -630/-624 in Jurkat cells, and to NF-κB (-187/-178) and SP1 (-59/-54) sites in HeLa cells. Unexpectedly, Tax2 did not activate the ICAM-1 promoter in any of the T cell or monocytic cell lines tested, but transactivated the promoter in HeLa cells through the NF-κB site at -187/-178. Tax1 expressing T cell lines contain nuclear protein complexes which specifically bind to the CRE-like site at -630/-624, but the identity of these proteins was not determined. Unlike Tax1, ectopic expression of Tax2 in stably transfected T cell lines did not upregulate ICAM-1 expression. ICAM-1 thus represents the first gene known to be differentially expressed in T cells by Tax1 and Tax2, and may relate to differences in pathogenicity and T cell subset tropism of HTLV-1 and HTLV-2.

ICAM-1 is the cellular receptor for 90% of rhinoviruses, as well as a key player in the recruitment of inflammatory cells to the airways in asthma and following rhinovirus infection. Rhinovirus infections are responsible for a large number of asthma exacerbations. Papi and Johnston[270] have shown that rhinovirus infection of primary bronchial epithelial cells and A549 respiratory epithelial cells markedly increased ICAM-1 cell surface expression. Interestingly, both major group and minor group rhinoviruses induced ICAM-1 expression; only major group rhinoviruses use ICAM-1 as their receptor. Soluble ICAM-1 blocked ICAM-1 induction by major group rhinoviruses, but not rhinovirus 2, a minor group virus. UV treatment of the viruses attenuated, but did not completely block ICAM-1 upregulation. Rhinovirus infection transcriptionally upregulated the ICAM-1 gene and drove expression of an ICAM-1 reporter construct, an effect that appeared to be dependent on the induction of RelA homodimer binding to the κB element at -187/-178. For the major group rhinoviruses, then, it would appear that binding to ICAM-1 on the cell surface induced outside-in signaling that results in NF-κB activation and more ICAM-1 expression. Since the ICAM-1 cytoplasmic domain is linked to ERM proteins, it will be of interest to determine whether virus activates NF-κB through a Rho kinase mechanism.

Lysophosphatidylcholine: Lysophosphatidylcholine (lyso-PC) is a major phospholipid component of lipoproteins (e.g., oxidized LDL and VLDL) found in atherosclerotic and inflammatory lesions. Lyso-PC at non-toxic concentrations selectively induced the expression of several genes in endothelial cells relevant to atherosclerosis and inflammation, including ICAM-1.[271] ICAM-1 mRNA induction by lyso-PC can be detected at 4hoursand continues to increase for up to 24 hours.[271,272] Lyso-PC-stimulated ICAM-1 mRNA accumulation is not dependent on message stabilization or *de novo* protein synthesis and appears to be transcriptionally regulated, although direct evidence for this conclusion has not been reported. Zhu et al[273] recently demonstrated that lyso-PC upregulated ICAM-1 mRNA in HUVECs and can modestly activate an ICAM-1 promoter-driven reporter gene construct. The proximal 282 bp of the ICAM-1 promoter appears to confer lyso-PC-responsiveness, but the element(s) responsible has not been determined. Lyso-PC weakly activated NF-κB in HUVECs, but a causal relationship between lyso-PC, NF-κB, and the ICAM-1 promoter has not yet been established. Lyso-PC signaling did not appear involve PKC, whereas genistein and lavendustin A reduced ICAM-1 mRNA levels and NF-κB activation elicited by lyso-PC, implicating protein tyrosine kinase(s). The pattern of gene expression in endothelial cells activated by lyso-PC is qualitatively different from endothelial cultures treated with TNF-α, LPS, or PMA. Unlike these agents, lyso-PC does not upregulate E-selectin, IL-8, or MCP-1, suggesting different inductive mechanisms.[272]

Inhibition of ICAM-1 Gene Expression

At least four endogenous substances, glucocorticoids, IL-10, and TGF-β, and nitric oxide (NO) possess potent immunomodulatory and anti-inflammatory activities, and have been shown to inhibit ICAM-1 expression.

Glucocorticoids: Glucocorticoids are probably the most physiologically important and effective anti-inflammatory agents known, yet their mechanism(s) of action *in vivo* remains ill-defined. Glucocorticoids have pleiotropic actions and their therapeutic effectiveness is most likely due to suppression of numerous proinflammatory mediators, including proteolytic enzymes, lipids, cytokines, chemokines and adhesion molecules. Although it is well-established that glucocorticoids suppress ICAM-1 induction *in vivo*, studies using cultured cells *in vitro* have yielded mixed results. Anti-inflammatory steroids have been shown to inhibit cytokine induced ICAM-1 expression on some cell types, including human synovial fibroblasts, and some monocytic and epithelial cell lines.[138,139,274-278] However, cytokine upregulation of ICAM-1 expression in other cell types, including renal tubular epithelial cells[279] and human alveolar macrophages,[280] was not inhibited by glucocorticoids. Inhibition of ICAM-1 expression on human endothelial cells has also been reported,[281,282] although there are conflicting reports.[283-287] It should be noted that in many of the studies where inhibitory effects of dexamethasone or other steroids on cytokine induced ICAM-1 expression were reported, the concentrations of steroid used were often suprapharmacological (10-100 μM) and the inhibitory effects incomplete (< 50%).

The biological effects of glucocorticoids are mediated through the glucocorticoid receptor (GR), a member of the steroid/thyroid hormone receptor superfamily. Upon ligand binding, the cytoplasmic GR dimerizes, sheds heat shock proteins and translocates to the nucleus where it binds to palindromic glucocorticoid response elements (GREs) in the promoters of target genes to activate, and in some cases inhibit, transcription. GR can also inhibit transcription by repressing the activity of other transcription factors through direct protein-protein interactions. This type of molecular cross-talk between transcription factors does not require *de novo* protein synthesis, and generally exhibits mutual repression such that the biological outcome depends on the ratio of the factors present. GR-mediated transrepression was first demonstrated for AP-1[288-290] and more recently, a similar interaction between GR and NF-κB was reported.[139,291] GR-mediated inhibition of ICAM-1 promoter activity was first explored in 293 cells by van de Stolpe et al.[139] Through deletion analysis of the promoter, these investigators demonstrated that the κB element at -187/-178 was responsible for glucocorticoid repression, and that liganded GR repressed RelA transactivation in cotransfection studies using a NF-κB responsive reporter construct.

The repression was mutual since overexpressed RelA strongly inhibited transactivation by activated GR.[292] The DNA binding domain of GR was critical for RelA repression, although GRE binding was not required. Although GR and RelA physically interact with each other, GR-mediated transrepression does not necessarily influence NF-κB nuclear binding complexes in gel shift assays.[292-294] Dexamethasone induced IκBα gene expression in HeLa cells, monocytic cells, and T lymphocytes, resulting in the suppression of NF-κB activity.[293,295] It should be noted that IκBα induction by glucocorticoids was not observed in normal rat kidney cells, endothelial cells, or A549/8 cells under conditions where NF-κB activity was repressed by glucocorticoids.[291,296,297] Since GR-mediated transcriptional interference has generally been investigated using ectopically expressed GR, the role of endogenous GR in this process needs to be clarified.

TGFβ: TGFβ inhibited ICAM-1 expression on glial cells in a stimulus-dependent manner.[298] TGFβ suppressed TNF-α- or IL-1β-induced ICAM-1 expression on primary rat astrocytes and human astroglioma cells, but did not inhibit IFNγ- or IFNγ/LPS-induced ICAM-1 expression in astrocytes, astroglioma cells, or microglial cells. Inhibition of TNF-α- and IL-1β-stimulated ICAM-1 mRNA and protein expression in astrocytes required pretreatment with TGFβ for 12 to 24 hours. Since TGFβ did not affect ICAM-1 message stability, and inhibited TNF-α-induced ICAM-1 promoter activity, it was suggested that the effects of TGFβ were mediated, at least in part, at the transcriptional level. The TGFβ responsive elements in the ICAM-1 promoter have not been reported. In endothelial cells, TGFβ inhibited basal and TNF-α-induced E-selectin mRNA and surface expression, but did not effect ICAM-1 or VCAM-1 expression.[299]

IL-10: IL-10 is a potent immunomodulatory cytokine which suppressed a wide range of immune and inflammatory responses (reviewed in [300, 301]). In addition to inhibiting the production of proinflammatory cytokines such as TNF-α, IL-1, IL-8, and IL-6, IL-10 suppressed the expression of cell surface proteins involved in antigen presentation and T cell costimulation, including MHC class II, B7, and ICAM-1. IL-10 inhibited ICAM-1 expression on IFNγ-stimulated human monocytes[302-304] and primary rat glial cells,[108] although the mechanisms were different. Song et al[304] reported that IL-10 inhibited IFNγ-induced ICAM-1 gene transcription with little effect on message stability. In this study, IL-10 did not affect IFNγ-induced STAT1α tyrosine phosphorylation or binding to the ICAM-1 pIRE, although it did induce immuno-reactive STAT3 binding activity. Unexpectedly, IFNγ induced nuclear binding activity to an ICAM-1 NF-κB probe, and IL-10 prevented formation of the complex. The relevance of these provocative results to the IL-10 mechanism of action awaits further characterization of the binding complex and functional studies on the ICAM-1 NF-κB site in monocytes. IL-10 has also been shown to inhibit the transcription of cytokine genes in LPS-stimulated macrophages,[305] an effect recently attributed to inhibition of NF-κB activation by IL-10.[306] In contrast, IL-10 inhibited the induction of ICAM-1 protein biosynthesis and cell surface expression in rat astrocytes and microglial cells without affecting ICAM-1 steady state mRNA levels.[108] Inhibition of ICAM-1 expression on glial cells was time-dependent, optimally requiring an 8 to 12 hour pretreatment period.

Nitric Oxide: Reperfusion of an ischemic vascular bed produces endothelial cell dysfunction, resulting in reduced basal and agonist-induced NO production, increased adhesion molecule expression, including ICAM-1, enhanced leukocyte adhesion, and tissue injury.[307,308] NO donors reduce ICAM-1 expression and leukocyte adherance. Takahashi et al[309] showed that IL-1β-induced ICAM-1 and VCAM-1 expression in HUVECs was significantly inhibited in the presence of the NO donor 3-morpholino-sydnonimine (SIN-1), but not 8-bromo-cGMP. The inhibitory effect of SIN-1 was abolished in the presence of hemoglobin, an NO scavenger. IL-1β-induced ICAM-1 and VCAM-1 mRNA expression was also inhibited by SIN-1 treatment. Gel shift experiments have shown that NO inhibited NF-κB activation.[310] NO did not appear to inhibit IKK activity, IκBα phosphorylation, or IκBα degradation. Rather, it is thought that NO transcriptionally induced IκBα expression.[311] These findings suggest that basal and inducible ICAM-1 expression may be regulated by endogenous NO.

Constitutive ICAM Regulation

As noted above, ICAM-1, ICAM-2 and ICAM-3 are constitutively expressed on distinctive, often overlapping cell types, whereas the expression of ICAM-4 and TLN is restricted to red blood cells and telencephalonic neurons, respectively. The molecular basis for the diverse expression patterns of these molecules remains obscure, but there has been recent progress.

ICAM-2 is constitutively expressed at relatively high levels on vascular endothelium in human tissues, and weakly expressed on monocytes, lymphocytes, platelets, and megakaryocytes.[48,59] Cowan et al[312] recently demonstrated that the 334-bp proximal promoter region of the ICAM-2 gene was sufficient to drive strong and uniform expression of a CD59 transgene throughout the vascular endothelium of transgenic mice. Little or no expression of CD59 was observed in other cell types with the exception of monocytes and neutrophils. Since ICAM-2 was not normally expressed on resting neutrophils, Cowan et al[312] suggest that the 334-bp promoter may lack negative regulatory features which would restrict myeloid expression. As shown in Figure 4-4, the human ICAM-2 promoter contains putative regulatory elements common to other genes expressed in endothelium (e.g., thrombomodulin, von Willebrand factor, preproendothelin, *Tie*-2, and P-Selectin genes), including potential binding sites for the GATA and Ets families of transcription factors.[312] In addition, the promoter contains potential binding sites for SP-1 and C/EBP, a CACCC motif frequently associated with GATA binding sites[313] and a NF-κB element. The ICAM-2 promoter does not contain a recognizable TATA box, and transcription in bovine endothelial cells initiates at four sites.[314] When cloned into a promoterless luciferase reporter vector, the 0.33 kb ICAM-2 promoter exhibited endothelial cell-specific activity, and inclusion of an additional 2.7 kb of upstream 5'-flanking sequence did not further increase promoter activity.[314] Of the putative transcription factor binding sites listed above, the Sp1 site and both GATA sites were shown to be positive regulatory elements in endothelial cells using mutational and gel shift analyses. Although the ICAM-2 promoter was minimally active in COS cells, ectopic expression of GATA-2 enhanced activity 3-4-fold. Mutation of an 8-bp palindromic element at -347/-341 (relative to ATG) also caused a substantial loss of ICAM-2 promoter activity in endothelial cells. This element is known to be a component of an intronic endothelial cell-specific enhancer in the murine *Tie2* gene.[315] Subsequently, McLaughlin et al[50] demonstrated that mutation of the NF-κB element and three out of the four putative Ets binding sites significantly reduced constitutive ICAM-2 promoter activity in HUVECs. In gel shift experiments, the Ets site at -122/-116 (relative to the ATG) and the tandem ets sites spanning -215 to -199 were shown to bind HUVECs nuclear factors in an Ets-specific manner, which was blocked with a pan Ets antibody. The Ets family member Erg was constitutively expressed in HUVECS, whereas the levels of Ets-1 and Nerf were considerably lower. Erg overexpression resulted in a modest transactivation of ICAM-2 promoter activity in HeLa cells and HUVECs. These investigators have shown that TNF-α downregulated the expression of ICAM-2 protein and mRNA in endothelial cells.[50] TNF-α treatment also reduced ICAM-2 promoter activity, an effect that was mediated through the tandem Ets sites at -215 to -199, and paralleled by a reduction in nuclear factor binding to these sites, and down regulation of Erg expression. These results suggested an important role for Ets in the regulation of ICAM-2 gene expression. Although the NF-κB element appears to be involved in basal ICAM-2 promoter activity, the nuclear factors mediating this effect have not been investigated. An NF-κB motif is also thought to regulate basal P-selectin expression in endothelial cells and megakaryocytes through a mechanism involving Bcl-3 and promoter bound p50 or p52 homodimers.[316] Like ICAM-2, P-selectin expression was not upregulated by a variety of inflammatory cytokines.

ICAM-3 expression is essentially restricted to the hemopoietic system. It was constitutively expressed on all lineages of resting leukocytes, particularly T lymphocytes. In contrast to ICAM-1 and ICAM-2, ICAM-3 was not expressed on resting or cytokine-stimulated endothelial cells. Like ICAM-2, however, ICAM-3 was strongly expressed on endothelial cells in malignant lymphoid tissues.[51,317] ICAM-3 expression on blood vessels does not appear to play a role in leukocyte recruitment during inflammation, but rather was associated with angiogenesis and tumor development.[318] Transcriptional regulation of the ICAM-3 gene and characterization of its 5'-regulatory region has not been reported yet.

ICAM-4/LW antigen is found only on red blood cells. The ICAM-4 promoter has not been characterized, and little is known about the regulation of expression of this gene in nucleated progenitor cells or cell lines.

A 1024-bp regulatory region upstream of the murine *Tlcn* gene has been sequenced, revealing a TATA-less promoter with a strong initiator element (Inr), and a number of putative transcription factor binding sites.[93] Four E-box and two N-box sequences were identified which are known binding sites for transcription factors belonging to the helix-loop-helix family and implicated in neural development.[319] It will be interesting to see whether analysis of this regulatory region will provide insights into the mechanisms governing the highly restricted spatial and developmental expression of the *Tlcn* gene.

SUMMARY AND FUTURE DIRECTIONS

From the discovery of ICAM-1 in the mid-1980's, the ICAM family has grown to five, and possibly six, members at the present time. With the human genome project nearing completion, database mining may yield additional ICAMs. The early ICAM work was propelled largely by immunologists seeking to understand leukocyte function, and armed principly with monoclonal antibodies. Targeted gene disruption and transgenic animals are now being used to extend earlier findings, and to identify new functions of ICAMs. An interesting turning point in the ICAM saga was the discovery of telencephalon, the first ICAM of non-hematopoietic origins. TLN may have foreshadowed the existence of additional ICAMs that exhibit very narrow cellular or developmental expression profiles, and may therefore have been overlooked.

The regulation of ICAM-1 gene expression has been extensively investigated by a number of groups, and progress is being made on some of the other ICAM family members. NF-κB has emerged as a dominant theme in ICAM-1 gene regulation. This transcription factor has been likened to a master switch of the inflammatory response, and undoubtedly plays a central role in ICAM-1 expression during acute inflammatory responses to injury and a host of environmental stresses, including infectious agents. The recent revelation that constituants of many microorganisms can elicit innate immune responses by acting on Toll-like receptors linked to the NF-κB pathway, underscores the importance of NF-κB and ICAM-1 in host defense. The induction of ICAM-1 by IFNγ through the JAK/STAT pathway likely contributes to adaptive immune responses and the chronicity of inflammatory disease. Although the regulation of ICAM-1 expression by the NF-κB and JAK/STAT pathways are relatively well characterized, our understanding of how other important regulators control gene expression is considerably less developed. Reactive oxygen species, for example, are ubiquitous, both under normal physiological conditions and in disease, and their effects are superimposed on other factors influencing ICAM-1 expression. A challenge of future work will be to integrate the signaling pathways and transcriptional responses of these regulators. Many cis-regulatory elements have been identified in the upstream regulatory region of the ICAM-1 gene, and in a number of cases, evidence has been presented demonstrating the functional significance of these sites and the factors that interact with them. There are likely many additional important regulatory regions, possibly in introns, the 3' UTR, or distant upstream and downstream sequences. Future studies will also need to address higher order transcriptional regulation, including interactions amongst multiple transcription factors, recruitment of coactivators (e.g., CBP/p300) and corepressors, and factors that influence chromatin structure. A sobering thought is that constitutive ICAM-1 expression is still poorly understood, a problem highlighted by the observation that transfection of a complete ICAM-1 gene into a cell that does not normally express ICAM-1 results in constitutive ICAM-1 expression.[114] The ICAM genes likely arose from duplication of an ancestral gene. Despite conservation of function as β_2-integrin ligands, the regulatory regions of these genes have diverged spectacularly. Future work will further examine why, for example, ICAM-2 is expressed constitutively on endothelium, ICAM-3 on T cells, and TLN on a subset of neurons. Another intriguing issue to resolve is the mechanism whereby endothelial cells express ICAM-3 during neovascularization in certain tumors.

ACKNOWLEDGMENTS

The authors gratefully acknowledge the support of many colleagues, who have contributed unpublished data, preprints of papers, and insightful discussions. As with many reviews on such a broad topic, there were inevitable omissions of many excellent studies.

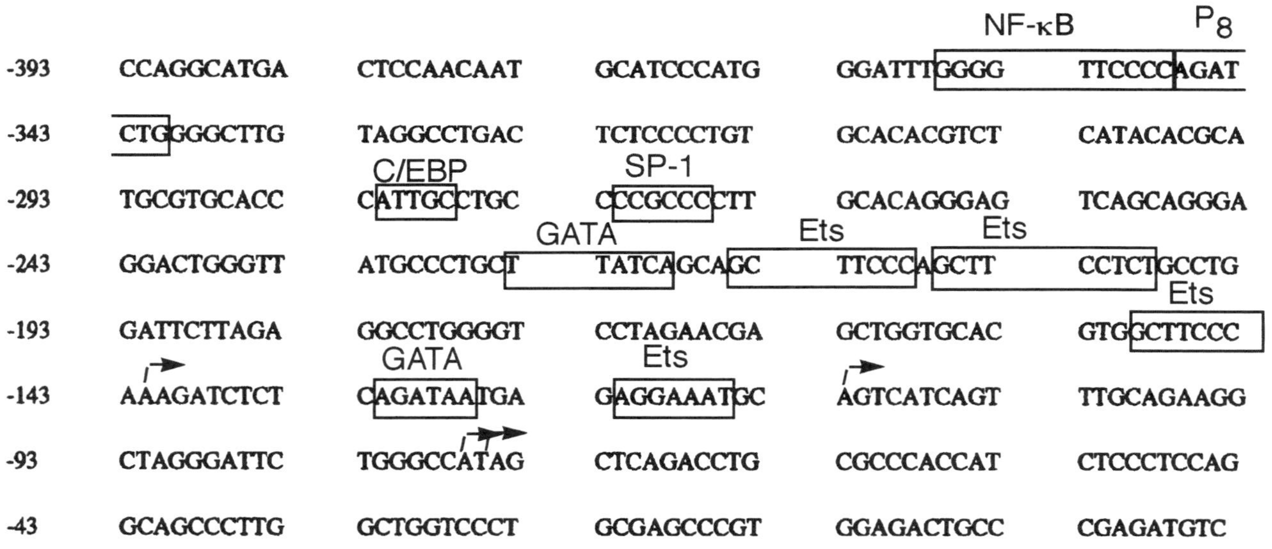

NF-κB P8

-393	CCAGGCATGA	CTCCAACAAT	GCATCCCATG	GGATTTGGGG	TTCCCCAGAT	-344
-343	CTGGGGCTTG	TAGGCCTGAC	TCTCCCCTGT	GCACACGTCT	CATACACGCA	-294
		C/EBP	SP-1			
-293	TGCGTGCACC	CATTGCCTGC	CCCGCCCCTT	GCACAGGGAG	TCAGCAGGGA	-244
		GATA		Ets		
-243	GGACTGGGTT	ATGCCCTGCT	TATCAGCAGC	TTCCCAGCTT	CCTCTGCCTG	-194
			Ets		Ets	
-193	GATTCTTAGA	GGCCTGGGGT	CCTAGAACGA	GCTGGTGCAC	GTGGCTTCCC	-144
		GATA	Ets		Ets	
-143	AAAGATCTCT	CAGATAATGA	GAGGAAATGC	AGTCATCAGT	TTGCAGAAGG	-94
			Ets			
-93	CTAGGGATTC	TGGGCCATAG	CTCAGACCTG	CGCCCACCAT	CTCCCTCCAG	-44
-43	GCAGCCCTTG	GCTGGTCCCT	GCGAGCCCGT	GGAGACTGCC	CGAGATGTC	4

Figure 4-4. The human ICAM-2 promoter. The sequence is derived from a prior publication.[74] Note that the sequence differs from that reported[74] by insertion of C at -349 and a G at -355 (TP Parks, unpublished data). Putative regulatory elements are indicated by the boxes, and the arrows indicate the transcriptional start sites of the human ICAM-2 gene. The numbering system is relative to the translation start site (ATG).

REFERENCES

1. Rothlein R, Dustin, ML, Marlin, SD, Springer, TA. A human intercellular adhesion molecule distinct from LFA-1. J Immunol 1986;137:1270-4.
2. Patarroyo M, Clark, EA, Prieto, J, Kantor, C, and Gahmberg, CG. Identification of a novel adhesion molecule in human leukocytes by monoclonal antibody LB-2. FEBS Lett 1987; 210:127-31.
3. Schulz TF, Mitterer, M, Vogetseder, W, Bock, G, Myones, BL, and Dierich, MP. Identification and characterization of a novel membrane activation antigen with wide cellular distribution. Eur J Immunol 1988; 18:7-11.
4. Johnson JP, Stade, BG, Hupke, U, Holzmann, B, and Riethmuller, G. The melanoma progression-associated antigen P3.58 is identical to the intercellular adhesion molecule, ICAM-1. Immunobiology 1988; 178:275-84.
5. Johnson J, and Shaw, S. Cluster report. CD54. In: *Leukocyte Typing IV:White Cell Differentiation Antigens*, edited by W. Knapp, B. Dorken, W. Gilks, E. Rieber, R. Schmidt, H. Stein and A. von dem borne. Oxford: Oxford University Press, 1989, p. 681-683.
6. Dustin ML, Rothlein, R, Bhan, AK, Dinarello, CA, and Springer, TA. Induction by IL 1 and interferon-gamma: tissue distribution, biochemistry, and function of a natural adherence molecule (ICAM-1). J Immunol 1986; 137:245-54.
7. Staunton DE, Marlin, SD, Stratowa, C, Dustin, ML, and Springer, TA. Primary structure of ICAM-1 demonstrates interaction between members of the immunoglobulin and integrin supergene families. Cell 1988; 52:925-33.
8. Simmons D, Makgoba, MW, Seed, B: ICAM, an adhesion ligand of LFA-1, is homologous to the neural cell adhesion molecule NCAM. Nature 1988; 331:624-7.
9. Dustin ML, and Springer, TA. Lymphocyte function-associated antigen-1 (LFA-1) interaction with intercellular adhesion molecule-1 (ICAM-1) is one of at least three mechanisms for lymphocyte adhesion to cultured endothelial cells. J Cell Biol 1988; 107:321-31.
10. Marlin SD, and Springer, TA. Purified intercellular adhesion molecule-1 (ICAM-1) is a ligand for lymphocyte function-associated antigen 1 (LFA-1). Cell 1987; 51:813-9.
11. Makgoba MW, Sanders, ME, Ginther Luce, GE, Dustin, ML, Springer, TA, Clark, EA, Mannoni, P, and Shaw, S. ICAM-1 a ligand for LFA-1-dependent adhesion of B, T and myeloid cells. Nature 1988; 331:86-8.
12. Diamond MS, Staunton, DE, Marlin, SD, and Springer, TA. Binding of the integrin Mac-1 (CD11b/CD18) to the third immunoglobulin-like domain of ICAM-1 (CD54) and its regulation by glycosylation. Cell 1991; 65:961-71.
13. Anderson D. The role of b2 integrins and intercellular adhesion molecule type 1 in inflmammation. In: *Physiology and Pathophysiology of Leukocyte Adhesion*, edited by D. Granger and G. Schmid-Schonbein. New York: Oxford University Press, 1995, p. 3-42.
14. Blackford J, Reid, HW, Pappin, DJ, Bowers, FS, and Wilkinson, JM. A monoclonal antibody, 3/22, to rabbit CD11c which induces homotypic T cell aggregation: evidence that ICAM-1 is a ligand for CD11c/CD18. Eur J Immunol 1996; 26:525-31.
15. Greve JM, Davis, G, Meyer, AM, Forte, CP, Yost, SC, Marlor, CW, Kamarck, ME, and McClelland, A. The major human rhinovirus receptor is ICAM-1. Cell 1989; 56:839-47.
16. Berendt AR, Simmons, DL, Tansey, J, Newbold, CI, and Marsh, K. Intercellular adhesion molecule-1 is an endothelial cell adhesion receptor for Plasmodium falciparum. Nature 1989; 341:57-9.
17. Languino LR, Plescia, J, Duperray, A, Brian, AA, Plow, EF, Geltosky, JE, and Altieri, DC. Fibrinogen mediates leukocyte adhesion to vascular endothelium through an ICAM-1-dependent pathway. Cell 1993; 73:1423-34.

156

18. Rosenstein Y, Park, JK, Hahn, WC, Rosen, FS, Bierer, BE, and Burakoff, SJ. CD43, a molecule defective in Wiskott-Aldrich syndrome, binds ICAM-1. Nature 1991; 354:233-5.

19. McCourt PA, Ek, B, Forsberg, N, and Gustafson, S. Intercellular adhesion molecule-1 is a cell surface receptor for hyaluronan. J Biol Chem 1994; 269:30081-4.

20. van de Stolpe A, and van der Saag, PT. Intercellular adhesion molecule-1. J Mol Med 1996; 74:13-33.

21. Kishimoto TK, and Rothlein, R. Integrins, ICAMs, and selectins: role and regulation of adhesion molecules in neutrophil recruitment to inflammatory sites. Adv Pharmacol 1994; 25:117-69.

22. Sligh JE, Jr., Ballantyne, CM, Rich, SS, Hawkins, HK, Smith, CW, Bradley, A, and Beaudet, AL. Inflammatory and immune responses are impaired in mice deficient in intercellular adhesion molecule 1. Proc Natl Acad Sci U S A 1993; 90:8529-33.

23. Xu H, Gonzalo, JA, St Pierre, Y, Williams, IR, Kupper, TS, Cotran, RS, Springer, TA, and Gutierrez-Ramos, JC. Leukocytosis and resistance to septic shock in intercellular adhesion molecule 1-deficient mice. J Exp Med 1994; 180:95-109.

24. Soriano SG, Lipton, SA, Wang, YF, Xiao, M, Springer, TA, Gutierrez-Ramos, JC, and Hickey, PR. Intercellular adhesion molecule-1-deficient mice are less susceptible to cerebral ischemia-reperfusion injury. Ann Neurol 1996; 39:618-24.

25. Hallahan DE, and Virudachalam, S. Ionizing radiation mediates expression of cell adhesion molecules in distinct histological patterns within the lung. Cancer Res 1997; 57:2096-9.

26. Bendjelloul F, Maly, P, Mandys, V, Jirkovska, M, Prokesova, L, Tuckova, L, and Tlaskalova-Hogenova, H. Intercellular adhesion molecule-1 (ICAM-1) deficiency protects mice against severe forms of experimentally induced colitis. Clin Exp Immunol 2000; 119:57-63.

27. Johnson JP, Stade, BG, Holzmann, B, Schwable, W, and Riethmuller, G. De novo expression of intercellular-adhesion molecule 1 in melanoma correlates with increased risk of metastasis. Proc Natl Acad Sci U S A 1989; 86:641-4.

28. Johnson JP. The role of ICAM-1 in tumor development. Chem Immunol 1991; 50:143-63.

29. Natali P, Nicotra, MR, Cavaliere, R, Bigotti, A, Romano, G, Temponi, M, and Ferrone, S. Differential expression of intercellular adhesion molecule 1 in primary and metastatic melanoma lesions. Cancer Res 1990; 50:1271-8.

30. Vogetseder W, Feichtinger, H, Schulz, TF, Schwaeble, W, Tabaczewski, P, Mitterer, M, Bock, G, Marth, C, Dapunt, O, Mikuz, G, and et al. Expression of 7F7-antigen, a human adhesion molecule identical to intercellular adhesion molecule-1 (ICAM-1) in human carcinomas and their stromal fibroblasts. Int J Cancer 1989; 43:768-73.

31. Johnson JP. Cell adhesion molecules in the development and progression of malignant melanoma. Cancer Metastasis Rev 1999; 18:345-57.

32. Harning R, Mainolfi, E, Bystryn, JC, Henn, M, Merluzzi, VJ, and Rothlein, R. Serum levels of circulating intercellular adhesion molecule 1 in human malignant melanoma. Cancer Res 1991; 51:5003-5.

33. Rothlein R, Mainolfi, EA, Czajkowski, M, and Marlin, SD. A form of circulating ICAM-1 in human serum. J Immunol 1991; 147:3788-93.

34. Pigott R, Dillon, LP, Hemingway, IH, and Gearing, AJ. Soluble forms of E-selectin, ICAM-1 and VCAM-1 are present in the supernatants of cytokine activated cultured endothelial cells. Biochem Biophys Res Commun 1992; 187:584-9.

35. Carley W, Ligon, G, Phan, S, Dziuba, J, Kelley, K, Perry, C, and Gerritsen, ME. Distinct ICAM-1 forms and expression pathways in synovial microvascular

endothelial cells. Cell Mol Biol (Noisy-le-grand) 1999; 45:79-88.

36. van Den Engel NK, Heidenthal, E, Vinke, A, Kolb, H, and Martin, S.Circulating forms of intercellular adhesion molecule (ICAM)-1 in mice lacking membranous ICAM-1. Blood 2000; 95:1350-5.

37. Adams DH, Mainolfi, E, Elias, E, Neuberger, JM, and Rothlein, R.Detection of circulating intercellular adhesion molecule-1 after liver transplantation--evidence of local release within the liver during graft rejection. Transplantation 1993; 55:83-7.

38. Cush JJ, Rothlein, R, Lindsley, HB, Mainolfi, EA, and Lipsky, PE.Increased levels of circulating intercellular adhesion molecule 1 in the sera of patients with rheumatoid arthritis. Arthritis Rheum 1993; 36:1098-102.

39. Koide M, Tokura, Y, Furukawa, F, and Takigawa, M.Soluble intercellular adhesion molecule-1 (sICAM-1) in atopic dermatitis. J Dermatol Sci 1994; 8:151-6.

40. Kojima T, Ono, A, Aoki, T, Kameda-Hayashi, N, and Kobayashi, Y.Circulating ICAM-1 levels in children with atopic dermatitis. Ann Allergy 1994; 73:351-5.

41. Staunton DE, Dustin, ML, and Springer, TA.Functional cloning of ICAM-2, a cell adhesion ligand for LFA-1 homologous to ICAM-1. Nature 1989; 339:61-4.

42. Nortamo P, Li, R, Renkonen, R, Timonen, T, Prieto, J, Patarroyo, M, and Gahmberg, CG. The expression of human intercellular adhesion molecule-2 is refractory to inflammatory cytokines. Eur J Immunol 1991; 21:2629-32.

43. de Fougerolles AR, Stacker, SA, Schwarting, R, and Springer, TA. Characterization of ICAM-2 and evidence for a third counter-receptor for LFA-1. J Exp Med 1991; 174:253-67.

44. Gahmberg CG, Nortamo, P, Zimmermann, D, and Ruoslahti, E. The human leukocyte-adhesion ligand, intercellular-adhesion molecule 2. Expression and characterization of the protein. Eur J Biochem 1991; 195:177-82.

45. Xie J, Li, R, Kotovuori, P, Vermot-Desroches, C, Wijdenes, J, Arnaout, MA, Nortamo, P, and Gahmberg, CG. Intercellular adhesion molecule-2 (CD102) binds to the leukocyte integrin CD11b/CD18 through the A domain. J Immunol 1995; 155:3619-28.

46. de Fougerolles AR, Klickstein, LB, and Springer, TA. Cloning and expression of intercellular adhesion molecule 3 reveals strong homology to other immunoglobulin family counter-receptors for lymphocyte function-associated antigen 1. J Exp Med 1993; 177:1187-92.

47. Nortamo P, Salcedo, R, Timonen, T, Patarroyo, M, and Gahmberg, CG. A monoclonal antibody to the human leukocyte adhesion molecule intercellular adhesion molecule-2. Cellular distribution and molecular characterization of the antigen. J Immunol 1991; 146:2530-5.

48. Diacovo TG, deFougerolles, AR, Bainton, DF, and Springer, TA. A functional integrin ligand on the surface of platelets. intercellular adhesion molecule-2. J Clin Invest 1994; 94:1243-51.

49. McLaughlin F, Hayes, BP, Horgan, CM, Beesley, JE, Campbell, CJ, and Randi, AM. Tumor necrosis factor (TNF)-alpha and interleukin (IL)-1beta down-regulate intercellular adhesion molecule (ICAM)-2 expression on the endothelium. Cell Adhes Commun 1998; 6:381-400.

50. McLaughlin F, Ludbrook, VJ, Kola, I, Campbell, CJ, and Randi, AM. Characterisation of the tumour necrosis factor (TNF)-(alpha) response elements in the human ICAM-2 promoter. J Cell Sci 1999; 112:4695-703.

51. Renkonen R, Paavonen, T, Nortamo, P, and Gahmberg, CG. Expression of endothelial adhesion molecules in vivo. Increased endothelial ICAM-2 expression in lymphoid malignancies. Am J Pathol 1992; 140:763-7.

52. Reiss Y, Hoch, G, Deutsch, U, and Engelhardt, B. T cell interaction with ICAM-1-deficient endothelium in vitro: essential role for ICAM-1 and ICAM-2 in transendothelial migration of T cells. Eur J Immunol 1998; 28:3086-99.

158

53.   Shang XZ, and Issekutz, AC. Contribution of CD11a/CD18, CD11b/CD18, ICAM-1 (CD54) and -2 (CD102) to human monocyte migration through endothelium and connective tissue fibroblast barriers. Eur J Immunol 1998; 28:1970-9.

54.   Issekutz AC, Rowter, D, and Springer, TA. Role of ICAM-1 and ICAM-2 and alternate CD11/CD18 ligands in neutrophil transendothelial migration. J Leukoc Biol 1999; 65:117-26.

55.   Gerwin N, Gonzalo, JA, Lloyd, C, Coyle, AJ, Reiss, Y, Banu, N, Wang, B, Xu, H, Avraham, H, Engelhardt, B, Springer, TA, and Gutierrez-Ramos, JC. Prolonged eosinophil accumulation in allergic lung interstitium of ICAM-2 deficient mice results in extended hyperresponsiveness. Immunity 1999; 10:9-19.

56.   Vazeux R, Hoffman, PA, Tomita, JK, Dickinson, ES, Jasman, RL, St. John, T, and Gallatin, WM. Cloning and characterization of a new intercellular adhesion molecule ICAM-R. Nature 1992; 360:485-8.

57.   Fawcett J, Holness, CL, Needham, LA, Turley, H, Gatter, KC, Mason, DY, and Simmons, DL. Molecular cloning of ICAM-3, a third ligand for LFA-1, constitutively expressed on resting leukocytes. Nature 1992; 360:481-4.

58.   Van der Vieren M, Le Trong, H, Wood, CL, Moore, PF, St. John, T, Staunton, DE, and Gallatin, WM. A novel leukointegrin, alpha d beta 2, binds preferentially to ICAM-3. Immunity 1995; 3:683-90.

59.   de Fougerolles AR, Qin, X, and Springer, TA. Characterization of the function of intercellular adhesion molecule (ICAM)-3 and comparison with ICAM-1 and ICAM-2 in immune responses. J Exp Med 1994; 179:619-29.

60.   Campanero MR, del Pozo, MA, Arroyo, AG, Sanchez-Mateos, P, Hernandez-Caselles, T, Craig, A, Pulido, R, and Sanchez-Madrid, F. ICAM-3 interacts with LFA-1 and regulates the LFA-1/ICAM-1 cell adhesion pathway. J Cell Biol 1993; 123:1007-16.

61.   Hernandez-Caselles T, Rubio, G, Campanero, MR, del Pozo, MA, Muro, M, Sanchez-Madrid, F, and Aparicio, P. ICAM-3, the third LFA-1 counterreceptor, is a co-stimulatory molecule for both resting and activated T lymphocytes. Eur J Immunol 1993; 23:2799-806.

62.   Hayflick JS, Kilgannon, P, and Gallatin, WM. The intercellular adhesion molecule (ICAM) family of proteins. New members and novel functions. Immunol Res 1998; 17:313-27.

63.   Bailly P, Hermand, P, Callebaut, I, Sonneborn, HH, Khamlichi, S, Mornon, JP, and Cartron, JP. The LW blood group glycoprotein is homologous to intercellular adhesion molecules. Proc Natl Acad Sci U S A 1994; 91:5306-10.

64.   Bailly P, Tontti, E, Hermand, P, Cartron, JP, and Gahmberg, CG. The red cell LW blood group protein is an intercellular adhesion molecule which binds to CD11/CD18 leukocyte integrins. Eur J Immunol 1995; 25:3316-20.

65.   Mori K, Fujita, SC, Watanabe, Y, Obata, K, and Hayaishi, O. Telencephalon-specific antigen identified by monoclonal antibody. Proc Natl Acad Sci U S A 1987; 84:3921-5.

66.   Murakami F, Tada, Y, Mori, K, Oka, S, and Katsumaru, H. Ultrastructural localization of telencephalon, a telencephalon-specific membrane glycoprotein, in rabbit olfactory bulb. Neurosci Res 1991; 11:141-5.

67.   Mizuno T, Yoshihara, Y, Inazawa, J, Kagamiyama, H, and Mori, K. cDNA cloning and chromosomal localization of the human telencephalon and its distinctive interaction with lymphocyte function-associated antigen-1. J Biol Chem 1997; 272:1156-63.

68.   Yoshihara Y, Oka, S, Nemoto, Y, Watanabe, Y, Nagata, S, Kagamiyama, H, and Mori, K. An ICAM-related neuronal glycoprotein, telencephalon, with brain segment-specific expression. Neuron 1994; 12:541-53.

69.   Tian L, Yoshihara, Y, Mizuno, T, Mori, K, and Gahmberg, CG. The neuronal glycoprotein telencephalon is a cellular ligand for the CD11a/CD18 leukocyte

integrin. J Immunol 1997; 158:928-36.

70. Ono M, Nomoto, K, and Nakazato, S. Gene structure of rat testicular cell adhesion molecule 1 (TCAM-1), and its physical linkage to genes coding for the growth hormone and BAF60b, a component of SWI/SNF complexes. Gene 1999; 226:95-102.

71. Casasnovas JM, Springer, TA, Liu, JH, Harrison, SC, and Wang, JH. Crystal structure of ICAM-2 reveals a distinctive integrin recognition surface. Nature 1997; 387:312-5.

72. Casasnovas JM, Stehle, T, Liu, JH, Wang, JH, and Springer, TA. A dimeric crystal structure for the N-terminal two domains of intercellular adhesion molecule-1. Proc Natl Acad Sci U S A 1998; 95:4134-9.

73. Bella J, Kolatkar, PR, Marlor, CW, Greve, JM, and Rossmann, MG. The structure of the two amino-terminal domains of human ICAM-1 suggests how it functions as a rhinovirus receptor and as an LFA-1 integrin ligand. Proc Natl Acad Sci U S A 1998; 95:4140-5.

74. Staunton DE, Merluzzi, VJ, Rothlein, R, Barton, R, Marlin, SD, and Springer, TA. A cell adhesion molecule, ICAM-1, is the major surface receptor for rhinoviruses. Cell 1989; 56:849-53.

75. Springer TA. Traffic signals for lymphocyte recirculation and leukocyte emigration: the multistep paradigm. Cell 1994; 76:301-14.

76. Staunton DE, Dustin, ML, Erickson, HP, and Springer, TA. The arrangement of the immunoglobulin-like domains of ICAM-1 and the binding sites for LFA-1 and rhinovirus [published errata appear in Cell 1990 Jun 15;61(2):1157 and 1991;66(6):following 1311]. Cell 1990; 61:243-54.

77. Li R, Nortamo, P, Valmu, L, Tolvanen, M, Huuskonen, J, Kantor, C, and Gahmberg, CG. A peptide from ICAM-2 binds to the leukocyte integrin CD11a/CD18 and inhibits endothelial cell adhesion. J Biol Chem 1993; 268:17513-8.

78. Sadhu C, Lipsky, B, Erickson, HP, Hayflick, J, Dick, KO, Gallatin, WM, and Staunton, DE. LFA-1 binding site in ICAM-3 contains a conserved motif and non-contiguous amino acids. Cell Adhes Commun 1994; 2:429-40.

79. Klickstein LB, York, MR, Fougerolles, AR, and Springer, TA. Localization of the binding site on intercellular adhesion molecule-3 (ICAM-3) for lymphocyte function-associated antigen 1 (LFA-1). J Biol Chem 1996; 271:23920-7.

80. Hermand P, Huet, M, Callebaut, I, Gane, P, Ihanus, E, Gahmberg, C, Cartron, JP, and Bailly, P. Binding sites of leukocyte (b)2 integrins (LFA-1, Mac-1) on the human ICAM4/LW blood group protein. J Biol Chem 2000; 275:26002-26010.

81. Berendt AR, McDowall, A, Craig, AG, Bates, PA, Sternberg, MJ, Marsh, K, Newbold, CI, and Hogg, N. The binding site on ICAM-1 for Plasmodium falciparum-infected erythrocytes overlaps, but is distinct from, the LFA-1-binding site. Cell 1992; 68:71-81.

82. Ockenhouse CF, Betageri, R, Springer, TA, and Staunton, DE. Plasmodium falciparum-infected erythrocytes bind ICAM-1 at a site distinct from LFA-1, Mac-1, and human rhinovirus [published erratum appears in Cell 1992 Mar 6;68(5):following 994]. Cell 1992; 68:63-9.

83. de Fougerolles AR, Diamond, MS, and Springer, TA. Heterogenous glycosylation of ICAM-3 and lack of interaction with Mac-1 and p150,95. Eur J Immunol 1995; 25:1008-12.

84. Tsukita S, and Yonemura, S. Cortical actin organization: lessons from ERM (ezrin/radixin/moesin) proteins. J Biol Chem 1999; 274:34507-10.

85. Ropers H, and Mohrenweiser, H. Report of the committee on the genetic constitution of chromosome 19. Genet Prior Rep 1993; 1:524-527.

86. Bossy D, Mattei, MG, and Simmons, DL. The human intercellular adhesion molecule 3 (ICAM3) gene is located in the 19p13.2-p13.3 region, close to the ICAM1 gene. Genomics 1994; 23:712-3.

160

87. Sistonen P. Linkage of the LW blood group locus with the complement C3 and Lutheran blood group loci. Ann Hum Genet 1984; 48:239-42.

88. Lewis M, Kaita, H, Coghlan, G, Philipps, S, Belcher, E, McAlpine, PJ, Coopland, GR, and Woods, RA. The chromosome 19 linkage group LDLR, C3, LW, APOC2, LU, SE in man. Ann Hum Genet 1988; 52:137-44.

89. Trask B, Fertitta, A, Christensen, M, Youngblom, J, Bergmann, A, Copeland, A, de Jong, P, Mohrenweiser, H, Olsen, A, Carrano, A, and et al. Fluorescence in situ hybridization mapping of human chromosome 19: cytogenetic band location of 540 cosmids and 70 genes or DNA markers. Genomics 1993; 15:133-45.

90. Kilgannon P, Turner, T, Meyer, J, Wisdom, W, and Gallatin, WM: Mapping of the ICAM-5 (telencephalon) gene, a neuronal member of the ICAM family, to a location between ICAM-1 and ICAM-3 on human chromosome 19p13.2. Genomics 1998; 54:328-30.

91. Sansom D, Borrow, J, Solomon, E, and Trowsdale, J. The human ICAM2 gene maps to 17q23-25. Genomics 1991; 11:462-4.

92. Ballantyne CM, Kozak, CA, O'Brien, WE, and Beaudet, AL. Assignment of the gene for intercellular adhesion molecule-1 (ICAM-1) to proximal mouse chromosome 9. Genomics 1991; 9:547-50.

93. Sugino H, Yoshihara, Y, Copeland, NG, Gilbert, DJ, Jenkins, NA, and Mori, K. Genomic organization and chromosomal localization of the mouse telencephalon gene, a neuronal member of the ICAM family. Genomics 1997; 43:209-15.

94. Williams AF, and Barclay, AN. The immunoglobulin superfamily--domains for cell surface recognition. Annu Rev Immunol 1988; 6:381-405.

95. Voraberger G, Schafer, R, and Stratowa, C. Cloning of the human gene for intercellular adhesion molecule 1 and analysis of its 5'-regulatory region. Induction by cytokines and phorbol ester. J Immunol 1991; 147:2777-86.

96. Ballantyne CM, Sligh, JE, Jr., Dai, XY, and Beaudet, AL. Characterization of the murine ICAM-1 gene. Genomics 1992; 14:1076-80.

97. King PD, Sandberg, ET, Selvakumar, A, Fang, P, Beaudet, AL, and Dupont, B. Novel isoforms of murine intercellular adhesion molecule-1 generated by alternative RNA splicing. J Immunol 1995; 154:6080-93.

98. Wakatsuki T, Kimura, K, Kimura, F, Shinomiya, N, Ohtsubo, M, Ishizawa, M, and Yamamoto, M. A distinct mRNA encoding a soluble form of ICAM-1 molecule expressed in human tissues. Cell Adhes Commun 1995; 3:283-92.

99. Xu H, Tong, IL, De Fougerolles, AR, and Springer, TA. Isolation, characterization, and expression of mouse ICAM-2 complementary and genomic DNA. J Immunol 1992; 149:2650-5.

100. Dean NM, McKay, R, Condon, TP, and Bennett, CF. Inhibition of protein kinase C-alpha expression in human A549 cells by antisense oligonucleotides inhibits induction of intercellular adhesion molecule 1 (ICAM-1) mRNA by phorbol esters. J Biol Chem 1994; 269:16416-24.

101. Wertheimer SJ, Myers, CL, Wallace, RW, and Parks, TP. Intercellular adhesion molecule-1 gene expression in human endothelial cells. Differential regulation by tumor necrosis factor-alpha and phorbol myristate acetate. J Biol Chem 1992; 267:12030-5.

102. Lane TA, Lamkin, GE, and Wancewicz, EV. Protein kinase C inhibitors block the enhanced expression of intercellular adhesion molecule-1 on endothelial cells activated by interleukin-1, lipopolysaccharide and tumor necrosis factor. Biochem Biophys Res Commun 1990; 172:1273-81.

103. Myers CL, Desai, SN, Schembri-King, J, Letts, GL, and Wallace, RW. Discriminatory effects of protein kinase inhibitors and calcium ionophore on endothelial ICAM-1 induction. Am J Physiol 1992; 262:C365-73.

104. Doukas J, and Pober, JS. IFN-gamma enhances endothelial activation induced by tumor necrosis factor but not IL-1. J Immunol 1990; 145:1727-33.

105.   Pober JS, Gimbrone, MA, Jr., Lapierre, LA, Mendrick, DL, Fiers, W, Rothlein, R, and Springer, TA. Overlapping patterns of activation of human endothelial cells by interleukin 1, tumor necrosis factor, and immune interferon. J Immunol 1986; 137:1893-6.

106.   Gerritsen ME, Kelley, KA, Ligon, G, Perry, CA, Shen, CP, Szczepanski, A, and Carley, WW. Regulation of the expression of intercellular adhesion molecule 1 in cultured human endothelial cells derived from rheumatoid synovium. Arthritis Rheum 1993; 36:593-602.

107.   Shrikant P, Chung, IY, Ballestas, ME, and Benveniste, EN. Regulation of intercellular adhesion molecule-1 gene expression by tumor necrosis factor-alpha, interleukin-1 beta, and interferon-gamma in astrocytes. J Neuroimmunol 1994; 51:209-20.

108.   Shrikant P, Weber, E, Jilling, T, and Benveniste, EN. Intercellular adhesion molecule-1 gene expression by glial cells. Differential mechanisms of inhibition by IL-10 and IL-6. J Immunol 1995; 155:1489-501.

109.   Cunningham AC, Zhang, JG, Moy, JV, Ali, S, and Kirby, JA. A comparison of the antigen-presenting capabilities of class II MHC-expressing human lung epithelial and endothelial cells. Immunology 1997; 91:458-63.

110.   Dang LH, and Rock, KL. Stimulation of B lymphocytes through surface Ig receptors induces LFA-1 and ICAM-1-dependent adhesion. J Immunol 1991; 146:3273-9.

111.   Branden H, and Lundgren, E. Differential regulation of LFA-1 and ICAM-1 on human primary B-lymphocytes. Cell Immunol 1993; 147:64-72

112.   Ohh M, and Takei, F.   Interferon-gamma- and phorbol myristate acetate-responsive elements involved in intercellular adhesion molecule-1 mRNA stabilization. J Biol Chem 1994; 269:30117-20.

113.   Ohh M, Smith, CA, Carpenito, C, and Takei, F. Regulation of intercellular adhesion molecule-1 gene expression involves multiple mRNA stabilization mechanisms. effects of interferon-gamma and phorbol myristate acetate. Blood 1994; 84:2632-9.

114.   Stade BG, Messer, G, Riethmuller, G, and Johnson, JP.   Structural characteristics of the 5' region of the human ICAM-1 gene. Immunobiology 1990; 182:79-87.

115.   Wawryk SO, Cockerill, PN, Wicks, IP, and Boyd, AW.   Isolation and characterization of the promoter region of the human intercellular adhesion molecule-1 gene. Int Immunol 1991; 3:83-93.

116.   Degitz K, Li, LJ, and Caughman, SW. Cloning and characterization of the 5'-transcriptional regulatory region of the human intercellular adhesion molecule 1 gene. J Biol Chem 1991; 266:14024-30.

117.   Maltzman JS, Carmen, JA, and Monroe, JG. Transcriptional regulation of the ICAM-1 gene in antigen receptor- and phorbol ester-stimulated B lymphocytes. role for transcription factor EGR1. J Exp Med 1996; 183:1747-59.

118.   Chiang MY, Chan, H, Zounes, MA, Freier, SM, Lima, WF, and Bennett, CF. Antisense oligonucleotides inhibit intercellular adhesion molecule 1 expression by two distinct mechanisms. J Biol Chem 1991; 266:18162-71

119.   Shaw G, and Kamen, R. A conserved AU sequence from the 3' untranslated region of GM-CSF mRNA mediates selective mRNA degradation. Cell 1986; 46:659-67.

120.   Ghersa P, Hooft van Huijsduijnen, R, Whelan, J, and DeLamarter, JF. Labile proteins play a dual role in the control of endothelial leukocyte adhesion molecule-1 (ELAM-1) gene regulation. J Biol Chem 1992; 267:19226-32

121.   Mahadevan LC, and Edwards, DR. Signaling and superinduction [letter]. Nature 1991, 349:747-8.

122.   Cano E, Hazzalin, CA, and Mahadevan, LC. Anisomycin-activated protein kinases p45 and p55 but not mitogen-activated protein kinases ERK-1 and -2 are implicated in the induction of c-fos and c-jun. Mol Cell Biol 1994;

14:7352-62.

123. Cano E, Doza, YN, Ben-Levy, R, Cohen, P, and Mahadevan, LC. Identification of anisomycin-activated kinases p45 and p55 in murine cells as MAPKAP kinase-2. Oncogene 1996; 12:805-12.

124. Hazzalin CA, Cuenda, A, Cano, E, Cohen, P, and Mahadevan, LC. Effects of the inhibition of p38/RK MAP kinase on induction of five fos and jun genes by diverse stimuli. Oncogene 1997; 15:2321-31.

125. Read MA, Whitley, MZ, Williams, AJ, and Collins, T. NF-kappa B and I kappa B alpha: an inducible regulatory system in endothelial activation. J Exp Med 1994; 179:503-12.

126. Read MA, Neish, AS, Gerritsen, ME, and Collins, T. Postinduction transcriptional repression of E-selectin and vascular cell adhesion molecule-1. J Immunol 1996; 157:3472-9.

127. Tartaglia LA, and Goeddel, DV. Two TNF receptors. Immunol Today 1992; 13:151-3.

128. Takeuchi M, Rothe, M, and Goeddel, DV. Anatomy of TRAF2. Distinct domains for nuclear factor-kappaB activation and association with tumor necrosis factor signaling proteins. J Biol Chem 1996, 271:19935-42

129. Fraser A, and Evan, G. A license to kill. Cell 1996; 85:781-4

130. Cleveland JL, and Ihle, JN. Contenders in FasL/TNF death signaling. Cell 1995, 81:479-82.

131. Yeh WC, Shahinian, A, Speiser, D, Kraunus, J, Billia, F, Wakeham, A, de la Pompa, JL, Ferrick, D, Hum, B, Iscove, N, Ohashi, P, Rothe, M, Goeddel, DV, and Mak, TW. Early lethality, functional NF-kappaB activation, and increased sensitivity to TNF-induced cell death in TRAF2-deficient mice. Immunity 1997; 7:715-25.

132. Lee SY, Reichlin, A, Santana, A, Sokol, KA, Nussenzweig, MC, and Choi, Y. TRAF2 is essential for JNK but not NF-kappaB activation and regulates lymphocyte proliferation and survival. Immunity 1997, 7:703-13.

133. Kelliher MA, Grimm, S, Ishida, Y, Kuo, F, Stanger, BZ, and Leder, P. The death domain kinase RIP mediates the TNF-induced NF-kappaB signal. Immunity 1998; 8:297-303.

134. Cheng G, and Baltimore, D. TANK, a co-inducer with TRAF2 of TNF- and CD 40L-mediated NF-kappaB activation. Genes Dev 1996; 10:963-73.

135. Rothe M, Xiong, J, Shu, HB, Williamson, K, Goddard, A, and Goeddel, DV. I-TRAF is a novel TRAF-interacting protein that regulates TRAF-mediated signal transduction. Proc Natl Acad Sci U S A 1996; 93:8241-6.

136. DiDonato JA, Hayakawa, M, Rothwarf, DM, Zandi, E, and Karin, M. A cytokine-responsive IkappaB kinase that activates the transcription factor NF-kappaB [see comments]. Nature 1997; 388:548-54.

137. Regnier CH, Song, HY, Gao, X, Goeddel, DV, Cao, Z, Rothe, M. Identification and characterization of an IkappaB kinase. Cell 1997; 90:373-83.

138. van de Stolpe A, Caldenhoven, E, Stade, BG, Koenderman, L, Raaijmakers, JA, Johnson, JP, and van der Saag, PT. 12-O-tetradecanoylphorbol-13-acetate- and tumor necrosis factor alpha-mediated induction of intercellular adhesion molecule-1 is inhibited by dexamethasone. Functional analysis of the human intercellular adhesion molecular-1 promoter. J Biol Chem 1994; 269:6185-92.

139. van de Stolpe A, Caldenhoven, E, Raaijmakers, JA, van der Saag, PT, and Koenderman, L. Glucocorticoid-mediated repression of intercellular adhesion molecule-1 expression in human monocytic and bronchial epithelial cell lines. Am J Respir Cell Mol Biol 1993; 8:340-7.

140. Jahnke A, and Johnson, JP. Synergistic activation of intercellular adhesion molecule 1 (ICAM-1) by TNF-alpha and IFN-gamma is mediated by p65/p50 and p65/c-Rel and interferon-responsive factor Stat1 alpha (p91) that can be activated by both IFN-gamma and IFN-alpha. FEBS Lett 1994; 354:220-6.

141. Hou J, Baichwal, V, and Cao, Z. Regulatory elements and transcription factors

controlling basal and cytokine-induced expression of the gene encoding intercellular adhesion molecule 1. Proc Natl Acad Sci U S A 1994; 91:11641-5.

142. Ledebur HC, and Parks, TP. Transcriptional regulation of the intercellular adhesion molecule-1 gene by inflammatory cytokines in human endothelial cells. Essential roles of a variant NF-kappa B site and p65 homodimers. J Biol Chem 1995; 270:933-43.

143. Shu HB, Agranoff, AB, Nabel, EG, Leung, K, Duckett, CS, Neish, AS, Collins, T, and Nabel, GJ. Differential regulation of vascular cell adhesion molecule 1 gene expression by specific NF-kappa B subunits in endothelial and epithelial cells. Mol Cell Biol 1993; 13:6283-9.

144. Kaszubska W, Hooft van Huijsduijnen, R, Ghersa, P, DeRaemy-Schenk, AM, Chen, BP, Hai, T, DeLamarter, JF, and Whelan, J. Cyclic AMP-independent ATF family members interact with NF-kappa B and function in the activation of the E-selectin promoter in response to cytokines. Mol Cell Biol 1993; 13:7180-90.

145. Catron KM, Brickwood, JR, Shang, C, Li, Y, Shannon, MF, and Parks, TP. Cooperative binding and synergistic activation by RelA and C/EBPbeta on the intercellular adhesion molecule-1 promoter. Cell Growth Differ 1998; 9:949-59.

146. Parry GC, and Mackman, N. A set of inducible genes expressed by activated human monocytic and endothelial cells contain kappa B-like sites that specifically bind c-Rel-p65 heterodimers. J Biol Chem 1994; 269:20823-5.

147. Kunsch C, Ruben, SM, and Rosen, CA. Selection of optimal kappa B/Rel DNA-binding motifs: interaction of both subunits of NF-kappa B with DNA is required for transcriptional activation. Mol Cell Biol 1992; 12:4412-21.

148. Paxton LL, Li, LJ, Secor, V, Duff, JL, Naik, SM, Shibagaki, N, and Caughman, SW. Flanking sequences for the human intercellular adhesion molecule-1 NF-kappaB response element are necessary for tumor necrosis factor alpha-induced gene expression. J Biol Chem 1997; 272:15928-35.

149. Wissink S, van de Stolpe, A, Caldenhoven, E, Koenderman, L, and van der Saag, PT. NF-kappa B/Rel family members regulating the ICAM-1 promoter in monocytic THP-1 cells. Immunobiology 1997; 198:50-64.

150. Cao Z, Henzel, WJ, and Gao, X. IRAK: a kinase associated with the interleukin-1 receptor. Science 1996, 271:1128-31.

151. Cao Z, Xiong, J, Takeuchi, M, Kurama, T, and Goeddel, DV. TRAF6 is a signal transducer for interleukin-1. Nature 1996; 383:443-6.

152. Yamaguchi K, Shirakabe, K, Shibuya, H, Irie, K, Oishi, I, Ueno, N, Taniguchi, T, Nishida, E, and Matsumoto, K. Identification of a member of the MAPKKK family as a potential mediator of TGF-beta signal transduction. Science 1995; 270:2008-11.

153. Ninomiya-Tsuji J, Kishimoto, K, Hiyama, A, Inoue, J, Cao, Z, and Matsumoto, K. The kinase TAK1 can activate the NIK-I kappaB as well as the MAP kinase cascade in the IL-1 signaling pathway. Nature 1999; 398:252-6.

154. Connelly MA, and Marcu, KB. CHUK, a new member of the helix-loop-helix and leucine zipper families of interacting proteins, contains a serine-threonine kinase catalytic domain. Cell Mol Biol Res 1995; 41:537-49.

155. Delhase M, Hayakawa, M, Chen, Y, and Karin, M. Positive and negative regulation of IkappaB kinase activity through IKKbeta subunit phosphorylation [see comments]. Science 1999; 284:309-13.

156. Lane TA, Lamkin, GE, and Wancewicz, E. Modulation of endothelial cell expression of intercellular adhesion molecule 1 by protein kinase C activation. Biochem Biophys Res Commun 1989; 161:945-52.

157. Mattila P, Majuri, ML, Mattila, PS, and Renkonen, R. TNF alpha-induced expression of endothelial adhesion molecules, ICAM-1 and VCAM-1, is linked to protein kinase C activation. Scand J Immunol 1992; 36:159-65.

158. Norris JG, Tang, LP, Sparacio, SM, and Benveniste, EN. Signal transduction

pathways mediating astrocyte IL-6 induction by IL-1 beta and tumor necrosis factor-alpha. J Immunol 1994; 152:841-50.

159. Nakanishi H, Brewer, KA, and Exton, JH. Activation of the zeta isozyme of protein kinase C by phosphatidylinositol 3,4,5-trisphosphate. J Biol Chem 1993; 268:13-6.

160. Cornelius LA, Taylor, JT, Degitz, K, Li, LJ, Lawley, TJ, and Caughman, SW. A 5' portion of the ICAM-1 gene confers tissue-specific differential expression levels and cytokine responsiveness. J Invest Dermatol 1993; 100:753-8.

161. Look DC, Pelletier, MR, and Holtzman, MJ. Selective interaction of a subset of interferon-gamma response element-binding proteins with the intercellular adhesion molecule-1 (ICAM-1) gene promoter controls the pattern of expression on epithelial cells. J Biol Chem 1994; 269:8952-8.

162. Sims SH, Cha, Y, Romine, MF, Gao, PQ, Gottlieb, K, and Deisseroth, AB. A novel interferon-inducible domain: structural and functional analysis of the human interferon regulatory factor 1 gene promoter. Mol Cell Biol 1993; 13:690-702.

163. Kanno Y, Kozak, CA, Schindler, C, Driggers, PH, Ennist, DL, Gleason, SL, Darnell, JE, Jr., and Ozato, K. The genomic structure of the murine ICSBP gene reveals the presence of the gamma interferon-responsive element, to which an ISGF3 alpha subunit (or similar) molecule binds. Mol Cell Biol 1993; 13:3951-63.

164. Naik SM, Shibagaki, N, Li, LJ, Quinlan, KL, Paxton, LL, and Caughman, SW. Interferon gamma-dependent induction of human intercellular adhesion molecule-1 gene expression involves activation of a distinct STAT protein complex. J Biol Chem 1997; 272:1283-90.

165. Walter MJ, Look, DC, Tidwell, RM, Roswit, WT, and Holtzman, MJ. Targeted inhibition of interferon-gamma-dependent intercellular adhesion molecule-1 (ICAM-1) expression using dominant-negative Stat1. J Biol Chem 1997; 272:28582-9.

166. Boulton TG, Zhong, Z, Wen, Z, Darnell, JE, Jr., Stahl, N, and Yancopoulos, GD. STAT3 activation by cytokines utilizing gp130 and related transducers involves a secondary modification requiring an H7-sensitive kinase. Proc Natl Acad Sci U S A 1995; 92:6915-9.

167. Zhang X, Blenis, J, Li, HC, Schindler, C, and Chen-Kiang, S. Requirement of serine phosphorylation for formation of STAT-promoter complexes. Science 1995; 267:1990-4.

168. Wen Z, Zhong, Z, and Darnell, JE, Jr.. Maximal activation of transcription by Stat1 and Stat3 requires both tyrosine and serine phosphorylation. Cell 1995; 82:241-50.

169. Schindler C, and Darnell, JE, Jr.. Transcriptional responses to polypeptide ligands: the JAK-STAT pathway. Annu Rev Biochem 1995; 64:621-51.

170. Tanaka N, Kawakami, T, and Taniguchi, T. Recognition DNA sequences of interferon regulatory factor 1 (IRF-1) and IRF-2, regulators of cell growth and the interferon system. Mol Cell Biol 1993; 13:4531-8.

171. Harada H, Fujita, T, Miyamoto, M, Kimura, Y, Maruyama, M, Furia, A, Miyata, T, and Taniguchi, T. Structurally similar but functionally distinct factors, IRF-1 and IRF-2, bind to the same regulatory elements of IFN and IFN-inducible genes. Cell 1989; 58:729-39.

172. Pine R, Decker T, Kessler DS, Levy DE, Darnell, JE. Purification and cloning of interferon-stimulated gene factor 2 (ISGF2): ISGF2 (IRF-1) can bind to the promoters of both beta interferon- and interferon-stimulated genes but is not a primary transcriptional activator of either. Mol Cell Biol 1990; 10:2448-57.

173. Miyamoto M, Fujita, T, Kimura, Y, Maruyama, M, Harada, H, Sudo, Y, Miyata, T, and Taniguchi, T. Regulated expression of a gene encoding a nuclear factor, IRF-1, that specifically binds to IFN-beta gene regulatory elements. Cell 1988; 54:903-13.

174. Fujita T, Shibuya, H, Hotta, H, Yamanishi, K, and Taniguchi, T. Interferon-beta gene regulation: tandemly repeated sequences of a synthetic 6 bp oligomer function as a virus-inducible enhancer. Cell 1987; 49:357-67.

175. Yu-Lee LY, Hrachovy, JA, Stevens, AM, and Schwarz, LA. Interferon-regulatory factor 1 is an immediate-early gene under transcriptional regulation by prolactin in Nb2 T cells. Mol Cell Biol 1990; 10:3087-94.

176. Abdollahi A, Lord, KA, Hoffman-Liebermann, B, and Liebermann, DA. Interferon regulatory factor 1 is a myeloid differentiation primary response gene induced by interleukin 6 and leukemia inhibitory factor: role in growth inhibition. Cell Growth Differ 1991; 2:401-7.

177. Tsukada J, Waterman, WR, Koyama, Y, Webb, AC, and Auron, PE. A novel STAT-like factor mediates lipopolysaccharide, interleukin 1 (IL-1), and IL-6 signaling and recognizes a gamma interferon activation site-like element in the IL1B gene [published erratum appears in Mol Cell Biol 1996 Jun;16(6):3233]. Mol Cell Biol 1996; 16:2183-94.

178. Ohmori Y, Schreiber, RD, and Hamilton, TA. Synergy between interferon-gamma and tumor necrosis factor-alpha in transcriptional activation is mediated by cooperation between signal transducer and activator of transcription 1 and nuclear factor kappaB. J Biol Chem 1997; 272:14899-907.

179. Caldenhoven E, Coffer, P, Yuan, J, Van de Stolpe, A, Horn, F, Kruijer, W, and Van der Saag, PT. Stimulation of the human intercellular adhesion molecule-1 promoter by interleukin-6 and interferon-gamma involves binding of distinct factors to a palindromic response element. J Biol Chem 1994; 269:21146-54.

180. Yuan J, Wegenka, UM, Lutticken, C, Buschmann, J, Decker, T, Schindler, C, Heinrich, PC, and Horn, F. The signaling pathways of interleukin-6 and gamma interferon converge by the activation of different transcription factors which bind to common responsive DNA elements. Mol Cell Biol 1994; 14:1657-68.

181. Stratowa C, and Audette, M. Transcriptional regulation of the human intercellular adhesion molecule-1 gene: a short overview. Immunobiology 1995; 193:293-304.

182. Romano M, Sironi, M, Toniatti, C, Polentarutti, N, Fruscella, P, Ghezzi, P, Faggioni, R, Luini, W, van Hinsbergh, V, Sozzani, S, Bussolino, F, Poli, V, Ciliberto, G, and Mantovani, A. Role of IL-6 and its soluble receptor in induction of chemokines and leukocyte recruitment. Immunity 1997; 6:315-25.

183. Muller S, Kammerbauer, C, Simons, U, Shibagaki, N, Li, LJ, Caughman, SW, Degitz, K. Transcriptional regulation of intercellular adhesion molecule-1: PMA-induction is mediated by NF kappa B. J Invest Dermatol 1995;104:970-5.

184. Chiu R, Imagawa, M, Imbra, RJ, Bockoven, JR, and Karin, M. Multiple cis-and trans-acting elements mediate the transcriptional response to phorbol esters. Nature 1987; 329:648-51.

185. Imagawa M, Chiu, R, and Karin, M. Transcription factor AP-2 mediates induction by two different signal-transduction pathways: protein kinase C and cAMP. Cell 1987; 51:251-60.

186. Baeuerle PA, Lenardo, M, Pierce, JW, and Baltimore, D. Phorbol-ester-induced activation of the NF-kappa B transcription factor involves dissociation of an apparently cytoplasmic NF-kappa B/inhibitor complex. Cold Spring Harb Symp Quant Biol 1988; 53:789-98.

187. McMahon SB, and Monroe, JG. The role of early growth response gene 1 (egr-1) in regulation of the immune response. J Leukoc Biol 1996; 60:159-66.

188. Nagel T, Resnick, N, Atkinson, WJ, Dewey, CF, Jr., and Gimbrone, MA, Jr.. Shear stress selectively upregulates intercellular adhesion molecule-1 expression in cultured human vascular endothelial cells. J Clin Invest 1994; 94:885-91.

189. Khachigian LM, Lindner, V, Williams, AJ, and Collins, T. Egr-1-induced endothelial gene expression: a common theme in vascular injury. Science 1996; 271:1427-31.

190. Schreck R, Meier, B, Mannel, DN, Droge, W, and Baeuerle, PA. Dithiocarbamates as potent inhibitors of nuclear factor kappa B activation in intact cells. J Exp Med 1992; 175:1181-94.

191. Traenckner EB, Pahl, HL, Henkel, T, Schmidt, KN, Wilk, S, and Baeuerle, PA. Phosphorylation of human I kappa B-alpha on serines 32 and 36 controls I kappa B-alpha proteolysis and NF-kappa B activation in response to diverse stimuli. EMBO J 1995; 14:2876-83.

192. Diaz-Meco MT, Dominguez, I, Sanz, L, Dent, P, Lozano, J, Municio, MM, Berra, E, Hay, RT, Sturgill, TW, and Moscat, J. zeta PKC induces phosphorylation and inactivation of I kappa B-alpha in vitro. EMBO J 1994; 13:2842-8.

193. Tojima Y, Fujimoto, A, Delhase, M, Chen, Y, Hatakeyama, S, Nakayama, K, Kaneko, Y, Nimura, Y, Motoyama, N, Ikeda, K, Karin, M, and Nakanishi, M. NAK is an IkappaB kinase-activating kinase. Nature 2000; 404:778-82.

194. Pomerantz JL, and Baltimore, D. NF-kappaB activation by a signaling complex containing TRAF2, TANK and TBK1, a novel IKK-related kinase. EMBO J 1999; 18:6694-704.

195. Kehry MR. CD40-mediated signaling in B cells. Balancing cell survival, growth, and death. J Immunol 1996; 156:2345-8.

196. Karmann K, Hughes, CC, Schechner, J, Fanslow, WC, and Pober, JS. CD40 on human endothelial cells: inducibility by cytokines and functional regulation of adhesion molecule expression. Proc Natl Acad Sci U S A 1995; 92:4342-6.

197. Hollenbaugh D, Mischel-Petty, N, Edwards, CP, Simon, JC, Denfeld, RW, Kiener, PA, and Aruffo, A. Expression of functional CD40 by vascular endothelial cells. J Exp Med 1995; 182:33-40.

198. Yellin MJ, Winikoff, S, Fortune, SM, Baum, D, Crow, MK, Lederman, S, and Chess, L. Ligation of CD40 on fibroblasts induces CD54 (ICAM-1) and CD106 (VCAM-1) up-regulation and IL-6 production and proliferation. J Leukoc Biol 1995; 58:209-16.

199. Hanissian SH, and Geha, RS. Jak3 is associated with CD40 and is critical for CD40 induction of gene expression in B cells. Immunity 1997; 6:379-87.

200. Karmann K, Min, W, Fanslow, WC, and Pober, JS. Activation and homologous desensitization of human endothelial cells by CD40 ligand, tumor necrosis factor, and interleukin 1. J Exp Med 1996; 184:173-82.

201. Lee HH, Dempsey, PW, Parks, TP, Zhu, X, Baltimore, D, and Cheng, G. Specificities of CD40 signaling: involvement of TRAF2 in CD40-induced NF-kappaB activation and intercellular adhesion molecule-1 up-regulation. Proc Natl Acad Sci U S A 1999; 96:1421-6.

202. Nolte D, Hecht, R, Schmid, P, Botzlar, A, Menger, MD, Neumueller, C, Sinowatz, F, Vestweber, D, and Messmer, K. Role of Mac-1 and ICAM-1 in ischemia-reperfusion injury in a microcirculation model of BALB/C mice. Am J Physiol 1994; 267:H1320-8.

203. Bradley JR, Johnson, DR, and Pober, JS. Endothelial activation by hydrogen peroxide. Selective increases of intercellular adhesion molecule-1 and major histocompatibility complex class I. Am J Pathol 1993; 142:1598-609.

204. Lo SK, Janakidevi, K, Lai, L, and Malik, AB. Hydrogen peroxide-induced increase in endothelial adhesiveness is dependent on ICAM-1 activation. Am J Physiol 1993; 264:L406-12.

205. Sellak H, Franzini, E, Hakim, J, and Pasquier, C. Reactive oxygen species rapidly increase endothelial ICAM-1 ability to bind neutrophils without detectable upregulation. Blood 1994; 83:2669-77.

206. Ikeda M, Schroeder, KK, Mosher, LB, Woods, CW, and Akeson, AL. Suppressive effect of antioxidants on intercellular adhesion molecule-1 (ICAM-1) expression in human epidermal keratinocytes. J Invest Dermatol 1994; 103:791-6.

207. Roebuck KA, Rahman, A, Lakshminarayanan, V, Janakidevi, K, and Malik, AB.

H2O2 and tumor necrosis factor-alpha activate intercellular adhesion molecule 1 (ICAM-1) gene transcription through distinct cis-regulatory elements within the ICAM-1 promoter. J Biol Chem 1995; 270:18966-74.

208. Wu H, Moulton, K, Horvai, A, Parik, S, and Glass, CK. Combinatorial interactions between AP-1 and ets domain proteins contribute to the developmental regulation of the macrophage scavenger receptor gene. Mol Cell Biol 1994; 14:2129-39.

209. Nose K, and Ohba, M. Functional activation of the egr-1 (early growth response-1) gene by hydrogen peroxide. Biochem J 1996; 316:381-3.

210. Meyer M, Schreck, R, and Baeuerle, PA. H2O2 and antioxidants have opposite effects on activation of NF-kappa B and AP-1 in intact cells: AP-1 as secondary antioxidant-responsive factor. EMBO J 1993; 12:2005-15.

211. Munoz C, Pascual-Salcedo, D, Castellanos, MC, Alfranca, A, Aragones, J, Vara, A, Redondo, MJ, and de Landazuri, MO. Pyrrolidine dithiocarbamate inhibits the production of interleukin-6, interleukin-8, and granulocyte-macrophage colony-stimulating factor by human endothelial cells in response to inflammatory mediators: modulation of NF-kappa B and AP-1 transcription factors activity. Blood 1996; 88:3482-90.

212. Marui N, Offermann, MK, Swerlick, R, Kunsch, C, Rosen, CA, Ahmad, M, Alexander, RW, and Medford, RM. Vascular cell adhesion molecule-1 (VCAM-1) gene transcription and expression are regulated through an antioxidant-sensitive mechanism in human vascular endothelial cells. J Clin Invest 1993; 92:1866-74.

213. Ferran C, Millan, MT, Csizmadia, V, Cooper, JT, Brostjan, C, Bach, FH, and Winkler, H. Inhibition of NF-kappa B by pyrrolidine dithiocarbamate blocks endothelial cell activation. Biochem Biophys Res Commun 1995; 214:212-23.

214. Bradley JR, Johnson, DR, and Pober, JS. Four different classes of inhibitors of receptor-mediated endocytosis decrease tumor necrosis factor-induced gene expression in human endothelial cells. J Immunol 1993; 150:5544-55.

215. Rushmore TH, Morton, MR, and Pickett, CB. The antioxidant responsive element. Activation by oxidative stress and identification of the DNA consensus sequence required for functional activity. J Biol Chem 1991; 266:11632-9.

216. Pinkus R, Weiner, LM, and Daniel, V. Role of quinone-mediated generation of hydroxyl radicals in the induction of glutathione S-transferase gene expression. Biochemistry 1995; 34:81-8.

217. Goldstein BD, Rozen, MG, Quintavalla, JC, and Amoruso, MA. Decrease in mouse lung and liver glutathione peroxidase activity and potentiation of the lethal effects of ozone and paraquat by the superoxide dismutase inhibitor diethyldithiocarbamate. Biochem Pharmacol 1979; 28:27-30

218. Munoz C, Castellanos, MC, Alfranca, A, Vara, A, Esteban, MA, Redondo, JM, and de Landazuri, MO. Transcriptional up-regulation of intracellular adhesion molecule-1 in human endothelial cells by the antioxidant pyrrolidine dithiocarbamate involves the activation of activating protein-1. J Immunol 1996; 157:3587-97.

219. Nobel CI, Kimland, M, Lind, B, Orrenius, S, and Slater, AF. Dithiocarbamates induce apoptosis in thymocytes by raising the intracellular level of redox-active copper. J Biol Chem 1995; 270:26202-8.

220. Hong JH, Chiang, CS, Campbell, IL, Sun, JR, Withers, HR, and McBride, WH. Induction of acute phase gene expression by brain irradiation. Int J Radiat Oncol Biol Phys 1995; 33:619-26.

221. Behrends U, Peter, RU, Hintermeier-Knabe, R, Eissner, G, Holler, E, Bornkamm, GW, Caughman, SW, and Degitz, K. Ionizing radiation induces human intercellular adhesion molecule-1 in vitro. J Invest Dermatol 1994; 103:726-30.

222. Hallahan DE, and Virudachalam, S. Intercellular adhesion molecule 1 knockout

168

abrogates radiation induced pulmonary inflammation. Proc Natl Acad Sci U S A 1997; 94:6432-7.

223. Hallahan D, Kuchibhotla, J, and Wyble, C. Cell adhesion molecules mediate radiation-induced leukocyte adhesion to the vascular endothelium. Cancer Res 1996; 56:5150-5.

224. Hallahan D, Clark, ET, Kuchibhotla, J, Gewertz, BL, and Collins, T. E-selectin gene induction by ionizing radiation is independent of cytokine induction. Biochem Biophys Res Commun 1995; 217:784-95

225. Sen R, and Baltimore, D. Inducibility of kappa immunoglobulin enhancer-binding protein Nf-kappa B by a posttranslational mechanism. Cell 1986; 47:921-8.

226. Brach MA, Gruss, HJ, Kaisho, T, Asano, Y, Hirano, T, Herrmann, F. Ionizing radiation induces expression of interleukin 6 by human fibroblasts involving activation of nuclear factor-kappa B. J Biol Chem 1993; 268:8466-72.

227. Brach MA, Hass, R, Sherman, ML, Gunji, H, Weichselbaum, R, and Kufe, D. Ionizing radiation induces expression and binding activity of the nuclear factor kappa B. J Clin Invest 1991; 88:691-5.

228. Resnick N, Collins, T, Atkinson, W, Bonthron, DT, Dewey, CF, Jr., and Gimbron, MA, Jr.. Platelet-derived growth factor B chain promoter contains a cis-acting fluid shear-stress-responsive element. Proc Natl Acad Sci U S A 1993; 90:7908.

229. Khachigian LM, Resnick, N, Gimbrone, MA, Jr., and Collins, T. Nuclear factor-kappa B interacts functionally with the platelet-derived growth factor B-chain shear-stress response element in vascular endothelial cells exposed to fluid shear stress. J Clin Invest 1995; 96:1169-75.

230. Bouillon M, Tessier, P, Boulianne, R, Destrempe, R, and Audette, M. Regulation by retinoic acid of ICAM-1 expression on human tumor cell lines. Biochim Biophys Acta 1991; 1097:95-102.

231. Wang Z, Cao, Y, D'Urso, CM, and Ferrone, S. Differential susceptibility of cultured human melanoma cell lines to enhancement by retinoic acid of intercellular adhesion molecule 1 expression. Cancer Res 1992; 52:4766-72.

232. Cilenti L, Toniato, E, Ruggiero, P, Fusco, C, Farina, AR, Tiberio, A, Hayday, AC, Gulino, A, Frati, L, and Martinotti, S. Transcriptional modulation of the human intercellular adhesion molecule gene I (ICAM-1) by retinoic acid in melanoma cells. Exp Cell Res 1995; 218:263-70.

233. Aoudjit F, Bosse, M, Stratowa, C, Voraberger, G, and Audette, M. Regulation of intercellular adhesion molecule-1 expression by retinoic acid: analysis of the 5' regulatory region of the gene. Int J Cancer 1994; 58:543-9.

234. Wang SY, and Gudas, LJ. Isolation of cDNA clones specific for collagen IV and laminin from mouse teratocarcinoma cells. Proc Natl Acad Sci U S A 1983; 80:5880-4.

235. Duester G, Shean, ML, McBride, MS, and Stewart, MJ. Retinoic acid response element in the human alcohol dehydrogenase gene ADH3: implications for regulation of retinoic acid synthesis. Mol Cell Biol 1991; 11:1638-46.

236. Imcke E, Ruszczak, Z, Mayer-da Silva, A, Detmar, M, and Orfanos, CE. Cultivation of human dermal microvascular endothelial cells in vitro: immunocytochemical and ultrastructural characterization and effect of treatment with three synthetic retinoids. Arch Dermatol Res 1991; 283:149-57.

237. Gille J, Paxton, LL, Lawley, TJ, Caughman, SW, Swerlick, RA. Retinoic acid inhibits regulated expression of vascular cell adhesion molecule-1 by cultured dermal microvascular endothelial cells. J Clin Invest 1997; 99:492-500.

238. Weber C, Calzada-Wack, JC, Goretzki, M, Pietsch, A, Johnson, JP, and Ziegler-Heitbrock, HW. Retinoic acid inhibits basal and interferon-gamma-induced expression of intercellular adhesion molecule 1 in monocytic cells. J Leukoc Biol 1995; 57:401-6.

239. Myers CL, Wertheimer, SJ, Schembri-King, J, Parks, T, and Wallace, RW.

Induction of ICAM-1 by TNF-alpha, IL-1 beta, and LPS in human endothelial cells after downregulation of PKC. Am J Physiol 1992; 263:C767-72.

240. Faure E, Equils, O, Sieling, PA, Thomas, L, Zhang, FX, Kirschning, CJ, Polentarutti, N, Muzio, M, and Arditi, M. Bacterial lipopolysaccharide activates NF-kappaB through toll-like receptor 4 (TLR-4) in cultured human dermal endothelial cells. Differential expression of TLR-4 and TLR-2 in endothelial cells. J Biol Chem 2000; 275:11058-63.

241. Zhang FX, Kirschning, CJ, Mancinelli, R, Xu, XP, Jin, Y, Faure, E, Mantovani, A, Rothe, M, Muzio, M, and Arditi, M. Bacterial lipopolysaccharide activates nuclear factor-kappaB through interleukin-1 signaling mediators in cultured human dermal endothelial cells and mononuclear phagocytes. J Biol Chem 1999; 274:7611-4.

242. Takeuchi S, Kawashima, S, Rikitake, Y, Ueyama, T, Inoue, N, Hirata, K, and Yokoyama, M. Cerivastatin suppresses lipopolysaccharide-induced ICAM-1 expression through inhibition of Rho GTPase in BAEC. Biochem Biophys Res Commun 2000; 269:97-102.

243. Poltorak A, He, X, Smirnova, I, Liu, MY, Huffel, CV, Du, X, Birdwell, D, Alejos, E, Silva, M, Galanos, C, Freudenberg, M, Ricciardi-Castagnoli, P, Layton, B, and Beutler, B. Defective LPS signaling in C3H/HeJ and C57BL/10ScCr mice: mutations in Tlr4 gene. Science 1998; 282:2085-8.

244. Boggemeyer E, Stehle, T, Schaible, UE, Hahne, M, Vestweber, D, and Simon, MM. Borrelia burgdorferi upregulates the adhesion molecules E-selectin, P-selectin, ICAM-1 and VCAM-1 on mouse endothelioma cells in vitro. Cell Adhes Commun 1994; 2:145-57.

245. Sellati TJ, Abrescia, LD, Radolf, JD, and Furie, MB. Outer surface lipoproteins of Borrelia burgdorferi activate vascular endothelium in vitro. Infect Immun 1996; 64:3180-7.

246. Wooten RM, Modur, VR, McIntyre, TM, and Weis, JJ. Borrelia burgdorferi outer membrane protein A induces nuclear translocation of nuclear factor-kappa B and inflammatory activation in human endothelial cells. J Immunol 1996; 157:4584-90.

247. Ebnet K, Brown, KD, Siebenlist, UK, Simon, MM, and Shaw, S. Borrelia burgdorferi activates nuclear factor-kappa B and is a potent inducer of chemokine and adhesion molecule gene expression in endothelial cells and fibroblasts. J Immunol 1997; 158:3285-92.

248. Hirschfeld M, Kirschning, CJ, Schwandner, R, Wesche, H, Weis, JH, Wooten, RM, and Weis, JJ. Cutting edge: inflammatory signaling by Borrelia burgdorferi lipoproteins is mediated by toll-like receptor 2. J Immunol 1999; 163:2382-6.

249. Lien E, Sellati, TJ, Yoshimura, A, Flo, TH, Rawadi, G, Finberg, RW, Carroll, JD, Espevik, T, Ingalls, RR, Radolf, JD, and Golenbock, DT. Toll-like receptor 2 functions as a pattern recognition receptor for diverse bacterial products. J Biol Chem 1999; 274:33419-25.

250. Medzhitov R, and Janeway, C, Jr.. Innate immune recognition: mechanisms and pathways. Immunol Rev 2000; 173:89-97.

251. Mori N, Wada, A, Hirayama, T, Parks, TP, Stratowa, C, and Yamamoto, N. Activation of intercellular adhesion molecule 1 expression by Helicobacter pylori is regulated by NF-kappaB in gastric epithelial cancer cells. Infect Immun 2000; 68:1806-14.

252. Huang GT, Eckmann, L, Savidge, TC, and Kagnoff, MF. Infection of human intestinal epithelial cells with invasive bacteria upregulates apical intercellular adhesion molecule-1 (ICAM)-1 expression and neutrophil adhesion. J Clin Invest 1996; 98:572-83.

253. Garofalo R, Sabry, M, Jamaluddin, M, Yu, RK, Casola, A, Ogra, PL, and Brasier, AR. Transcriptional activation of the interleukin-8 gene by respiratory syncytial virus infection in alveolar epithelial cells: nuclear

170

translocation of the RelA transcription factor as a mechanism producing airway mucosal inflammation. J Virol 1996; 70:8773-81.

254. Ferran C, Cooper, JT, Brostjan, C, Stroka, DM, Millan, MT, Goodman, DJ, and Bach, FH. Expression of a truncated form of the human p55 TNF-receptor in bovine aortic endothelial cells renders them resistant to human TNF. Transplant Proc 1996; 28:618-9.

255. Fiedler MA, Wernke-Dollries, K, and Stark, JM. Mechanism of RSV-induced IL-8 gene expression in A549 cells before viral replication. Am J Physiol 1996; 271:L963-71.

256. Maran A, Maitra, RK, Kumar, A, Dong, B, Xiao, W, Li, G, Williams, BR, Torrence, PF, and Silverman, RH. Blockage of NF-kappa B signaling by selective ablation of an mRNA target by 2-5A antisense chimeras. Science 1994; 265:789-92.

257. Doukas J, Cutler, AH, and Mordes, JP. Polyinosinic:polycytidylic acid is a potent activator of endothelial cells. Am J Pathol 1994; 145:137-47.

258. Hildreth JE, and Orentas, RJ. Involvement of a leukocyte adhesion receptor (LFA-1) in HIV-induced syncytium formation. Science 1989; 244:1075-8.

259. Pantaleo G, Butini, L, Graziosi, C, Poli, G, Schnittman, SM, Greenhouse, JJ, Gallin, JI, and Fauci, AS. Human immunodeficiency virus (HIV) infection in CD4+ T lymphocytes genetically deficient in LFA-1:LFA-1 is required for HIV-mediated cell fusion but not for viral transmission. J Exp Med 1991;173:511-4.

260. Valentin A, Lundin, K, Patarroyo, M, and Asjo, B. The leukocyte adhesion glycoprotein CD18 participates in HIV-1-induced syncytia formation in monocytoid and T cells. J Immunol 1990; 144:934-7.

261. Dhawan S, Weeks, BS, Soderland, C, Schnaper, HW, Toro, LA, Asthana, SP, Hewlett, IK, Stetler-Stevenson, WG, Yamada, SS, Yamada, KM, and et al. HIV-1 infection alters monocyte interactions with .human microvascular endothelial cells. J Immunol 1995; 154:422-32

262. Shrikant P, Benos, DJ, Tang, LP, and Benveniste, EN. HIV glycoprotein 120 enhances intercellular adhesion molecule-1 gene expression in glial cells. Involvement of Janus kinase/signal transducer and activator of transcription and protein kinase C signaling pathways. J Immunol 1996; 156:1307-14.

263. Fukudome K, Furuse M, Fukuhara N, Orita S, Imai T, Takagi S, Nagira M, Hinuma Y, Yoshie O. Strong induction of ICAM-1 in human T cells transformed by human T-cell-leukemia virus type 1 and depression of ICAM-1 or LFA-1 in adult T-cell-leukemia-derived cell lines. Int J Cancer 1992; 52:418-27.

264. Imai T, Tanaka, Y, Fukudome, K, Takagi, S, Araki, K, and Yoshie, O. Enhanced expression of LFA-3 on human T-cell lines and leukemic cells carrying human T-cell-leukemia virus type 1. Int J Cancer 1993; 55:811-6.

265. Sugamura K, and Hinuma, Y. Human retroviruses: HTLV-1 and HTLV-II. In: *The Retroviridae 2*, edited by J. Levy. New York: Plenum Press, 1993.

266. Lindholm P, Kashanchi, F, and Brady, H. Transcriptional regulation in the human retrovirus HTLV-1. Semin virol 1993; 4:53-60.

267. Mori N, Murakami, S, Oda, S, and Eto, S. Human T-cell leukemia virus type I tax induces intracellular adhesion molecule-1 expression in T cells [letter]. Blood 1994; 84:350-1.

268. Tanaka Y, Fukudome, K, Hayashi, M, Takagi, S, and Yoshie, O. Induction of ICAM-1 and LFA-3 by Tax1 of human T-cell leukemia virus type 1 and mechanism of down-regulation of ICAM-1 or LFA-1 in adult-T-cell-leukemia cell lines. Int J Cancer 1995; 60:554-61.

269. Tanaka Y, Hayashi, M, Takagi, S, and Yoshie, O. Differential transactivation of the intercellular adhesion molecule 1 gene promoter by Tax1 and Tax2 of human T-cell leukemia viruses. J Virol 1996; 70:8508-17.

270. Papi A, and Johnston, SL. Rhinovirus infection induces expression of its own receptor intercellular adhesion molecule 1 (ICAM-1) via increased NF-kappaB-mediated transcription. J Biol Chem 1999; 274:9707-20.

271. Kume N, Cybulsky, MI, and Gimbrone, MA, Jr.. Lysophosphatidylcholine, a component of atherogenic lipoproteins, induces mononuclear leukocyte adhesion molecules in cultured human and rabbit arterial endothelial cells. J Clin Invest 1992; 90:1138-44.

272. Kume N, and Gimbrone, MA, Jr.. Lysophosphatidylcholine transcriptionally induces growth factor gene expression in cultured human endothelial cells. J Clin Invest 1994; 93:907-11.

273. Zhu Y, Lin, JH, Liao, HL, Verna, L, and Stemerman, MB. Activation of ICAM-1 promoter by lysophosphatidylcholine: possible involvement of protein tyrosine kinases. Biochim Biophys Acta 1997; 1345:93-8.

274. Colic M, and Drabek, D. Expression and function of intercellular adhesion molecule 1 (ICAM-1) on rat thymic macrophages in culture. Immunol Lett 1991; 28:251-7.

275. Tessier P, Audette, M, Cattaruzzi, P, and McColl, SR. Up-regulation by tumor necrosis factor alpha of intercellular adhesion molecule 1 expression and function in synovial fibroblasts and its inhibition by glucocorticoids. Arthritis Rheum 1993; 36:1528-39.

276. Yasuda M, Kokubu, F, Izumi, H, Matsukura, S, Tokunaga, H, Yamamoto, T, Kuroiwa, Y, and Adachi, M. [Effect of dexamethasone on intercellular adhesion molecule-1 expression on cultured bronchial epithelial cells stimulated by inflammatory cytokines]. Arerugi 1995; 44:100-3.

277. Perretti M, Wheller, SK, Harris, JG, and Flower, RJ. Modulation of ICAM-1 levels on U-937 cells and mouse macrophages by interleukin-1 beta and dexamethasone. Biochem Biophys Res Commun 1996; 223:112-7.

278. Rothlein R, Czajkowski, M, O'Neill, MM, Marlin, SD, Mainolfi, E, Merluzzi, VJ. Induction of intercellular adhesion molecule 1 on primary and continuous cell lines by pro-inflammatory cytokines. Regulation by pharmacologic agents and neutralizing antibodies. J Immunol 1988; 141:1665-9.

279. Wuthrich RP, and Sekar, P. Effect of dexamethasone, 6-mercaptopurine and cyclosporine A on intercellular adhesion molecule-1 and vascular cell adhesion molecule-1 expression. Biochem Pharmacol 1993; 46:1349-53.

280. Fattal-German M, Ladurie, FL, Cerrina, J, Lecerf, F, and Berrih-Aknin, S. Modulation of ICAM-1 expression in human alveolar macrophages in vitro. Eur Respir J 1996; 9:463-71.

281. Cronstein BN, Kimmel, SC, Levin, RI, Martiniuk, F, and Weissmann, G. A mechanism for the antiinflammatory effects of corticosteroids: the glucocorticoid receptor regulates leukocyte adhesion to endothelial cells and expression of endothelial-leukocyte adhesion molecule 1 and intercellular adhesion molecule 1. Proc Natl Acad Sci U S A 1992; 89:9991-5.

282. Aziz KE, and Wakefield, D. Modulation of endothelial cell expression of ICAM-1, E-selectin, and VCAM-1 by beta-estradiol, progesterone, and dexamethasone. Cell Immunol 1996; 167:79-85.

283. Swerlick RA, Garcia-Gonzalez, E, Kubota, Y, Xu, YL, Lawley, TJ. Studies of the modulation of MHC antigen and cell adhesion molecule expression on human dermal microvascular endothelial cells. J Invest Dermatol 1991; 97:190-6.

284. Detmar M, Tenorio, S, Hettmannsperger, U, Ruszczak, Z, and Orfanos, CE. Cytokine regulation of proliferation and ICAM-1 expression of human dermal microvascular endothelial cells in vitro. J Invest Dermatol 1992; 98:147-53.

285. Hettmannsperger U, Tenorio, S, Orfanos, CE, and Detmar, M. Corticosteroids induce proliferation but do not influence TNF- or IL-1 beta-induced ICAM-1 expression of human dermal microvascular endothelial cells in vitro. Arch Dermatol Res 1993; 285:347-51.

286. Hess DC, Bhutwala, T, Sheppard, JC, Zhao, W, and Smith, J. ICAM-1 expression on human brain microvascular endothelial cells. Neurosci Lett 1994; 168:201-4.

287. Burke-Gaffney A, and Hellewell, PG. Regulation of ICAM-1 by dexamethasone

in a human vascular endothelial cell line EAhy926. Am J Physiol 1996; 270:C552-61.

288.   Jonat C, Rahmsdorf, HJ, Park, KK, Cato, AC, Gebel, S, Ponta, H, and Herrlich, P. Antitumor promotion and antiinflammation: down-modulation of AP-1 (Fos/Jun) activity by glucocorticoid hormone. Cell 1990; 62:1189-204.

289.   Diamond MI, Miner, JN, Yoshinaga, SK, and Yamamoto, KR. Transcription factor interactions: selectors of positive or negative regulation from a single DNA element. Science 1990; 249:1266-72.

290.   Yang-Yen HF, Chambard, JC, Sun, YL, Smeal, T, Schmidt, TJ, Drouin, J, and Karin, M. Transcriptional interference between c-Jun and the glucocorticoid receptor: mutual inhibition of DNA binding due to direct protein-protein interaction. Cell 1990; 62:1205-15.

291.   Kleinert H, Euchenhofer, C, Ihrig-Biedert, I, and Forstermann, U. Glucocorticoids inhibit the induction of nitric oxide synthase II by down-regulating cytokine-induced activity of transcription factor nuclear factor-kappa B. Mol Pharmacol 1996; 49:15-21.

292.   Caldenhoven E, Liden, J, Wissink, S, Van de Stolpe, A, Raaijmakers, J, Koenderman, L, Okret, S, Gustafsson, JA, and Van der Saag, PT. Negative cross-talk between RelA and the glucocorticoid receptor: a possible mechanism for the antiinflammatory action of glucocorticoids. Mol Endocrinol 1995; 9:401-12.

293.   Scheinman RI, Gualberto, A, Jewell, CM, Cidlowski, JA, and Baldwin, AS, Jr.. Characterization of mechanisms involved in transrepression of NF-kappa B by activated glucocorticoid receptors. Mol Cell Biol 1995; 15:943-53.

294.   Ray A, and Prefontaine, KE. Physical association and functional antagonism between the p65 subunit of transcription factor NF-kappa B and the glucocorticoid receptor. Proc Natl Acad Sci U S A 1994; 91:752-6.

295.   Auphan N, DiDonato, JA, Rosette, C, Helmberg, A, and Karin, M. Immunosuppression by glucocorticoids: inhibition of NF-kappa B activity through induction of I kappa B synthesis [see comments]. Science 1995; 270:286-90.

296.   Ohtsuka T, Kubota, A, Hirano, T, Watanabe, K, Yoshida, H, Tsurufuji, M, Iizuka, Y, Konishi, K, and Tsurufuji, S.   Glucocorticoid-mediated gene suppression of rat cytokine-induced neutrophil chemoattractant CINC/gro, a member of the interleukin-8 family, through impairment of NF-kappa B activation. J Biol Chem 1996; 271:1651-9.

297.   Brostjan C, Anrather, J, Csizmadia, V, Natarajan, G, and Winkler, H. Glucocorticoids inhibit E-selectin expression by targeting NF-kappaB and not ATF/c-Jun. J Immunol 1997; 158:3836-44.

298.   Shrikant P, Lee, SJ, Kalvakolanu, I, Ransohoff, RM, and Benveniste, EN. Stimulus-specific inhibition of intracellular adhesion molecule-1 gene expression by TGF-beta. J Immunol 1996; 157:892-900.

299.   Gamble JR, Khew-Goodall, Y, and Vadas, MA. Transforming growth factor-beta inhibits E-selectin expression on human endothelial cells. J Immunol 1993; 150:4494-503.

300.   Moore KW, O'Garra, A, de Waal Malefyt, R, Vieira, P, and Mosmann, TR. Interleukin-10. Annu Rev Immunol 1993; 11:165-90.

301.   Mosmann TR. Properties and functions of interleukin-10. Adv Immunol 1994; 56:1-26.

302.   Most J, Schwaeble, W, Drach, J, Sommerauer, A, and Dierich, MP. Regulation of the expression of ICAM-1 on human monocytes and monocytic tumor cell lines. J Immunol 1992; 148:1635-42.

303.   Willems F, Marchant, A, Delville, JP, Gerard, C, Delvaux, A, Velu, T, de Boer, M, and Goldman, M. Interleukin-10 inhibits B7 and intercellular adhesion molecule-1 expression on human monocytes. Eur J Immunol 1994; 24:1007-9.

304.   Song S, Ling-Hu, H, Roebuck, KA, Rabbi, MF, Donnelly, RP, and Finnegan, A.

Interleukin-10 inhibits interferon-gamma-induced intercellular adhesion molecule-1 gene transcription in human monocytes. Blood 1997; 89:4461-9

305.   Wang P, Wu, P, Anthes, JC, Siegel, MI, Egan, RW, and Billah, MM. Interleukin-10 inhibits interleukin-8 production in human neutrophils. Blood 1994; 83:2678-83.

306.   Wang P, Wu, P, Siegel, MI, Egan, RW, and Billah, MM. Interleukin (IL)-10 inhibits nuclear factor kappa B (NF kappa B) activation in human monocytes. IL-10 and IL-4 suppress cytokine synthesis by different mechanisms. J Biol Chem 1995; 270:9558-63.

307.   Lefer AM, Lefer, DJ. The role of nitric oxide and cell adhesion molecules on the microcirculation in ischaemia-reperfusion. Cardiovasc Res 1996; 32:743-51.

308.   Lindemann S, Sharafi, M, Spiecker, M, Buerke, M, Fisch, A, Grosser, T, Veit, K, Gierer, C, Ibe, W, Meyer, J, and Darius, H. NO reduces PMN adhesion to human vascular endothelial cells due to downregulation of ICAM-1 mRNA and surface expression. Thromb Res 2000; 97:113-23.

309.   Takahashi M, Ikeda, U, Masuyama, J, Funayama, H, Kano, S, and Shimada, K. Nitric oxide attenuates adhesion molecule expression in human endothelial cells. Cytokine 1996; 8:817-21.

310.   Peng HB, Libby, P, and Liao, JK. Induction and stabilization of I kappa B alpha by nitric oxide mediates inhibition of NF-kappa B. J Biol Chem 1995; 270:14214-9.

311.   Spiecker M, Darius, H, Kaboth, K, Hubner, F, and Liao, JK. Differential regulation of endothelial cell adhesion molecule expression by nitric oxide donors and antioxidants. J Leukoc Biol 1998; 63:732-9.

312.   Cowan PJ, Shinkel, TA, Witort, EJ, Barlow, H, Pearse, MJ, and d'Apice, AJ. Targeting gene expression to endothelial cells in transgenic mice using the human intercellular adhesion molecule 2 promoter. Transplantation 1996; 62:155-60.

313.   Pan J, and McEver, RP. Characterization of the promoter for the human P-selectin gene. J Biol Chem 1993; 268:22600-8.

314.   Cowan PJ, Tsang, D, Pedic, CM, Abbott, LR, Shinkel, TA, d'Apice, AJ, and Pearse, MJ. The human ICAM-2 promoter is endothelial cell-specific in vitro and in vivo and contains critical Sp1 and GATA binding sites. J Biol Chem 1998; 273:11737-44.

315.   Schlaeger TM, Bartunkova, S, Lawitts, JA, Teichmann, G, Risau, W, Deutsch, U, and Sato, TN. Uniform vascular-endothelial-cell-specific gene expression in both embryonic and adult transgenic mice. Proc Natl Acad Sci U S A 1997; 94:3058-63.

316.   Pan J, and McEver, RP. Regulation of the human P-selectin promoter by Bcl-3 and specific homodimeric members of the NF-kappa B/Rel family. J Biol Chem 1995; 270:23077-83.

317.   Doussis-Anagnostopoulou I, Kaklamanis, L, Cordell, J, Jones, M, Turley, H, Pulford, K, Simmons, D, Mason, D, and Gatter, K. ICAM-3 expression on endothelium in lymphoid malignancy. Am J Pathol 1993; 143:1040-3.

318.   Patey N, Vazeux, R, Canioni, D, Potter, T, Gallatin, WM, and Brousse, N. Intercellular adhesion molecule-3 on endothelial cells. Expression in tumors but not in inflammatory responses [see comments]. Am J Pathol 1996; 148:465-72.

319.   Kageyama R, Sasai, Y, Akazawa, C, Ishibashi, M, Takebayashi, K, Shimizu, C, Tomita, K, and Nakanishi, S. Regulation of mammalian neural development by helix-loop-helix transcription factors. Crit Rev Neurobiol 1995; 9:177-88.

Chapter 5

TRANSCRIPTIONAL REGULATION OF VCAM-1

Andrew S. Neish,# Sarita Aggarwal,≠ and Tucker Collins†

#Department of Pathology
Emory University
1369 Pierce Street, Atlanta, GA 30322
≠Department of Medicine
Beth Israel Deaconess Medical Center
330 Brookline Avenue, Boston, MA 02215
†Vascular Research Division, Department of Pathology
Brigham and Women's Hospital and Harvard Medical School
221 Longwood Avenue, Boston MA 02115

INTRODUCTION

The recruitment of circulating leukocytes into the extravascular space is critical for immune/inflammatory responses and repair of tissue injury. The process of leukocyte adhesion and transmigration involves an ordered sequence of events involving specific endothelial-leukocyte adhesion molecules (reviewed in refs. 1 and 2 and Chapters 1 and 3 of this volume). The initial interaction between leukocytes and endothelium appears to be a transient, reversible rolling of leukocytes along the vessel wall. This process is mediated by members of the selectin family present on the endothelial surface. Rolling leukocytes become activated by local factors generated by the endothelium. The activated leukocytes then stably arrest and adhere firmly to the vessel wall, an event mediated by the interaction of integrins on the surface of leukocytes with immunoglobulin gene superfamily members expressed by endothelial cells. Among these is vascular cell adhesion molecule 1 (VCAM-1), a 110 kD endothelial surface glycoprotein.[3,4] VCAM-1 specifically interacts with the integrins α4β1 (VLA4) and α4β7, via its first and fourth immunoglobulin domain.[5,6] The integrin counter-receptors are found on circulating monocytic leukocytes, eosinophils, and basophils, but not neutrophils.[5,7] Because VCAM-1 binds a distinct subset of leukocytes, endothelial VCAM-1 expression *in vivo* influences the composition of leukocyte populations recruited into inflamed tissue.

VCAM-1 is considered to be a critical mediator of the chronic inflammatory and immune responses (reviewed in 8). Immunohistochemical surveys of pathologic human tissues consistently detect endothelial surface VCAM-1 associated with chronic inflammatory reactions, including rheumatoid arthritis,[9-11] chronic cardiac graft rejection,[12-14] and inflammatory dermatoses and sarcoidosis.[9] Endothelial expression of VCAM-1 has been described in experimental disease models. VCAM-1 immunoreactivity has been described in large arterial vessels[15] and hepatic sinusoids in rodent models of septic shock,[16] and paratracheal endothelial cells in a murine antigen inhalation model of asthma.[17] Endothelial VCAM-1 expression has been associated with natural and experimental atherosclerotic lesions.[18-24] It is postulated that VCAM-1-mediated monocyte adhesion to aortic endothelial cells may represent an initial event in the pathogenesis of atherosclerosis.

VCAM-1 is involved in cell-cell interactions involving cells other than leukocytes. VCAM-1 binds tumor cells and may participate in early events of metastasis.[25,26] VCAM-1 expression on cerebral endothelium may mediate the binding of *Plasmodium falciparum*-infected red cells.[27] VCAM-1/VLA4 interactions are involved in developmental processes. VCAM-1 is expressed on embryonic skeletal muscle, heart, and extra-embryonic tissues.[28-30] Mice homozygous null for VCAM-1 (and α4 integrin) die at 9.5 days of development secondary to failure of chorioallantoic fusion and subsequent vascular development.[28]

VCAM-1 expression is not restricted to endothelial cells. Immunohistochemical expression of VCAM-1 is present *in vivo* on smooth muscle cells in atherosclerotic vessels[19,21,31] and on macrophages and synovial lining layer cells in rheumatoid articular tissues.[10,11,32] VCAM-1 is present on follicular dendritic cells in lymphoid tissues,[9] mediating interactions with B cells.[33,34] VCAM-1 is expressed on macrophages[9] and bone marrow stromal cells[35,36] and may play a role in lymphopoiesis. VCAM-1 may act as a T-cell costimulatory molecule and contribute to T cell extravasation at sites of inflammation.[37]

Endothelial expression of VCAM-1 and other endothelial leukocyte adhesion molecules is dynamically regulated. Quiescent endothelial cells are not adhesive for monocytic leukocytes. Exposure of the endothelial cells to soluble inflammatory mediators results in upregulation of VCAM-1 and hyperadhesivity for circulating monocytic leukocytes. Therefore, control over VCAM-1 expression on the endothelial surface is critical for modulating the leukocytic composition, spatial distribution, and temporal extent of an inflammatory reaction. In this review, we will present the current knowledge about the mechanism of VCAM-1 regulation on endothelial cells. Evidence will be summarized suggesting that cytokine-induced gene expression of VCAM-1 is a transcriptional activation process mediated by the specific interactions of a small group of DNA-binding transcription factors.

VCAM-1 Is An Inducible Molecule

Cytokine Activation

Upregulation of endothelial surface VCAM-1 expression can be observed *in vivo*. In primates, subcutaneous administration of TNF-α alone or in combination with IL-4 induces VCAM-1 in endothelial cells and vascular smooth muscle cells.[38] In murine models, surface endothelial VCAM-1 expression is absent or trace positive in the aorta and pulmonary and renal vascular beds. Administration of intraperitoneal lipopolysaccharide (LPS) results in a striking upregulation of surface protein expression in these vessels, peaking in six hours. A concomitant increase in VCAM-1 mRNA is also seen in these tissues, suggesting induction is the result of new transcription.[15]

The use of cultured endothelial cells has allowed much more detailed studies of VCAM-1 induction. The expression of endothelial surface VCAM-1 immunoreactivity is induced by the classic triad of inflammatory mediators: interleukin-1β (IL-1β), tumor necrosis factor-α (TNF-α), and LPS.[39] Use of *in vitro* adhesion assays confirms that cytokine induction of surface VCAM-1 immunoreactivity on cultured endothelial cells correlates with increased adhesion of monocytes, lymphocytes, eosinosphils, and basophils.[4,7,39-42] The kinetics of VCAM-1 induction are consistent with a role in modulating chronic inflammation. Low levels are detected on unstimulated endothelial cells. Poststimulation, surface expression is first seen within two hours, peaks at 6-12 hours, and remains high for several days.[39] VCAM-1 expression is delayed and sustained relative to the expression of the neutrophil-specific adhesion molecule E-selectin. In contrast, E-selectin surface expression is induced by TNF-α within one hour, peaks at four hours, and rapidly declines by 24 hours, consistent with its role in modulating acute inflammatory reactions[43] (see Chapters 1 and 2). VCAM-1 expression thus corresponds with the preferential adhesion and infiltration of mononuclear leukocytes typical of chronic inflammatory processes.[4]

Induction of VCAM-1 messenger RNA in cultured endothelial cells parallels the kinetics of surface protein. VCAM-1 mRNA is virtually absent from unstimulated endothelial cells, but activation with inflammatory mediators results in detectable accumulation of steady state mRNA levels within one hour. In the continuous presence of cytokine, cultured endothelial VCAM-1 message levels reach a sustained high level by 2-3 hours, and then gradually diminish over several days.[3] Again, the *de novo* appearance of VCAM-1 mRNA following activation with cytokine is consistent with a transcriptional activation event. Additionally, the concordance between transcript levels and appearance of functional surface protein implies that the dominant regulatory point in VCAM-1 induction occurs at the level of RNA synthesis/stabilization.

Interestingly, the upregulation of VCAM-1 messenger RNA is partially dependent on new protein synthesis. Endothelial cells treated with cycloheximide prior to cytokine induction exhibit markedly reduced accumulation of VCAM-1 mRNA and a reduced rate of new transcript synthesis.[44,45] This observation is in striking contrast to the effects of protein synthesis blockade on other inducible endothelial adhesion molecules in that both E-selectin and ICAM mRNAs are highly superinduced by cycloheximide pretreatment.[45,46] The selective sensitivity of VCAM-1 mRNA to protein synthesis blockade implies newly synthesized factors are involved in VCAM-1 upregulation.

Combinations of IL-4 and TNF-α /IL-1 induce VCAM-1 surface expression and endothelial adhesivity to levels far greater than the effects of TNF-α /IL-1 alone.[41,42] Treatment of cultured endothelial cells with IL-4 results in a weak but selective induction of VCAM-1 mRNA and surface protein.[7,47] This observation is also seen *in vivo*.[38] Studies undertaken by Iadamarco et al[48] demonstrated that the synergistic effects of IL-4 on TNF-α-activated transcription is in part secondary to increased VCAM-1 message stabilization, which allows for greater accumulation of steady state mRNA levels. A similar phenomenonon has been described during activation of ICAM-1 by TNF-α and phorbol esters.[46] Finally, in endothelial cells, interferon-α and -γ pretreatment augment TNF-α-induced activation of VCAM-1 message and protein; however, no message stabilization was observed, suggesting effects at the transcriptional level.[49]

Other activators of VCAM-1 expression *in vitro* have been described, including such diverse stimuli as ionizing radiation,[50] the matrix constituent hyaluronan,[51] and the hormone angiotensin II.[52] IL-13, which shares many biologic properties with IL-4, stimulates new synthesis of VCAM-1 mRNA and surface protein.[53-55] The neuropeptide Substance P has been shown to be an inducer of endothelial VCAM-1 expression *in vitro* and *in vivo*.[56-58] Poly(I)•poly(C), a double strand RNA analog utilized as a model of viral infection, induces VCAM-1 mRNA in cultured cells to levels higher and more prolonged than that achieved by cytokine.[59] Consistently, VCAM-1 expression has been detected in T-cells infected with HTLV-I,[60] respiratory epithelial cells infected with rhinovirus,[61] and endothelial cells infected with *Borrelia Burgdorferi*.[62]

Endothelial cells from distinct vascular beds may exhibit heterogeneous responses to cytokine. Cultured endothelial cells derived from human dermal microvasculature can express VCAM-1 surface protein and mRNA in response to TNF-α but do not respond to IL-1,[63-65] though IL-1 induces other adhesion molecules in this cell population. Additionally, cultured human iliac vein endothelial cells can express VCAM-1 in response to TNF-α, however, human arterial endothelial cells can not.[60]

Oxidant Stress and Atherogenic Stimuli

As mentioned earlier, VCAM-1 expression on arterial endothelial cells binds monocytes and thus may mediate leukocyte recruitment in early atherosclerosis. A variety of agents and stimuli implicated in the pathogenesis of natural and experimental atherosclerosis have been reported to induce expression of VCAM-1 protein and mRNA in cultured endothelial cells and are discussed below.

Oxidative stress has been implicated in the pathogenesis of atherosclerosis.[67] For example, oxidized lipid species such as oxidized LDL and its components lysophosphatidylcholine (Lyso-PC) and 13-HPODE are associated with atherosclerotic lesions.[68] Lyso-PC induces VCAM-1 (and ICAM-1) surface expression and mRNA levels in cultured arterial endothelial cells.[69] Similarly, oxidized LDL can augment cytokine-induced surface VCAM-1 expression in aortic endothelial cell cultures.[70,71] Significantly, neither lipid species activates E-selectin synthesis. The reasons for this relative specificity are unknown. Conversely, the antioxidants pyrrolidine dithiocarbmate (PDTC) and N-acetylcysteine (NAC) selectively block the cytokine-mediated upregulation of VCAM-1, while having little effect on ICAM-1 or E-selectin activation. These compounds diminish the cytokine-induced nuclear translocation of NF-κB, consistent with a postulated redox sensitive mechanism of NF-κB activation.[72,73]

Nitric oxide (NO) is known to have anti-adhesive effects *in vivo*.[74] In cultured cells, NO donors diethylamine-NO (DETA-NO) and S-nitroso-glutathione (GNSO) reduce the surface expression of VCAM-1 and other adhesion molecules, as well as VCAM-1 mRNA accumulation and transcriptional induction.[75,76] NO can inhibit the activation of NF-κB by both increasing the level and stabilizing the inhibitor IκB-α.[77,78]

Other pro-atherogenic stimuli also affect VCAM-1 expression. Advanced glycation endproducts (AGEs) induce VCAM-1 mRNA and surface protein on cultured HUVEC,[79] and induce aortic VCAM-1 surface expression *in vivo* in a rabbit model of AGE-induced atherosclerosis.[80] Cultured endothelial cells subjected to fluid mechanical forces designed to mimic physiologic blood flow downregulate VCAM-1 expression at both the mRNA and surface protein levels.[81] VCAM-1 is expressed in endothelial cells following balloon injury in the rabbit aorta[82] and rat carotid.[31,83] Expression of protein and mRNA localizes to the region of injury.[83] Taken together, these studies suggest that pro-atherogenic factors upregulate the level of VCAM-1 expression. The relationship between VCAM-1 and atherosclerosis is reviewed in detail elsewhere.[84]

REGULATORY SEQUENCES CONTROLLING CYTOKINE-INDUCED VCAM-1 EXPRESSION

Mechanisms of Activation

Increased levels of VCAM-1 mRNA seen with both proinflammatory and pro-atherogenic stimuli suggest that these agents affect the rate of VCAM-1 transcription. Experimentally, assays of nascent transcript formation are necessary to prove this. Nuclear run-off analysis reveals that active transcription of the VCAM-1 gene does not occur in quiescent cultured endothelial cells and that new transcript

180

formation is rapidly detectable within one hour after TNF-α induction.[45,85] These latter experiments indicate that synthesis of messenger RNA in response to the inflammatory cytokines is indeed the result of transcriptional activation. This finding has led us to focus on transcriptional control of the VCAM-1 locus as the dominant point of regulation in the control of expression of the gene.

To understand the molecular mechanisms controlling the characteristic VCAM-1 response to cytokine, the genomic organization of the human and mouse VCAM-1 genes were determined[86,87] and the VCAM-1 5' flanking region characterized.[85] The human VCAM-1 gene is located on chromosome 1q and spans about 25 kilobases of DNA. The human gene contains 9 exons which correlate with the functional immunoglobulin domains in the mature protein.[86] A diagramatic representation of the VCAM-1 genomic architecture is shown in Figure 5-1. The mature surface protein is shown in Figure 5-2. Alternative splicing of exon five results in transcripts containing six or seven Ig domains, though the seven domain form appears most abundant *in vivo*.[88] Both forms support adhesion of VLA-4-bearing cells.[89]

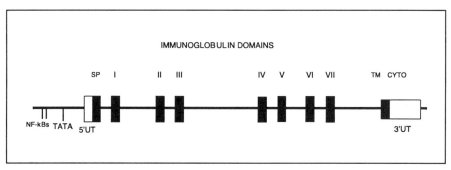

Figure 5-1. Structural organization of the VCAM-1 gene. The positions of the exon-intron boundaries correlates with the immunoglobulin domain structure of the protein. Exons are indicated by boxes. Introns as well as 5' and 3' flanking sequences are represented by lines. The 5' and 3' untranslated regions are represented by open boxes. The positions of the consensus transcriptional elements TATAA and NF-κB are indicated. SP, signal peptide; TM, transmembrane region; CYTO, cytoplasmic domain and UT, untranslated region. Immunoglobulin-like domains are indicated with Roman numerals.

Primer extension analysis of mRNA from cytokine-induced endothelial cells reveals a single transcriptional start site located 24 base pairs downstream of a consensus TATA box. The regulatory sequences controlling the VCAM-1 response to cytokine reside within the 100 base pairs upstream of the start site. This region was mapped using a series of promoter-reporter constructs containing segments of VCAM-1 genomic sequence fused to a CAT reporter gene. These constructs were assayed for ability to mediate cytokine-responsive gene transcription by transient transfection into cultured bovine aortic endothelial cells and subsequent stimulation with recombinant TNF-α. A fragment spanning 2190 bp upstream of the start site could indeed induce transcription of the reporter gene. A series of 5' deletion end point constructs demonstrated that a reporter containing only 98 bp of upstream flank functioned as a minimal promoter, capable of directing full cytokine-induced gene expression.[85,90] We have utilized additional fine mutational analysis of this minimal promoter to define discrete functional elements mediating response to cytokine. A diagram of the 2190 base pair 5' flank with the putative regulatory elements discussed below is shown in Figure 5-3.

The Cytokine-Inducible Enhancer

The cytokine activation of the VCAM-1 promoter requires a contiguous 45 bp region of DNA which functions as a cytokine-responsive enhancer, located between -42 and -86 bp upstream of the transcriptional start site (see Fig. 5-3). This region was mapped by saturation mutagenesis (linker scan) of the entire minimal promoter.[90] In these experiments, 10 independent 10 base pair block mutations were generated, systematically introducing mutations into the entire VCAM-1 minimal promoter sequence upstream of the TATA box, and were assayed by transient transfection in endothelial cells. The sequence thus mapped can be mobilized to a noncytokine-inducible promoter, such as the SV-40 viral promoter, and can confer cytokine responsiveness on the previously noninducible promoter construct. Additionally, the cytokine-responsive enhancer is functional in either orientation and can act at a distance, fulfilling the criteria for an enhancer element (Neish and Collins, unpublished data).

Inspection of the cytokine-responsive enhancer at positions -73 and -58 reveals tandem 10 base pair sequence motifs, separated by five base pairs, matching the consensus binding site for the inducible transcription factor NF-κB (GGGUNNYCC).[91] Both of these sites are absolutely necessary for cytokine-mediated transcriptional response; point mutations of either of these elements totally abolishes cytokine responsiveness in mutant promoter constructs.[85,92] Furthermore, when both sites are mobilized to a noncytokine-responsive promoter, the resulting construct acquires TNF-α inducibility.[93] Mobilization of either site in isolation results in an only marginally inducible construct (Neish and Collins, unpublished data). Thus, the integrity of these motifs appears central to the inducible function of the VCAM-1 promoter.

The cytokine-inducible enhancer also exhibits a GC rich motif immediately 3' of the NF-κB motifs, bearing the sequence TCCGCCTC. GC rich motifs are common in the basal promoters of both constitutively active and inducible genes. Disruption of the GC core in the VCAM-1 promoter results in a strong reduction, though not abolition, of cytokine-induced reporter gene activity.[90] Mobilization of this motif to a heterologous promoter results in increased basal activity of the hybrid reporter, but confers no ability to respond to cytokine (Neish and Collins, unpublished data).

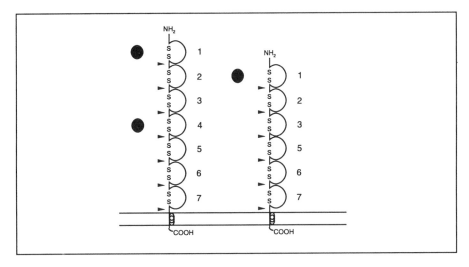

Figure 5-2. VCAM-1 exists as a 6 or 7 domain form. Schematic illustration of the different forms of VCAM1 generated by alternate splicing. Binding sites of α4 integrins to Ig-like domains 1 and 4 are indicated by solid circles. Loops represent disulfide-linked Ig-like domains (numbered) and arrowheads point to exon splice junctions.

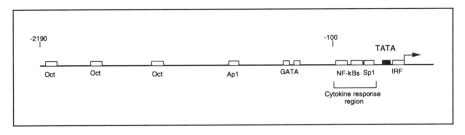

Figure 5-3. Diagram of the VCAM-1 promoter from +4 to –2190, indicating the relative position of regulatory elements. The TATA box is in bold. Transcriptional start site is designated with an arrow. Defined and putative regulatory elements are boxed. See text.

Other Positively Activating Domains

Our saturation mutagenesis experiments extended from -100 to the TATA box of the VCAM-1 basal promoter. Further inspection of the VCAM-1 core promoter downstream of the TATA box reveals the motif GAAATAGAAA, located one base pair upstream from the transcriptional start site. This site comprises an interferon-stimulated response element (ISRE), a well-defined motif found in the promoters of a number of other inducible genes.[94] These sequences interact with the interferon regulatory factor (IRF) family of transcriptional activators and will be discussed in a subsequent section. Mutagenesis experiments demonstrate that disruption of the IRF binding site decreases TNF-α-induced reporter gene expression in transfected endothelial cells.[44] This binding motif in the VCAM-1 promoter was independently characterized by Iademarco et al in myoblast C2C12 cells.[95] These investigators demonstrated that a 5' deletion construct spanning only -34 base pairs upstream of the transcriptional start site, containing only the TATA box and encompassing this 3' motif (but without NF-κB sites), was fully active in transiently transfected C2C12 cells, while completely inactive in HUVEC cells. Further mutational analysis by this group localized activity to the ISRE and demonstrated that this motif was necessary for activity in myoblastic cells.[30]

Upstream Silencer

Initial characterizations of the 5' flanking region of the VCAM-1 gene in our and other laboratories showed that while a fragment spanning greater than two kilobases of 5' upstream sequence functioned as a inducible promoter, fragments of less than 933 base pairs resulted in a promoter construct with greater levels of both basal and cytokine-induced activity.[85,92] The transcriptional silencing activity present in this region has been ascribed to "octamer" sequence elements present at positions -1554 and -1180, along with several other sites of more divergent sequence. The VCAM-1 octamer motifs have the sequence ATTTACAT, similar to the classically described octamer motif ATTTGCAT originally characterized as a transcriptional silencing motif in immunoglobulin heavy chain gene expression.[96] Intrinsic silencing activity of the VCAM-1 octamers was demonstrated by Iademarco et al.[95] Positioning selected VCAM-1 octamers upstream of a constitutively active heterologous promoter resulted in octamer site-dependent repression of basal transcription, though these octamer motifs were unable to repress NF-κB-mediated inducible expression. A specific DNA-binding activity is present in endothelial cells, though its identity and relationship to the known octamer-binding factors is unknown.

Interestingly, the ISRE and the entire cytokine-responsive region are completely conserved in the murine VCAM-1 promoter.[87] Such conservation of putative motifs is strongly suggestive of the functional importance of these sequences.

TRANSCRIPTIONAL ACTIVATORS CONTROLLING CYTOKINE-INDUCED VCAM-1 EXPRESSION

Mutational mapping of a functional promoter element is inferred to represent discovery of a binding site for a site-specific DNA-binding protein. Transcription factors themselves are low abundance nuclear proteins, typically with separable domains that mediate DNA binding and transcriptional activation. Often, homology between transcription factors defines "families," with potentially similar binding sites (which collectively define the "consensus") and modes of activation. Certain transcription factor families can homo- and heterodimerize between members of their own (or extended) families, forming complex combinations of DNA-binding specificities and transactivational properties. In this section we will describe the transcriptional activators currently thought to interact with the functional sequence motifs present in the VCAM-1 promoter.

NF-κB

As with many rapidly inducible genes involved in immune/inflammatory function, NF-κB plays a central role in the inducible expression of VCAM-1. The biology of the Rel family of transcription factors and their mechanism of rapid post-translational activation are described in detail in Chapters 6 and 9.

Rel family members can form combinations of homo- and heterodimers, each with potentially different specificities of binding and activation. Various κB motifs, both synthetic and natural, bind the activating p65/p50 heterodimers, p65 homodimers, and the nonactivating p50 homodimers with strikingly different affinities.[97,98] Furthermore, different cell types (and presumably different tissues) express different Rel proteins,[99] which may have distinct properties. For these reasons it is necessary to determine the composition of the dimer associating with each functional κB element in each specific cell type.

In TNF-α-activated endothelial cells, heterodimeric p50 and p65 appear to be the predominant NF-κB species binding to the VCAM-1 promoter, though homodimeric forms of both proteins are present, and c-Rel and RelB are detectable in trace amounts.[44,100] Multiple lines of evidence support this. First, gel shift and supershift analyses reveal that the dominant induced nucleoprotein complexes binding to both NF-κB sites contain both p50 and p65 immunoreactivity. Second, UV crosslinking studies also reveal that both the 5' and 3' VCAM-1 NF-κB sites bind predominantly heterodimeric p50 and p65.[44] In transactivation experiments utilizing overexpressed Rel proteins, p65 acts as a powerful activator of VCAM-1 promoter reporter constructs, and the addition of p50, while having no transactivating potential of its own, further increases the level of expression (Neish and Collins, unpublished data and 100). Interestingly, higher levels of overexpressed p50 result in transcriptional repression. This phenomenon has been observed with other promoters and may reflect competition between homodimeric p50 and heterodimeric p50/p65 for NF-κB sites.[101-102] The in vivo significance of this observation is unclear; nuclear levels of p50 do not increase coincident with VCAM-1 transcriptional downregulation.[45,100]

Although the endothelial NF-κB transcription factor system is necessary, it is not sufficient to mediate appropriate cytokine-induced VCAM-1 expression. Simply put, activation of other genes, such as E-selectin and ICAM-1, is critically dependent on NF-κB, yet each gene's kinetics and distribution of expression are distinct from those of VCAM-1. While the number of binding sites and composition of Rel dimers may account for some differences in expression patterns, it is clear that a relatively small number of other transcription factors must cooperate with NF-κB to appropriately orchestrate activation of the VCAM-1 promoter.

Sp Family

Sp1 was one of the earliest transcription factors biochemically characterized and cloned (reviewed in 103). It is a large (95-110 kD) phosphoprotein containing three DNA-binding zinc fingers and a glutamine rich activation domain. The protein is also glycosylated, an unusual modification of a nuclear transcription factor. Sp1 binds to the characteristic symmetrical recognition motif (G/T)(G/A)GG(C/A)G (G/T)(G/A)(G/A)(C/T), often found in the promoters of basally expressed "housekeeping genes." In endothelial cells, Sp1 is abundant, constitutively present, and noncytokine-inducible at the RNA, protein, and DNA-binding level.[90,104] Sp1 was implicated in the transcriptional control of VCAM-1 when saturation mutagenesis studies assigned a functional role to the GC rich motif immediately 3' of the NF-κB site. Sp1 is the dominant immunoreactivity interacting with this motif in electrophoretic mobility shift assays with endothelial cell extracts. Recombinant Sp1 specifically interacts with the VCAM-1 promoter binding motif and can co-occupy the cytokine-responsive enhancer with NF-κB.[90] Overexpression of Sp1 in cells devoid of the native protein is a powerful and site specific transactivator of VCAM-1 promoter constructs (Neish and Collins, unpublished data).

Recently, it has been shown that Sp1 represents the first member of a family of proteins with at least four members; Sp2, Sp3, and Sp4.[105,106] As with other families of transcription factors, different members may have discrete properties of DNA site recognition or transcriptional stimulating ability. Electrophoretic mobility shift assays using endothelial nuclear extracts and the VCAM-1 cytokine response region as a probe reveal that the Sp3 family member interacts with the GC rich motif (Neish and Collins, unpublished data). The role of this related species in VCAM-1 regulation is under investigation.

IRF-1

As mentioned earlier, stimulated levels of VCAM-1 transcription are reduced in the presence of cycloheximide, indicating that protein synthesis is necessary for maximal transcriptional activation. This finding is in striking contrast to E-selectin and ICAM-1, both of which are superinduced in the presence of the protein synthesis inhibitor. The sensitivity of VCAM-1 transcription to protein synthesis blockade is consistent with factors other than NF-κB contributing to cytokine-induced VCAM-1 expression. A possible candidate for such an activator is IRF-1, which indeed is a protein synthesis-dependent, cytokine-inducible transcriptional activator.[107-110] This protein belongs to a family of DNA-binding proteins which recognize the ISRE sequence and includes IRF-2, a repressor of activation,[111] and p48, the binding subunit of the interferon-inducible factor ISGF-3.[112] IRF-1 was discovered as an inducible activator of IFN-β transcription[109,113] and has been implicated in the activation of other basally quiescent genes, such as the inducible form of nitric oxide synthase,[114] the major histocompatibility class-1 B-27 gene,[115] and the transcriptional repressor IRF-2.[108] Interestingly, all these promoters also have functional NF-κB motifs.[108,114,115]

EMSA with endothelial nuclear extracts reveals that IRF-1 is TNF-α inducible within 2 hours and binds specifically to the VCAM-1 IRF motif.[44] Interestingly, DNase I footprint analysis reveals that binding of NF-κB to the tandem sites on the VCAM-1 promoter increases the binding affinity of IRF-1 to its cognate site. IRF-1 is a weak transactivation of the native VCAM-1 promoter, but acts in synergy with overexpressed NF-κB.[44] Working in C2C12 myoblastic cells, Iadamarco et al demonstrated a specific interaction between the IRF motif and IRF-2.[30,95] In these cells, IRF-2 was constitutively present and acted as a transactivator.

Other Proteins

Several other transcription factors have been implicated in the regulation of VCAM-1.

HMGI(Y): The high mobility group (HMG) proteins are low molecular weight nonhistone chromosomal proteins that have been placed into three groups: HMG 1/2, HMG 14/17 and HMG I(Y).[116] These proteins do not have transcriptional activating capacity. HMGI(Y) binds to the minor groove of DNA, induces DNA bending, and can increase the affinity of transcription factors binding to the adjacent or overlying major groove.[117] HMGI(Y) can facilitate binding of NF-κB to a subset of binding motifs characterized by an extended AT rich core sequence.[98] Such sites have been described in the promoters of IFN-β and E-selectin[117-119] (Chapter 2). HMG I(Y) interacts with the AT rich core of the VCAM IRF site and facilitates binding of IRF-1.[44] The role, if any, of this protein in mediating NF-κB binding to the tandem VCAM-1 sites is under investigation.

GATA: The GATA proteins are zinc finger DNA-binding proteins, first described as factors restricted to cells of hematopoietic lineage, though subsequent family members have been detected in a variety of cell types, including endothelial cells.[120,121] Tandem sequence motifs matching the binding consensus for these transcriptional activators are present in the VCAM-1 promoter at positions -258 and -249. A single GATA motif is present at the homologous region of the murine VCAM-1 promoter.[87] Mutations or deletions of these sites result in a reduction of TNF-α-induced reporter gene activity in transfected endothelial cells. Additionally, over-expression of the GATA family member GATA2 can transactivate the VCAM-1 promoter in a site-specific manner.[85] Papi and Johnston demonstrated that mutations in the VCAM-1 GATA binding site attenuate promoter activity in transfected epithelial cells stimulated by rhinovirus.[61] These authors were also able to identify an inducible DNA-binding activity specific to the GATA motif, though its identity is unknown.

AP-1: The transcription factor AP-1 was originally described as a heterodimer of the basic leucine zipper oncoproteins, cJun and cFos.[122] Subsequent work has shown that there are multiple bZip family members, including jun B, jun D, and Fra 1, which are capable of homo- and heterodimerization.[123]

The AP-1 factors recognize the consensus binding motif, TGACTCA.[122] Such motif is present at -573 of the VCAM-1 promoter, and interestingly, tandem consensus motifs are present at the homologous locations of the murine VCAM-1 promoter.[85,87] Nevertheless, mutation and transactivation studies to date have not demonstrated a functional role for the AP-1 site in TNF-α-mediated activation of the VCAM-1 promoter in endothelial cells. The possibility exists, indeed is probable, that this motif may play a functional role in alternate signal transduction pathways, in different cell types, or in developmental contexts. Interestingly, Ahmed et al have shown that AP-1 subunits, cJun and cFos, can transactivate a VCAM-1 reporter bearing the tandem NF-κB motifs, but not the AP-1 motif.[124] This example of a potential physical interaction between two families of transcription factors will be discussed further in a subsequent section.

NEGATIVE TRANSCRIPTIONAL REGULATION

Downregulation of cytokine-induced gene expression, though of great interest, is less well understood than activation. New transcription of the VCAM-1 and E-selectin genes is rapidly shut down once cytokine is removed from the media of cultured endothelial cells. This shutdown occurs at the transcriptional level and requires new protein synthesis.[45] One possibility is that newly synthesized IκB-α may have a role in this postactivation repression of the VCAM-1 and other NF-κB-dependent gene(s) (see discussion in Chapter 2).

In the nuclei of transcriptionally quiescent endothelial cells, Western blotting and EMSA experiments indicate that abundant Sp1 and small amounts of homodimeric p50 are present and capable of interacting with the VCAM-1 promoter.[90] In the cytoplasm, relatively large amounts of p50, p65, and IκB-α are detectable, presumably in the form of inactive heterotrimer.[100] When cultured endothelial cells are exposed to TNF-α, IκB-α is rapidly degraded within 15 minutes via the proteasome pathway, coincident with nuclear translocation of p50/p65.[93,100] The IκB-α promoter itself is κB-dependent and is activated by newly translocated NF-κB. This leads to rapid increases in the cellular levels of IκB-α protein.[125,126] Increased IκB-α sequesters free cytoplasmic NF-κB and diminishes expression of κB-dependent genes once the activating stimulus has ceased.[100,126,127] Additionally, nuclear localized IκB-α may contribute to negative regulation of VCAM-1. Overexpressed IκB-α potently and specifically abolishes NF-κB-mediated transactivation of VCAM-1 promoter-reporter constructs (Neish and Collins, unpublished). *In vitro*, protein binding assays reveal IκB-α is capable of specifically displacing heterodimeric p50/p65 from the E-selectin and VCAM-1 κB elements, though not homodimeric p50.[45] Treatment of cytokine-stimulated cultured cells with proteosomal inhibitors post-activation blocks the rapid degradation of IκB-α, allowing IκB-α to be visualized in the nucleus coincident with the shutdown of VCAM-1 transcription.[45] Taken together, these data imply that a nuclear IκB-α-mediated displacement process may play a role in post-induction repression of VCAM-1 (and other NF-κB-dependent genes) and help prevent inappropriate expression of these proteins.

COMBINATORIAL INTERACTIONS BETWEEN TRANSCRIPTION FACTORS

We have noticed that a relatively small number of transcription factors appear to be involved in the regulation of multiple cytokine-inducible genes. Beyond cataloging the transcription factors interacting with a given promoter, a further goal is to explain how this select set of transcription cooperate to appropriately regulate a target gene. Certain combinations of transcription factors may represent a regulatory unit with a common function. For example, combinations of tandem NF-κB motifs in close proximity to Sp1 sites may form a "cassette" with the ability to powerfully and rapidly activate transcription in response to proinflammatory stimuli. Such arrangements are found in the regulatory regions of highly cytokine-inducible genes such as the HIV-LTR[128] and the IκB-α promoter[126] (Table 5-1). Combinations of functional IRF-1 and NF-κB motifs are found in several other inducible genes: VCAM-1, inducible nitric oxide synthase (NOS),[114] major histocompatibility complex (MHC) class 1 genes,[115] and even in bactericidal proteins expressed in Drosophila.[129] This combination of factors may regulate genes involved in innate immunity.

Experimentally, the interactions of transcription factors are inferred by demonstration of protein-protein interactions and the ability of two factors to synergistically transactivate a synthetic or natural promoter bearing the appropriate binding sites. Numerous recent studies demonstrate such interactions between different families of transcription factors. For example, NF-κB has been demonstrated to both physically and functionally interact with C/EBP-β,[130,131] c-Jun,[132] and

VCAM-1

TGGCTCTGCCCT**GGGTTTCCCC**TTGAA**GGGATTTCCC**TCCGCCTCTGCAAGACC
CTT**TATAA**
ACCGAGACGGGA**CCCAAAGGGG**AACTT**CCCTAAAGGG**AGGCGGAGACGTTGTGG
GAA**ATATT**

HIV LTR

ACAA**GGGACTTTCC**GCTG**GGGACTTTCC**AGGGAGGCGTGGCCTGGGCGGGACTG
GGGAGTGGCGAGCCCTCAGATGGCTGCA**TATAA**
TGTT**CCCTGAAAGG**CGAC**CCCTGAAAGG**TCCCTCCGCACCGGACCCGCCCTGAC
CCCTCACCGCTCGGGAGTCTACCGACGT**ATATT**

MAdCAM-1

GGGTCGGGCTG**GGAAAGCCCC**CTG**GGAAAGTCCC**ACAGAGCCGG
CCCAGCCCGAC**CCTTTCGGGG**GAC**CCTTTCAGGG**TGTCTCGGCC

IκB-α

ACTGGCTT**GGAAATTCCCC**CGAGCTTGACCCGCCCAG**GAGAAATCCC**CTGCCAG
CG
TGACCGAA**CCTTTAAGGG**GCTCGAACTGGGGCGGGTC**CTCTTTAGGG**GACGGTC
GC

Mouse KC (MIP-1)

TTCCGGTTGCA**GGGAAACACCC**TGTACTCC**GGGAATTTCCC**TGGCCCGGAGCT
AAGGCCAACGT**CCCTTTGTGGG**ACATGAGG**CCCTTAAAGGG**ACCGGGCCTCGA

RANTES

CTATTTT**GGAAACTCCC**CTTA**GGGGATGCCC**CTCAA
GATAAAA**CCTTTGAGGG**GAAT**CCCCTACGGG**GAGTT

Table 5-1. Sp1 (underlined) and tandem NF-κB (bold) motifs are present in the promoters of several inducible promoters.

ATF-2.[133] In the context of the VCAM-1 promoter, we have demonstrated that transactivation by overexpressed NF-κB can be increased further by overexpressing IRF-1.[44] This effect can be duplicated with heterologous promoter constructs bearing isolated VCAM-1 NF-κB and IRF binding elements. Additionally, IRF-1 physically interacts with the p50 subunit of NF-κB in *in vitro* assays of protein-protein interactions. This observation is consistent with the data of Lechleitner et al., showing additive effects of TNF-α (an NF-κB inducer) and IFN-γ (a classical inducer of IRF-1) on induction of VCAM-1 protein, message, and reporter constructs.[49] Nabel et al have shown physical and functional interaction between p65 and Sp1 in the context of the HIV-LTR.[128,134] Preliminary experiments demonstrate a striking synergistic transactivation of the VCAM-1 promoter by NF-κB and Sp1 (Neish and Collins, unpublished data). Thus, in the context of the VCAM-1 promoter, positive combinatorial interactions are seen with NF-κB and Sp1, and with NF-κB and IRF-1. The observations of Ahmad et al[124], showing transactivation of the VCAM NF-κB motifs by overexpressed AP-1 subunits, may imply the existence of an interaction between non-DNA bound cJun and cFos with NF-κB subunits.

Interestingly, multiple NF-κB dimers (p50/p65) may functionally interact in the proper promoter context. Arrangements of multiple NF-κB motifs have been reported in several genes. The E-selectin promoter contains three closely spaced NF-κB sites.[119,135] Two closely spaced functional κB elements have been identified in the promoters of MAdCAM-1,[136] class I MHC,[137] and IκB-α.[125] An intriguing observation in the VCAM-1 promoter is the absolute requirement for both NF-κB motifs to be intact, in the proper orientation and proper relative position. Cooperative binding of NF-κB molecules does not take place (Neish and Collins, unpublished data). The mechanisms by which multiple NF-κB binding motifs, or other combinations of binding sites, influence factor binding and promoter activation is of great interest.

Regulatory sequences may stimulate transcription by positioning bound factors for optimal interactions with each other and with coactivators/basal factors. As discussed in Chapter 2, the net result of multiple bound transcription factors is a specific higher order protein structure, or "enhanceosome," which presumably interacts with proteins of the basal transcriptional apparatus.[133,138,139] This model, in part, explains the phenomena of transcriptional "synergy," where combinations of transactivating factors result in a greater than additive activation of a given promoter. Transcriptional synergy between the factors bound to κB elements and a small set of other transcriptional activators may be a common theme in cytokine-induced gene expression.[140]

Beyond glimpses into the mechanisms of inducible gene expression, studies of transcription factor interactions on specific promoters may offer insights into how appropriate regulation is integrated from cytoplasmic signals. The promoter with its bound proteins may be viewed as the final integrator/effector of multiple signal transduction pathways. To date, we have characterized three different proteins involved in the induction of the VCAM-1 promoter with three distinct mechanisms of activation. The first, NF-κB, is activated by rapid post-translational events; the second, IRF-1, requires new protein-synthesis; and the third, Sp1, is constitutively active. By comparison, the E-selectin promoter utilizes NF-κB and c-Jun/ATF-2, both of which are rapidly post-translationally activated.[119,140] E-selectin activation is rapid and protein synthesis-independent, while VCAM-1 induction is slower, more prolonged, and partially protein synthesis-dependent. Therefore, analysis of the VCAM-1 and E-selectin promoters and their interacting proteins may help mechanistically explain the respective kinetics of expression of these two adhesion molecules.

SUMMARY AND FUTURE DIRECTIONS

This review has focused on the mechanisms controlling inducible expression of VCAM-1 in endothelial cells. As described earlier, VCAM-1 expression is present in extra-embryonic tissues, developing skeletal and cardiac muscle, and vascular smooth muscle. Antigen-presenting cells and some tissue macrophages also constitutively express VCAM-1. The mechanisms controlling this developmentally regulated, constitutive expression are totally unknown. Future work may uncover novel regulatory elements and/or previously uncharacterized proteins necessary for the nonendothelial functions of VCAM-1.

Based on our analysis, we suggest a model for the transcriptional regulation of cytokine-activated VCAM-1 gene expression in endothelial cells (Fig. 5-4). In the uninduced state, the VCAM-1 promoter may be occupied by homodimeric p50 and Sp1. Conceivably, these bound proteins could displace bound chromatin elements maintaining the promoter in an "open" state. A similar arrangement has been postulated in the context of the HIV-LTR.[141] Cytokine stimulation of the endothelial cell results in the nuclear translocation of NF-κB. The transcriptionally active heterodimeric p50/p65 displaces homodimeric p50 and interacts with Sp1 to maximally activate transcription. In the continued presence of an activating stimulus, IRF-1 synthesis is induced, cooperating with NF-κB to allow transcription of the promoter. These transcription factors interact with architectural proteins to form a complex that recruits coactivators and components of the basal transcription machinery. IRF-1 may allow transcription of VCAM-1 to continue longer than relatively transiently expressed molecules such as E-selectin and to potentially respond to signals that do not (or weakly) activate NF-κB. With the cessation of activating signals, IκB-α degradation stops, allowing cytoplasmic levels to reaccumulate and end nuclear translocation of active heterodimer. Concurrently, newly synthesized IκBα may enter the nucleus and selectively displace bound heterodimer, shutting off

192

transcription. Finally, the system would reset once homodimeric p50 rebound the promoter. It must be kept in mind that this model is based on experiments with transiently transfected reporter constructs and binding assays with cell extracts. The transfected promoters under study are in the form of large numbers of supercoiled, episomal plasmids. Such promoter sequences are outside their natural context, enmeshed in local chromatin structure and in numeric excess of the physiologic two alleles per cell. Similarly, binding assays involve short fragments of DNA interacting with recombinant proteins or nuclear extracts. The recent development of *in vivo* footprinting methods and cell free transcription systems may allow more direct testing of this hypothesis in the near future.

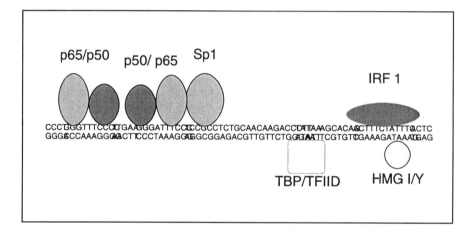

Figure 5-4. Model of the cytokine-induced VCAM-1 enhancer.

ACKNOWLEDGEMENTS

This work was supported by NIH grants HL03011 to A.S.N., as well as grants HL45462, HL35716, and P0136028 to T.C. S.A. was a Fellow of the Stanley L. Sarnoff endowment.

REFERENCES

1.   Springer T.   Traffic signals for lymphocyte recirculation and leukocyte emigration: the multistep paradigm.   Cell 1994; 76:301-314.
2.   Carlos TM, Harlan JM. Leukocyte-endothelial adhesion molecules.   Blood 1994; 84:2068-2101.
3.   Osborn L, Hession C, Tizard R, et al   Direct expression cloning of vascular cell adhesion molecule 1, a cytokine-induced endothelial protein that binds to lymphocytes.   Cell   1989; 59:1203-1211.

4. Rice GE, Munro JM, Bevilacqua MP. Inducible cell adhesion molecule 110 (INCAM-110) is an endothelial receptor for lymphocytes. J Exp Med 1990; 171:1369-1374.

5. Elices M, Osborn L, Takada Y, et al. VCAM-1 on activated endothelium interacts with the leukocyte integrin VLA-4 at a site distinct from the VLA-4/fibronectin binding site. Cell 1990; 60:577-584.

6. Vonderheide R, Springer TA. Lymphocyte adhesion through very late antigen 4: evidence for a novel binding site in the alternatively spliced domain of vascular cell adhesion molecule 1 and an additional α4 integrin counter-receptor on stimulated endothelium. J Exp Med 1992; 175:1433-1442.

7. Schleimer RP, Sterbinsky SA, Kaiser J, et al. IL-4 induces adherence of human eosinophils and basophils but not neutrophils to endothelium: association with expression of VCAM-1. J Immunol 1992; 148:1086-1092.

8. Postigo AA, Teixido J, Sanchez-Madrid F. The α4β1/VCAM1 adhesion pathway in physiology and disease. Res Immunol 1993; 144:723-735.

9. Rice GE, Munro JM, Corless C, et al. Vascular and nonvascular expression of INCAM-110. Am J Pathol 1991; 138:385-393.

10. Koch AE, Burrows JC, Haines GK, et al. Immunolocalization of endothelial and leukocyte adhesion molecules in human rheumatoid and osteoarthritic synovial tissues. Lab Invest 1991; 64:313-320.

11. Morales-Ducret JE, Wayner MJ, Elices M, et al. α4/β1 integrin (VLA-4) ligands in arthritis: vascular cell adhesion molecule-1 expression on synovium and on fibroblast-like synoviocytes. J Immunol 1992; 149:1583-1593.

12. Briscoe DM, Schoen FJ, Rice GE, et al. Induced expression of endothelial-leukocyte adhesion molecules in human cardiac allografts. Transplantation 1991; 51:537-547.

13. Pelletier RP, Ohye GR, Vanbuskirk A, et al. Importance of endothelial VCAM-1 for inflammatory leukocytic infiltration in vivo. J Immunol 1992; 149:2473-2481.

14. Bergese S, Pelletier R, Vallera D, et al. Regulation of endothelial VCAM-1 expression in murine cardiac grafts: roles for TNF and IL4. Am J Pathol 1995; 146:989-997.

15. Fries JWU, Williams AJ, Atkins RC, et al. Expression of VCAM-1 and E-selectin in an in vivo model of endothelial activation. Am J Pathol 1993; 143:725-737.

16. van Oosten M, van de Bilt E, De Vries HE, et al. Vascular adhesion molecule-1 and intercellular adhesion molecule-1 expression on rat liver cells after lipopolysaccharide administration in vivo. Hepatology 1995; 22:1538-1546.

17. Nakajima H, Sano H, Nishimura T, et al. Role of vascular cell adhesion molecule 1/very late activation antigen 4 and intercellular adhesion molecule 1/lymphocyte function-associated antigen 1 interactions in antigen-induced eosinophil and T cell recruitment into the tissue. J Exp Med 1994; 179:1145-1154.

18. Cybulsky M, Gimbrone M. Endothelial expression of a mononuclear leukocyte adhesion molecule during atherogenesis. Science 1991; 251:788-791.

19. Li H, Cybulski MI, Gimbrone MA, et al. Inducible expression of vascular cell adhesion molecule-1 by vascular smooth muscle cells in vitro and within rabbit atheroma. Am J Pathol 1993; 143:1551-1559.

20.	Li H, Cybulsky MI, Gimbrone MA Jr, et al. An atherogenic diet rapidly induces VCAM-1, a cytokine-regulatable mononuclear leukocyte adhesion molecule in rabbit aortic endothelium. Arterioscler Thromb Vasc Biol 1993; 13:197-204.

21.	O'Brien K, Allen M, McDonald A, et al. Vascular cell adhesion molecule-1 is expressed in human coronary atherosclerotic plaques: implications for the mode of progression of advanced coronary atherosclerosis. J Clin Invest 1993; 92:945-951.

22.	Iiyama K, Hajra L, Iiyama M, Li H, DiChiara M, Medoff BD, and Cybulsky MI. Patterns of vascular cell adhesion molecule-1 and intercellular adhesion molecule-1 expression in rabbit and mouse atherosclerotic lesions and at sites predisposed to lesion formation. Circ Res 1999; 85:199-207.

23.	Nakashima Y, Raines EW, Plump AS, Breslow JL, Ross R. Upregulation of VCAM-1 and ICAM-1 at atherosclerosis-prone sites on the endothelium in the ApoE-deficient mouse. Arterioscler Thromb Vasc Biol 1998; 18:842-851.

24.	Hajra L, Evans AI, Chen M, Hyduk SJ, Collins T, Cybulsky MI. The NF-kappa B signal transduction pathway in aortic endothelial cells is primed for activation in regions predisposed to atherosclerotic lesion formation. Proc Natl Acad Sci USA 2000; 97:9052-9057.

25.	Rice GE, Gimbrone MA, Bevilacqua MP. Tumor cell-endothelial cell interactions: increased adhesion of human melanoma cells to activated vascular endothelium. Am J Pathol 1988; 133:204-210.

26.	Rice GE, Bevilacqua MP. An inducible endothelial cell surface glycoprotein mediates melanoma adhesion. Science 1989; 246:1303-1306.

27.	Ockenhouse CF, Tegoshi T, Maeno Y, et al. Human vascular endothelial cell adhesion receptors for Plasmodium falciparum-infected erythrocytes: roles for endothelial leukocyte adhesion molecule 1 and vascular cell adhesion molecule 1. J Exp Med 1992; 176:1183-1189.

28.	Gurtner GC, Davis V, Li H, et al. Targeted disruption of the murine VCAM1 gene: essential role of VCAM-1 in chorioallantoic fusion and placentation. Genes Dev 1995; 9:1-14.

29.	Rosen GD, Sanes JR, LaChance R, et al. Roles for the integrin VLA-4 and its counter receptor VCAM-1 in myogenesis. Cell 1992; 69:1107-1119.

30.	Jesse TL, LaChance R, Iademarco MF, et al. Interferon regulatory factor-2 is a transcriptional activator in muscle where it regulates expression of vascular cell adhesion molecule-1. J Cell Biol 1998; 140:1265-1276.

31.	Landry DB, Couper LL, Bryant SR, et al. Activation of the NF-kappaB and I kappa B system in smooth muscle after rat arterial injury. Induction of vascular cell adhesion molecule-1 and monocyte chemoattractant protein-1. Am J Path 1997; 151:1085-1095.

32.	Kienzle G, von Kempis J. Vascular cell adhesion molecule 1 (CD106) on primary human articular chondrocytes: functional regulation of expression by cytokines and comparison with intercellular adhesion molecule 1 (CD54) and very late activation antigen 2. Arthritis Rheum 1998; 41:1296-1305.

33.	Koopman G, Parmentier HK, Schuurman H-J, et al. Adhesion of human B cells to follicular dendritic cells involves both the lymphocyte function-associated antigen/intercellular adhesion molecule 1 and very late antigen 4/vascular cell adhesion molecule 1 pathways. J Exp Med 1991; 173:1297-1304.

34.   Freedman AS, Munro JM, Rice GE, et al. Adhesion of human B cells to germinal centers *in vitro* involves VLA-4 and INCAM-110. Science 1990; 249:1030-1033.

35.   Miyake K, Medina K, Ishahara H, et al. A VCAM-like adhesion molecule on murine bone marrow stromal cells mediates binding of lymphocyte precursors in culture. J Cell Biol 1991; 114:557-565.

36.   Ryan DH, Nuccie BL, Abboud CN, et al. Vascular cell adhesion molecule-1 and the intergin VLA-4 mediate adhesion of human B cell precursors to cultured bone marrow adherent cells. J Clin Invest 1991; 88:995-1004.

37.   van Seventer GA, Newman W, Shimizu Y, et al. Analysis of T-cell stimulation by superantigen plus major histocompatibility complex II molecules or by CD3 monoclonal antibody: costimulation by purified adhesion ligands VCAM-1, ICAM-1, but not ELAM-1. J Exp Med 1991; 174:901-913.

38.   Briscoe DM, Cotran RS, Pober JS. Effects of tumor necrosis factor, lipopolysaccharide, and IL-4 on the expression of vascular cell adhesion molecule-1 *in vivo*. J Immunol 1992; 149:2954-2960.

39.   Carlos TM, Schwartz BR, Korach NL, et al. Vascular cell adhesion molecule-1 mediates lymphocyte adherence to cytokine-activated cultured human endothelial cells. Blood 1990; 76:965-970.

40.   Bochner BS, Luscinskas FW, Gimbrone MA, et al. Adhesion of human basophils, eosinophils, and neutrophils to interleukin 1-activated human vascular endothelial cells: contributions of endothelial adhesion molecules. J Exp Med 1991; 173:1553-1556.

41.   Thornhill MH, Wellicome SM, Mahiouz DL et al. Tumor necrosis factor combines with IL-4 or IFN-γ to selectively enhance endothelial cell adhesiveness for T-cells. J Immunol 1991; 146:592-598.

42.   Masinovsky B, Urdal D, Gallatin WM. IL-4 acts synergistically with IL-1 to promote lymphocyte adhesion to microvascular endothelium by induction of vascular cell adhesion molecule-1. J Immunol 1990; 145:2886-2895.

43.   Bevilacqua MP, Stengelin S, Gimbrone MA, et al. Endothelial leukocyte adhesion molecule 1: an inducible receptor for neutrophils related to complement regulatory proteins and lectins. Science 1989; 243:1160-1165.

44.   Neish AS, Read MA, Thanos D, et al. Endothelial IRF-1 cooperates with NF-κB as a transcriptional activator of vascular cell adhesion molecule-1. Mol Cell Biol 1995; 15:2558-2569.

45.   Read MA, Neish AS, Gerritsen ME, et al. Post-induction transcriptional repression of the E-selectin and VCAM-1 genes. J Immunol 1996; 157:3472-3479.

46.   Wertheimer SJ, Myers CL, Wallace RW, et al. Intercellular adhesion molecule-1 gene expression in human endothelial cells. J Biol Chem 1992; 267:12030-12035.

47.   Luscinskas FW, Kansas GS, Ding H, et al. Monocyte rolling, arrest and spreading on IL-4-activated vascular endothelium under flow is mediated via sequential action of L-selectin, β1-integrins, and β2-integrins. J Cell Biol 1994; 125:1417-1427.

48.   Iademarco MF, Barks JL, Dean DC. Regulation of vascular cell adhesion molecule-1 expression by IL-4 and TNF-α in cultured endothelial cells. J Clin Invest 1995; 95:264-271.

49. Lechleitner S, Gille J, Johnson DR, et al. Interferon enhances tumor necrosis factor-induced vascular cell adhesion molecule 1 (CD106) expression in human endothelial cells by an interferon-related factor 1-related pathway. J Exp Med 1998; 187:2023-2030.

50. Heckmann M, Douwes K, Peter R, et al. Vascular activation of adhesion molecule mRNA and cell surface expression by ionizing radiation. Exp Cell Res 1998; 238:148-154.

51. Oertli B, Beck-Schimmer B, Fan X, et al. Mechanisms of hyaluronan-induced up-regulation of ICAM-1 and VCAM-1 expression by murine kidney tubular epithelial cells: hyaluronan triggers cell adhesion molecule expression through a mechanism involving activation of nuclear factor-kappa B and activating protein-1. J Immunol 1998; 161:3431-3437.

52. Tummala PE, Chen XL, Sundell CL, et al. Angiotensin II induces vascular cell adhesion molecule-1 expression in rat vasculature: A potential link between the renin-angiotensin system and atherosclerosis. Circulation 1999; 100:1223-1229.

53. Sironi M, Sciacca FL, Matteucci C, et al. Regulation of endothelial and mesothelial cell function by interleukin-13: selective induction of vascular cell adhesion molecule-1 and amplification of interleukin-6 production. Blood 1994; 84:1913-1921.

54. Bochner BS, Klunk DA, Sterbinsky SA, et al. IL-13 selectively induces vascular cell adhesion molecule-1 expression in human endothelial cells. J Immunol 1995; 154:799-803.

55. Doucet C, Brouty-Boye D, Pottin-Clemenceau C, et al. IL-4 and IL-13 specifically increase adhesion molecule and inflammatory cytokine expression in human lung fibroblasts. Int Immunol 1998; 10:1421-1433.

56. Quinlan KL, Song IS, Naik SM et al. VCAM-1 expression on human dermal microvascular endothelial cells is directly and specifically up-regulated by substance P. J Immunol 1999;162:1656-1661.

57. Quinlan KL, Naik SM, Cannon G, et al. Substance P activates coincident NF-AT- and NF-kappa B-dependent adhesion molecule gene expression in microvascular endothelial cells through intracellular calcium mobilization. J Immunol 1999; 163:5656-5665.

58. Lambert N, Lescoulie PL, Yassine-Diab B, et al. Substance P enhances cytokine-induced vascular cell adhesion molecule-1 (VCAM-1) expression on cultured rheumatoid fibroblast-like synoviocytes. Clin Exp Immunol 1998; 113:269-275.

59. Offermann MK, Zimring J, Mellits KH, et al. Activation of the double-stranded-RNA-activated protein kinase and induction of vascular cell adhesion molecule-1 by poly(I)•poly(C) in endothelial cells. Eur J Biochem 1995; 232:28-36.

60. Valentin H, Lemasson I, Hamaia S, et al. Transcriptional activation of the vascular cell adhesion molecule −1 gene in T lymphocytes expressing human T-cell leukemia virus type 1 Tax protein. J Virol 1997; 71:8522-8530.

61. Papi A, Johnston SL. Respiratory epithelial cell expression of vascular cell adhesion molecule-1 and its upregulation by rhinovirus infection via NF-κB and GATA transcription factors. J Biol Chem 1999; 274:30041-30051.

62. Ebnet K, Brown KD, Siebenlist UK, et al. Borrelia Burgdorferi activates nuclear factor-kappa B and is a potent inducer of chemokine and adhesion molecule gene

expression in endothelial cells and fibroblasts. J Immunol 1997; 158:3285-3292.

63. Gille J, Swerlick RA, Lawley TJ, et al. Differential regulation of vascular cell adhesion molecule-1 gene transcription by tumor necrosis factor α and interleukin-1α in dermal microvascular endothelial cells. Blood 1996; 87:211-217.

64. Swerlick RA, Lee KH, Li L-J, et al. Regulation of vascular cell adhesion molecule 1 on human dermal microvascular endothelial cells. J Immunol 1992; 149:698-705.

65. Petzelbauer P, Bender JR, Wilson J, et al. Heterogeneity of dermal microvascular endothelial cell antigen expression and cytokine responsiveness *in situ* and in cell culture. J Immunol 1993; 151:5062-5072.

66. Hauser IA, Johnson DR, Madri JA. Differential induction of VCAM-1 on human iliac venous and arterial endothelial cells and its role in adhesion. J Immunol 1993; 151:5172-5185.

67. Kunsch C, Medford RM. Oxidative stress as a regulator of gene expression in the vasculature. Circ Res 1999; 85:753-766.

68. Witztum JL, Steinberg D. Role of oxidized low density lipoprotein in atherosclerosis. J Clin Invest 1991; 88:1785-1792.

69. Kume N, Cybulsky MI, Gimbrone MA. Lysophosphatidylcholine, a component of atherogenic lipoproteins, induces mononuclear leukocyte adhesion molecules in cultured human and rabbit arterial endothelial cells. J Clin Invest 1992; 90:1138-1144

70. Khan BV, Parthasarathy SS, Alexander RW, et al. Modified low density lipoprotein and its constituents augment cytokine-activated vascular cell adhesion molecule-1 gene expression in human vascular cells. J Clin Invest 1995; 95:1262-1270.

71. Lin JH, Zhu Y, Liao HL, et al. Induction of vascular cell adhesion molecule-1 by low-density lipoprotein. Atherosclerosis 1996; 127:185-194.

72. Weber C, Erl W, Pietsch A, et al. Antioxidants inhibit monocyte adhesion by suppressing nuclear factor-κB mobilization and induction of vascular cell adhesion molecule-1 in endothelial cells stimulated to generate radical. Arterioscler Thromb 1994; 14:1665-1673.

73. Marui N, Offerman MK, Swerlick R, et al. Vascular cell adhesion molecule-1 (VCAM-1) gene transcription and expression are regulated through an antioxidant-sensitive mechanism in human vascular endothelial cells. J Clin Invest 1993; 92:1866-1874.

74. Kubes P, Suzuki M, Granger DN. Nitric oxide: an endogenous modulator of leukocyte adhesion. Proc Natl Acad Sci U S A 1991; 88:4651-4655.

75. DeCaterina R, Libby P, Peng H-B, et al. Nitric oxide decreases cytokine-induced endothelial activation. J Clin Invest 1995; 96:60-68.

76. Khan BV, Harrison DG, Olbrych MT, et al. Nitric oxide regulates vascular cell adhesion molecule 1 gene expression and redox-sensitive transcriptional events in human vascular endothelial cells. Proc Natl Acad Sci U S A 1996; 93:9114-9119.

77. Peng H-B, Libby P, Liao JK. Induction and stabilization of IκBα by nitric oxide mediates inhibition of NF-κB. J Biol Chem 1995; 270:14214-14219.

78. Peng HB, Spiecker M, Liao JK. Inducible nitric oxide: an autoregulatory feedback inhibitor of vascular inflammation. J Immunol 1998; 161:1970-1976.

79. Schmidt AM, Hori O, Chen JX et al. Advanced glycation endproducts interacting with their endothelial receptor induce expression of vascular cell adhesion molecule-1 (VCAM-1) in cultured human endothelial cells and mice. J Clin Invest 1995; 96:1395-1403.

80. Vlassara H, Fuh H, Donnelly T, et al. Advanced glycation endproducts promote adhesion molecule (VCAM-1, ICAM-1) expression and atheroma formation in normal rabbits. Mol Med 1995; 1:447-456.

81. Ando J, Tsuboi H, Korenaga R, et al. Shear stress inhibits adhesion of cultured mouse endothelial cells to lymphocytes by downregulating VCAM-1 expression. Am Physiol Soc 1994; 267:C679--C687.

82. Tanaka H, Sukhova GK, Swanson SJ, et al. Sustained activation of vascular cells and leukocytes in the rabbit aorta after balloon injury. Circulation 1993; 88:1788-1803.

83. Linder V, Collins T. Expression of NF-κB and IκB-α by aortic endothelium in an arterial injury model. 1996; 148:427-438.

84. Collins T, Cybulsky MI. Nuclear factor-κB: pivotal mediator or innocent bystander in atherogenesis. J Clin Invest, in press.

85. Neish AS, Williams AJ, Palmer HJ, et al. Functional analysis of the human vascular cell adhesion molecule 1 promoter. J Exp Med 1992; 176:1583-1593.

86. Cybulsky M, Fries J, Williams A, et al Gene structure, chromosomal location, and basis for alternative mRNA splicing of the human VCAM1 gene. Proc Natl Acad Sci U S A 1991; 88:7859-7863.

87. Cybulsky MI, Allan-Motamed M, Collins T. Structure of the murine VCAM1 gene. Genomics 1993; 18:387-391.

88. Cybulsky M, Fries J, Williams A, et al. Alternative splicing of human VCAM-1 in activated vascular endothelium. Am J Pathol 1991; 138:815-820.

89. Hession C, Tizard R, Vassalo C, et al. Cloning of an alternate form of vascular cell adhesion molecule-1 (VCAM1). J Biol Chem 1991; 266:6682-6685.

90. Neish AS, Khachigian LM, Park A, et al. Sp1 is a component of the cytokine-inducible enhancer in the promoter of vascular cell adhesion molecule-1. J Biol Chem 1995; 270:28903-28909.

91. Grilli M, Chiu J-S, Lenardo M. NF-κB and Rel, participants in a multiform transcriptional regulatory system. In: Jeon KW, Friedlander M, Jarvik J, eds. International Review of Cytology: A Survey of Cell Biology. San Diego: Academic Press, Inc., Harcourt Brace Jovanovich, 1993:1-62.

92. Iademarco MF, McQuillan JJ, Rosen GD, et al. Characterization of the promoter for vascular cell adhesion molecule-1 (VCAM-1). J Biol Chem 1992; 267:16323-16329.

93. Read MA, Neish AS, Luscinskas FW, et al. The proteasome pathway is required for cytokine-induced endothelial leukocyte adhesion molecule expression. Immunity 1995; 2:493-506.

94. Tanaka N, Kawakami T, Taniguchi T. Recognition DNA sequences of interferon regulatory factor 1 (IRF-1) and IRF-2, regulators of cell growth and the interferon system. Mol Cell Biol 1993; 13:4531-4538.

95. Iademarco MF, McQuillan JJ, Dean DC. Vascular cell adhesion molecule-1: contrasting transcriptional control mechanisms in muscle and endothelium. Proc Natl Acad Sci U S A 1993; 90:3943-3947.

96. Lenardo MJ, Staudt L, Robbins P, et al. Repression of the IgH enhancer in teratocarcinoma cells associated with a novel octamer factor. Science 1989; 243:544-545.

97. Kunsch C, Ruben SM, Rosen CA. Selection of optimal κB/Rel DNA-binding motifs: interaction of both subunits of NF-κB with DNA is required for transcriptional activation. Mol Cell Biol 1992; 12:4412-4421.

98. Thanos D, Maniatis T. Identification of the rel family members required for virus induction of the human beta interferon gene. Mol Cell Biol 1995; 15:152-164.

99. Carrasco D, Ryseck R-P, Bravo R. Expression of relB transcripts during lymphoid organ development: specific expression in dendritic antigen-presenting cells. Development 1993; 118:1221-1231.

100. Read MA, Whitley MZ, Williams AJ, et al. NF-κB and IκB-α: an inducible regulatory system in endothelial activation. J Exp Med 1994; 179:503-512.

101. Shu HB, Agranoff AB, Nabel EG, et al. Differential regulation of vascular cell adhesion molecule 1 gene expression by specific NF-κB subunits in endothelial and epithelial cells. Mol Cell Biol 1993; 13:6283-6289.

102. Franzoso G, Bour V, Park S, et al. The candidate oncoprotein Bcl-3 is an antagonist of p50/ NF-κB-mediated inhibition. Nature 1992; 359:339-342.

103. Courey AJ, Tjian R. Mechanisms of transcriptional control as revealed by studies of human transcription factor Sp1. In: Transcriptional Regulation. Cold Spring Harbor, NY: Cold Spring Harbor Laboratory Press, 1992:743-769.

104. Khachigan LM, Lindner V, Williams AJ, et al. Egr-1-induced endothelial gene expression: a common theme in vascular injury. Science 1996; 271:1427-1431.

105. Hagen G, Muller S, Beato M, et al. Sp1-mediated transcriptional activation is repressed by Sp3. EMBO J 1994; 13:3843-3851.

106. Majello B, DeLuca P, Hagan G, et al. Different members of the Sp1 multigene family exert opposite transcriptional regulation of the long terminal repeat of HIV-1. Nucleic Acids Res 1994; 22:4914-4921.

107. Fujita T, Reis LFL, Watanabe N, et al. Induction of the transcription factor IRF-1 and interferon-β mRNAs by cytokines and activators of second messenger pathways. Proc Natl Acad Sci U S A 1989; 86:9936-9940.

108. Harada H, Takahashi E-I, Itoh S, et al. Structure and regulation of the human interferon regulatory factor 1 (IRF-1) and IRF-2 genes: implications for a gene network in the interferon system. Mol Cell Biol 1994; 14:1500-1509.

109. Miyamoto M, Fujita T, Kimura Y, et al. Regulated expression of a gene encoding a nuclear factor, IRF-1, that specifically binds to IFN-β gene regulatory elements. Cell 1988; 54:903-913.

110. Pine R, Decker T, Kessler DS, et al. Purification and cloning of interferon-stimulated gene factor 2 (ISGF2): ISGF2 (IRF-1) can bind to the promoters of both beta interferon and interferon-stimulated genes but is not a primary transcriptional activator of either. Mol Cell Biol 1990; 10:2448-2457.

111. Harada H, Fujita T, Miyamoto M, et al. Structurally similar but functionally distinct factors, IRF-1 and IRF-2, bind to the same regulatory elements IFN and IFN-inducible genes. Cell 1989; 58:729-739.

112. Veals SA, Schindler C, Leonard D, et al. Subunit of alpha-interferon-responsive transcription factor is related to interferon regulatory factor and Myb families of DNA-binding proteins. Mol Cell Biol 1992; 12:3315-3324.

113. Reis LFL, Harada H, Wolchok JD, et al. Critical role of a common transcription factor, IRF-1, in the regulation of the IFN-β and IFN-inducible genes. EMBO J 1992; 11:185-193.

114. Xie Q, Whisnant R, Nathan C. Promoter of the mouse gene encoding calcium independent nitric oxide synthetase confers inducibility by interferon-γ and bacterial lipopolysaccharide. J Exp Med 1993; 177:1779-1784.

115. Johnson DR, Pober JS. HLA class 1 heavy chain gene promoter elements mediating synergy between tumor necrosis factor and interferons. Mol Cell Biol 1994; 14:1322-1332.

116. Crothers DM. Architectural elements in nucleoprotein complexes. Curr Biol 1993; 3:675-676.

117. Thanos D, Maniatis T. The high mobility group protein HMG I(Y) is required for NF-κB-dependent virus induction of the human IFN-β gene. Cell 1992; 71:777-789.

118. Lewis H, Kaszubska W, DeLamarter JF, et al. Cooperativity between two NF-κB complexes, mediated by high-mobility-group protein I(Y), is essential for cytokine-induced expression of the E-selectin promoter. Mol Cell Biol 1994; 14:5701-5709.

119. Whitley MZ, Thanos D, Read MA, et al. A striking similarity in the organization of the E-selectin and beta interferon gene promoters. Mol Cell Biol 1994; 14:6464-6475.

120. Dorfman DM, Wilson DB, Bruns GAP, et al. Human transcription factor GATA-2: evidence for regulation of preproendothelin-1 gene expression in endothelial cells. J Biol Chem 1992; 267:1279-1285.

121. Lee M-E, Temizer DH, Clifford JA, et al. Cloning of the GATA-binding protein that regulates endothelin-1 gene expression in endothelial cells. J Biol Chem 1991; 24:16188-16192.

122. Ransone LJ, Verma IM. Nuclear proto-oncogenes fos and jun. Annu Rev Cell Biol 1990; 6:539-557.

123. Lee W, Mitchell P, Tjian R. Purified transcription factor AP-1 interacts with TPA-inducible enhancer elements. Cell 1987; 49:741-752.

124. Ahmed M, Theofanidis P, Medford RM. Role of activating protein-1 in the regulation of vascular cell adhesion molecule-1 gene expression by tumor necrosis factor-α. J Biol Chem 1998; 273:4616-4621.

125. Cheng Q, Cant CA, Moll T, et al. NF-κB subunit-specific regulation of the IκBα promoter. J Biol Chem 1994; 269:13551-13557.

126. Sun S-C, Ganchi P, Ballard D, et al. NF-κB controls expression of inhibitor IκBα: evidence for an inducible autoregulatory pathway. Science 1993; 259:1912-1915.

127. Rice NR, Ernst MK. In vivo control of NF-κB and activation by IκBα. EMBO J 1993; 12:4685-4695.

128. Perkins ND, Edwards NL, Duckett CS, et al. A cooperative interaction between NF-κB and Sp1 is required for HIV-1 enhancer activation. EMBO J 1993; 12:3551-3558.

129. Georgel P, Kappler C, Langley E, et al. Drosophila immunity: a sequence homologous to mammalian interferon consensus response element enhances the activity of the diptercin promoter. Nucleic Acids Res 1995; 23;1140-1145.

130. Matsusaka T, Fujikawa K, Nishio Y, et al. Transcription factors NF-IL6 and NF-κB synergistically activate transcription of the inflammatory cytokines, interleukin 6 and interleukin 8. Proc Natl Acad Sci U S A 1993; 90:10193-10197.

131. Stein B, Cogswell PC, Baldwin AS. Functional and physical associations between NF-κB and C/EBP family members: a Rel domain-bZIP interaction. Mol Cell Biol 1993; 13:3964-3974.

132. Stein B, Baldwin AS. Cross-coupling of the NF-κB p65 and Fos/Jun transcription factors produces potentiated biological function. EMBO J 1993; 12:3879-3891.

133. Du W, Thanos D, Maniatis T. Mechanisms of transcriptional synergism between distinct virus-inducible enhancer elements. Cell 1993; 74:887-898.

134. Perkins ND, Agranoff AB, Pascal E, et al. An interaction between the DNA-binding domains of RelA(p65) and Sp1 mediates human immunodeficiency virus gene activation. Mol Cell Biol 1994; 14:6570-6583.

135. Schindler U, Baichwal VR. Three NF-κB binding sites in the human E-selectin gene required for maximal tumor necrosis factor alpha-induced expression. Mol Cell Biol 1994; 14:5820-5831.

136. Takeuchi M, Baichwal VR. Induction of the gene encoding mucosal vascular addressin cell adhesion molecule 1 by tumor necrosis factor-α is mediated by NF-κB proteins. Proc Natl Acad Sci U S A 1995; 92:3561-3565.

137. Mansky P, Brown WM, Park J-H, et al. The second κB element, κB2, of the HLA-α class I regulatory complex is an essential part of the promoter. J Immunol 1994; 153:5082-5090.

138. Thanos D, Maniatis T. Virus induction of human IFNβ gene expression requires the assembly of an enhanceosome. Cell 1995; 83:1091-1100.

139. Tjian R, Maniatis T. Transcriptional activation: a complex puzzle with few easy pieces. Cell 1994; 77:5-8.

140. Collins T, Read MA, Neish AS et al. Transcriptional regulation of endothelial cell adhesion molecules: NF-κB and cytokine-inducible enhancers. FASEB J 1995; 9:899-909.

141. Pazin MJ, Sheridan PL, Cannon K, et al. NF-κB-mediated chromatin reconfiguration and transcriptional activation of the HIV-1 enhancer in vitro. Genes Dev 1996; 10:37-49.

Chapter 6

ACTIVATION OF NUCLEAR FACTOR-κB

Frank S. Lee,# Robert T. Peters,≠ Zhijian J. Chen,† and Tom Maniatis*

#Department of Pathology and Laboratory Medicine
University of Pennsylvania School of Medicine
218 John Morgan Bldg., Philadelphia, PA 19104
≠Syntonix Pharmaceuticals, Inc.
9 Fourth Avenue, Waltham, MA 02451-7506
†Department of Molecular Biology and Oncology
UT Southwestern Medical Center
5323 Harry Hines Boulevard, Dallas, TX 75235-9148
*Department of Molecular and Cellular Biology
Harvard University, 7 Divinity Avenue
Cambridge, Massachusetts 02138

INTRODUCTION

NF-κB comprises a family of transcription factors that play a central role in the transcriptional regulation of a large number of diverse genes involved in inflammatory and immune responses. These genes include those encoding for cytokines such as IFNβ and IL-2, transcription factors such as c-myc, immunoreceptors such as class I and II MHC, and acute-phase proteins such as angiotensinogen. The various chapters of this volume highlight the critical importance of NF-κB in the regulation of an additional important class of genes, namely those encoding adhesion molecules such as E-selectin, ICAM-1, and VCAM-1. NF-κB binding sites have also been identified in the transcriptional enhancers of a number of different viruses, the best known example being HIV. This review provides a brief overview of the proteins of the NF-κB regulatory system, and then focuses on recent advances in understanding the mechanisms by which NF-κB is activated. The regulation of NF-κB involves the phosphorylation and proteolysis of its inhibitor protein IκB and the nuclear translocation of NF-κB. Aspects of NF-κB not covered here, such as its involvement in oncogenesis and apoptosis, and its relationship to the Drosophila immune response, have been addressed in several recent reviews.[1-4]

THE NF-κB AND IκB FAMILIES OF PROTEINS

Members of the mammalian NF-κB family of proteins include p65 (RelA), RelB, c-Rel, p50/p105, and p52/p100 (Fig. 6-1).[5-11] These members can either hetero- or homodimerize; prototypical NF-κB consists of a p50/p65 heterodimer. A central feature of the NF-κB family of transcription factors is the presence of the Rel homology domain (RHD). This domain, originally identified in the avian oncoprotein v-Rel,[12] contains about 300 amino acids, is the DNA binding and dimerization domain, possesses a nuclear localization sequence (NLS), and is the domain that interacts with IκB. X-ray crystallographic studies of the p50 homodimer have revealed that this domain actually consists of two structural domains, each with immunoglobulin-like folds.[13,14] Although both domains contact DNA, most of the critical contacts are made by the N-terminal domain, whereas the C-terminal domain contains the dimerization interface. p65, RelB, and c-Rel are all synthesized as mature proteins. In contrast, mature p50 arises from processing of the precursor protein p105[15] in a manner detailed later. Likewise, p52 is processed from p100. All precursor and mature proteins just described contain RHDs; hence, all are NF-κB family members.

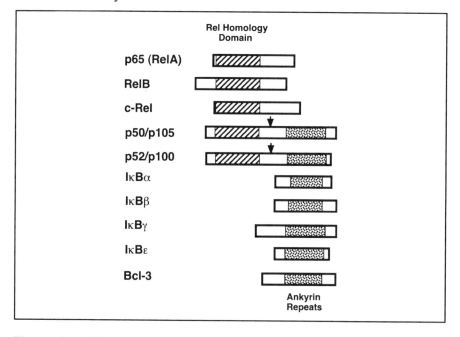

Figure 6-1. Mammalian members of the NF-κB and IκB families of proteins. Members of the NF-κB family are denoted by the presence of the Rel homology domain (*hatched*). Members of the IκB family are denoted by the presence of the ankyrin repeat domain (*stippled*). p105 and p100 are members of both families. Proteolytic cleavage (denoted by *arrows*) and degradation of their C-termini results in mature p50 and p52, respectively.

NF-κB binds to decameric DNA sequences typically of the consensus GGGRNNYYCC (R = A,G; N = A,C,G,T; Y = C,T) with dissociation constants approaching 10^{-12} M, unusually low for a eukaryotic transcription factor.[16] Extensive protein:DNA contacts present in the crystal structure of the p50 homodimer bound to DNA correlates with this tight binding.[13,14] Among the mammalian NF-κB family of proteins, p65, RelB, and c-Rel possess transcriptional activation domains and exhibit potent transcriptional activity when transiently transfected.[17,18] p65 binds the transcriptional coactivator CBP/p300 in vitro and in vivo.[19-21] CBP/p300 contains histone acetylase activity[22,23] and is also found associated with the mammalian RNA polymerase II holoenzyme.[24] Thus, NF-κB functions at least in part by recruiting CBP/p300 and the holoenzyme to promoters. p65 and c-Rel have also been shown to bind the basal transcription factors TBP and TFIIB in vitro, providing additional potential mechanisms for their transcriptional activity.[25-27] Other family members, such as p50 and p52, lack transcriptional activation domains, and thus possess weak transcriptional activity or are transcriptionally inert in vivo.[17] The transcriptional activity of NF-κB therefore depends on its subunit composition. For example, p50/p65 is a potent transcriptional activator, whereas p50 homodimers are not. In fact, p50 homodimers in some situations are thought to serve as transcriptional repressors.[28]

The mammalian IκB family of proteins include IκBα, IκBβ, IκBε, IκBγ/p105, p100, and Bcl-3 (Fig. 6-1).[29-34] An essential feature of the IκB proteins is the presence of a domain consisting of six to seven ankyrin repeats, each approximately 33 residues. This ankyrin repeat domain is responsible for interaction with and inhibition of NF-κB. Insights into these activities have been provided by reports on the structure of the IκBα/NF-κB complex,[35,36] detailed below. IκBα, β, and ε bind preferentially to p65 and c-Rel,[29,33,34] and all contain an N-terminal domain with critical regulatory functions detailed below. At the C-terminal domains of IκBα and β reside PEST domains, rich in the amino acids proline, glutamic acid, serine, and threonine.[29,33] This domain has been implicated in the turnover of proteins with short half-lives.[38]

IκBγ is the product of an alternatively spliced message from the p105 gene. Identified in pre-B cells, it comprises the 70-kDa C-terminal fragment of p105 and preferentially binds p50 and p52 homodimers.[30] This fragment contains the ankyrin repeat domain of p105 and thus it, as well as p105 itself, is a member of the IκB family. Likewise, p100 is also a member. Bcl-3 is an IκB molecule, but its activity is distinct from other members of the family. Bcl-3 preferentially binds to either p50 or p52 homodimers[31] and has been proposed to potentiate transcription by two different mechanisms. Bcl-3 inhibits binding of p50 homodimers to DNA and thus relieves $(p50)_2$-mediated inhibition of transcription.[39] Bcl-3 can also bind to either p50 or p52 homodimers bound to DNA, and in this capacity acts as a transcriptional coactivator.[40,41] It is unclear which mechanism predominates *in vivo*. IκBβ can bind to NF-κB bound to DNA, thus potentially shielding NF-κB from IκBα that would facilitate its nuclear export (see below).[42]

IMPORTANCE OF THE UBIQUITIN-PROTEASOME PATHWAY AND PHOSPHORYLATION IN THE ACTIVATION OF NF-κB

NF-κB serves as a paradigm for a transcription factor whose activity is regulated by nuclear translocation. For the sake of simplicity, NF-κB will refer to the p50/p65 heterodimer for the remainder of this chapter. In resting cells, NF-κB is sequestered in the cytoplasm by virtue of two related mechanisms (Fig. 6-2). In one mechanism, the ankyrin repeat domain of p105 in the p105/p65 heterodimer masks NLSs in the RHDs, thereby preventing the nuclear transport of NF-κB.[32,43] In the other mechanism, after processing from p105/p65, mature p50/p65 then associates with IκB, with the ankyrin repeat domain of IκB then serving a role analogous to that in the C-terminus of p105.[8,44]

Activation of NF-κB in both mechanisms involves the proteolytic degradation of the inhibitory ankyrin repeat domain – the C-terminus of p105 in the former case, IκB in the latter – by the ubiquitin-proteasome pathway.[45-47] Ubiquitination is catalyzed by a minimum of two, and in most cases, three, enzymes. Thus, ubiquitin activating enzyme (E1) initially forms a high-energy thioester bond with ubiquitin in a reaction that requires ATP. E1 then transfers ubiquitin to ubiquitin conjugating enzyme (E2). E2, either alone or in collaboration with ubiquitin protein ligase (E3), can then catalyze the covalent attachment of ubiquitin to the ε-amino group of lysine residues on the protein substrates. Ubiquitin itself can serve as a substrate for ubiquitination; in fact, protein ubiquitination generally results in the covalent attachment of long ubiquitin chains. In general, targeting of the ubiquitination machinery to specific proteins is conferred by E2s and E3s, each of which comprises a family of proteins. The E2s involved in ubiquitination of IκB are members of the Ubc4/Ubc5 class,[48,49] which have been implicated in the ubiquitination of proteins involved in the stress response and of proteins with rapid turnover, such as the tumor suppressor p53.[50,51] Ubc4/Ubc5 has also been implicated in the ubiquitination of p105 by some groups[52] but not by others (A. Goldberg, personal communication).

The specific E3 involved in the recognition of phosphorylated IκBα (see below) has recently been identified as the F-box protein β-TrCP.[53-56] F-box proteins in general have been identified as one of the core proteins of a multicomponent ubiquitin ligase complex. The other known components are the proteins Skp1, Roc1/Rbx1, and Cdc53/Cul1, which together with an F-box protein make up an SCF ubiquitin ligase complex. The Cdc53/Cul1 protein serves as a scaffold.

It interacts with Roc1/Rbx1, and both of these proteins then interact with a specific E2. Cdc53/Cul1 simultaneously interacts with Skp1, which in turn interacts with β-TrCP through the F-box motif of the latter. β-TrCP then interacts with the specific target for the SCF complex through another protein-protein interaction motif, which in β-TrCP is a WD repeat domain. Thus, the protein β-TrCP specifically brings the phosphorylated IκBα into an SCF complex, which in turn recruits the E2 proteins Ubc4/Ubc5 to carry out the actual ubiquitination of IκBα. It is not yet known if a similar mechanism is involved in the ubiquitination of p105. Ubiquitination, in turn, targets these protein/protein domains for degradation by the 26S proteasome, an ATP-dependent multisubunit complex.[57]

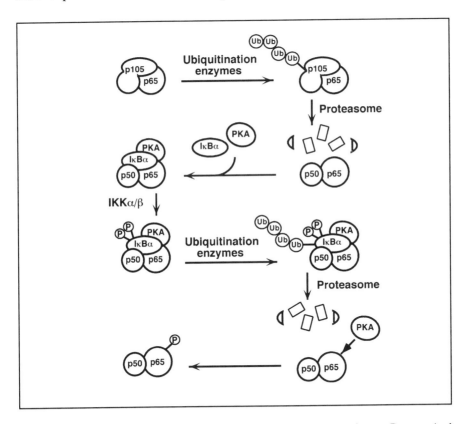

Figure 6-2. Activation of NF-κB by the ubiquitin-proteasome pathway. Prototypical NF-κB is initially synthesized as a p105/p65 heterodimer, with the C-terminal ankyrin repeat domain of p105 masking the NLS. Ubiquitination- and proteasome-mediated degradation of the C-terminus of p105 yields p50/p65, which then associates with IκBα and PKA. IκBα blocks both the NLS of NF-κB and the catalytic activity of PKA. After signal-induced phosphorylation of IκBα at ser-32 and -36, IκBα is ubiquitinated and degraded by the proteasome while still bound to NF-κB. This allows both the activation of PKA, which phosphorylates ser-276 of p65, and the nuclear translocation of NF-κB.

An important distinction between IκB degradation and p105 processing should be noted. Whereas processing of p105 to p50, which requires a glycine rich region at the C-terminus of p50,[58] is largely a constitutive process (for possible exceptions, see ref. 3), degradation of IκB is tightly regulated. Indeed, the induced degradation of IκB is the primary mechanism by which NF-κB is activated in response to a wide variety of stimuli, such as the proinflammatory cytokines TNF-α and IL-1, X irradiation, lipopolysaccharide, virus infection, oxidants, and certain phosphatase inhibitors such as okadaic acid.[3,59] Thus, in most physiologic settings, NF-κB is largely regulated by the inducible degradation of its inhibitor IκB.

Among the IκB family of proteins, IκBα has been the most intensely studied member, and thus most of the discussion to follow focuses mainly on this protein (Fig. 6-2). Numerous stimuli that activate NF-κB, such as TNF-α, phorbol myristate acetate (PMA)/ionomycin, and the HTLV-1 tax protein, have all been shown to induce phosphorylation of IκBα at ser-32 and -36.[60-63] This phosphorylation targets IκBα for ubiquitination at lys-21 and/or -22, and as mentioned previously, ubiquitination is the signal for subsequent degradation by the proteasome.[47,64,65] Thus, site-specific phosphorylation of the N-terminus of IκBα is the event that eventually results in its proteolytic degradation.

Similar events probably occur with IκBβ and ε. Thus, inducible phosphorylation occurs at the IκBβ residues (ser-19 and -23) that are homologous to ser-32 and -36 of IκBα.[63,66] The corresponding residues in IκBε (ser-18 and -22) are similarly likely to be inducibly phosphorylated.[34] It should be noted, however, that the IκBβ and ε residues (lys-9 and −6, respectively) that are homologous to lys-21/22 of IκBα may not be the preferred sites of inducible ubiquitination.[34,67] A Drosophila homologue of IκB, Cactus, which binds to the Drosophila NF-κB homologue, Dorsal, also undergoes signal-induced degradation.[68,69] This degradation is dependent on the N-terminus of Cactus, which contains potential phosphorylation sites that may be functionally homologous to ser-32 and -36 of IκBα.[68,69]

Remarkably, both IκBα and β not only associate with NF-κB but also with protein kinase A (PKA) (Fig. 6-2).[70] In this manner, IκB serves a role analogous to that of the regulatory subunit of PKA in inhibiting PKA. Thus, a quaternary complex of p50/p65: IκB:PKA exists in the cell. Upon signal-induced degradation of IκBα and β by stimuli such as LPS, active PKA then phosphorylates ser-276 in the RHD of the p65 subunit of NF-κB, thereby potentiating its transcriptional activity by promoting its interaction with the coactivator CBP/p300.[21] PKA inhibitors block NF-κB activation. Conversely, proteasome inhibitors block LPS-dependent but not cAMP-dependent PKA activation. It remains to be seen what fraction of the total cytoplasmic pool of NF-κB: IκB is associated with PKA and whether this mechanism more generally applies to other NF-κB-inducing stimuli, such as exposure to TNF-α. It also remains to be seen whether other NF-κB subunits with potential PKA phosphorylation sites, such as RelB, c-rel, and p50/p105, are activated by the same mechanism.

A distinctive feature of both the processing of p105 and the degradation of IκBα is their selectivity. In the case of p105, selective proteolysis of its C-terminal ankyrin repeat domain occurs.[15] Both p65 and the N-terminal portion of p105 containing the p50 RHD domain are spared. In the case of IκBα, phosphorylation, ubiquitination, and proteolysis all occur while it is bound to NF-κB;[47,71-73] p50/p65 and PKA are spared from proteolysis, and NF-κB is then free to translocate into the nucleus. This phenomenon of subunit-specific destruction has been elegantly demonstrated in model substrates (β-galactosidase),[74] and has been observed with other natural substrates such as cyclin/CDK complexes. However, the partial proteolysis of p105 to p50 by the proteasome is a novel finding, and its mechanism remains unresolved. The proteasome could function as an exoprotease, progressively digesting p105 from its C-terminus to eventually generate p50. Alternatively, an endoprotease activity associated with the proteasome may cleave off the C-terminus of p105, which is then rapidly degraded.

Shortly after the degradation of IκBα and the subsequent activation of NF-κB, transcription of the IκBα gene is rapidly induced and new IκBα protein is synthesized.[75,76] This upregulation of IκBα synthesis is due in large part to the presence of functional NF-κB binding sites in the promoter of the IκBα gene. Newly synthesized IκBα then enters the nucleus, displaces NF-κB from its target genes, and then transports the NF-κB: IκBα complex back to the cytoplasm, thus returning the cells to a quiescent state.[77,78] This autoregulatory loop, which may apply to IκBε but not IκBβ,[33,34] ensures the rapid and transient nature of NF-κB activation by most stimuli, such as TNF-α. Other stimuli, such as LPS and IL-1β, lead to a persistent activation of NF-κB. In this regard, it should be noted that the induced degradation of IκBβ and ε appears to occur with slower kinetics than that of IκBα;[33,34,63] thus, the degree of persistence of NF-κB activation may be related to the efficiency with which various stimuli induce degradation of the various IκB proteins.

Further insight into the regulatory functions of the IκB molecules was recently provided by two reports on the co-crystallization of an IκBα/NF-κB complex using truncated versions of these proteins.[35,36] Both reports find a similar arrangement of the six ankyrin repeats of IκBα, each composed of a β loop followed by two antiparallel α-helices, which form stacked layers with a slight overall twist. A large contact area, comprised of a series of discrete patches, is formed between repeats 3-5 of IκBα and the dimerization domain of the p50/p65 heterodimer. Such an interface allows for the independent development of interactions between these contacts, and may account for the different specificity of the various IκB proteins for various rel family members.

Additional interactions occur between ankyrin repeats 1-2 and the NLS of p65. This interaction imposes an α-helical structure on the NLS that is distinctly different from the extended chain conformation found in a related NLS when bound to the nuclear importin karyopherin-α, or the disordered structure found in this region in the structures of NF-κB when either bound to DNA or free in solution. Thus, the NLS of p65 seems to have no intrinsic secondary structure on its own, but adopts specific conformations dependent on its interacting proteins. This also demonstrates how the NLS is sequestered in the IκBα /NF-κB complex, both by forming specific interactions with IκBα, and by adopting a structure that is incompatible with recognition by a nuclear importin.

The structure of the p50/p65 dimerization domain in the IκBα/NF-κB complex, composed of the C-terminal portions of their RHDs, is virtually identical to that found in the DNA-bound forms of p50/p65. However, the N-terminal portion of the p65 RHD, while exhibiting similar structure to the DNA-bound form, is rotated almost 180° with respect to the rest of the heterodimer, such that NF-κB is unable to bind DNA.[35] Furthermore, it appears that the C-terminus of full length IκBα would extend into the cleft between p50 and p65 and thereby sterically hinder NF-κB from interacting with DNA as well.

Thus, the information provided by these structures elucidates mechanisms by which IκBα carries out its regulatory functions, namely the binding of NF-κB, the inhibition of nuclear translocation, and the inability of the NF-κB/ IκBα complex to bind DNA. Furthermore, the amino terminus of IκBα is positioned such that the regulatory serines-32 and -36 and lysines-21 and -22 contained within this region are readily accessible. The putative nuclear export signal present in the carboxy terminus of IκBα is also exposed in this structure, allowing for the clearance of nuclear NF-κB/ IκBα complexes necessary for the transient induction of genes by NF-κB. Overall, these structures demonstrate the elegant and complex manner in which the IκB proteins control activation by the NF-κB/rel family of proteins.

The inducible N-terminal phosphorylation and subsequent degradation of IκBα is the primary mechanism by which NF-κB is regulated. One notable exception to this is seen in mature B cells, in which NF-κB is present as a constitutive nuclear activity.[79] It has been suggested that the basal turnover of IκBα is accelerated in these cells, which may in part account for this.[80] Another notable exception is seen in the setting of certain select situations, such as reoxygenation after hypoxia treatment of T cells. In this case, IκBα is tyrosine phosphorylated in a manner dependent on the T cell tyrosine kinase Lck.[81] IκBα subsequently dissociates from the NF-κB: IκBα complex but is not degraded. NF-κB is thus activated in a manner independent of proteolytic degradation. Tyr-42 of IκBα has been suggested as the phosphoacceptor site in this situation; it remains to be seen whether mutation of this site inhibits reoxygenation-induced NF-κB activation. It should be noted that mutation of this site does not block

TNF-α-induced degradation of IκBα *in vivo*, unlike the ser-32/36 to ala mutation;[60,61] thus, this mechanism does not appear to be operative in TNF-α-induced activation. It may also be noted that tyr-42 of IκBα does not appear to be conserved in IκBβ or ε; thus, if this mechanism also applies to IκBβ or ε, different phosphoacceptor sites would presumably be employed.

It should also be recognized that a basal level of IκBα phosphorylation exists.[2] The sites of basal phosphorylation reside in the PEST domain of IκBα, and one enzyme responsible for this phosphorylation has been identified as casein kinase II.[82-87] Sites of phosphorylation that have been identified include ser-283, -288, and -293, and thr-291 and -299. Mutations of various combinations of these residues modestly prolong the half-life of IκBα *in vivo*, indicating that these residues play a role in the basal turnover of IκBα.[84,85] However, these mutations, as well as deletions of the entire PEST region, do not prevent signal-induced phosphorylation and degradation of IκBα.[61,85,87,88] Thus, phosphorylation of the PEST domain appears to play a role mainly in the constitutive, basal, turnover of IκBα. That being said, evidence has also been presented that this PEST domain, although not essential for signal-induced degradation of IκBα, might be important for optimal kinetics of the process.[89] Other kinases may also participate in the C-terminal phosphorylation of IκBα. For example, kinases distinct from casein kinase II have been identified and partially characterized in endothelial cells that phosphorylate the C-terminus of IκBα, and these kinases are inducible by TNF-α, IL-1β, and LPS.[90]

IκB KINASE COMPLEX(ES) AS A CRITICAL REGULATOR OF NF-κB ACTIVATION

Once the degradation of IκBα was found to be dependent on the inducible phosphorylation of the critical serines in its N-terminus, a central area of study became the identification of the kinase or kinases responsible for this modification. The first major advance was the identification of a 700 kDa kinase complex that was able to specifically phosphorylate ser-32 and ser-36 of IκBα.[49] Subsequently, this same high molecular weight complex was isolated from TNF-α-treated HeLa cells and found to contain two kinases, designated IκB kinase-α (IKKα) and IκB kinase-β (IKKβ) (also known as IKK1 and IKK2).[91-93] The cDNA for IKKα was cloned independently in a two-hybrid screen, as encoding a protein that interacted with the NF-κB inducing kinase (NIK), which had previously been implicated in the NF-κB signaling pathway.[94] This same group also isolated the cDNA for IKKβ after identifying it as an expressed sequence tag (EST) homologous to IKKα.[95] Both protein kinases possess an N-terminal kinase domain, a central leucine zipper dimerization domain, and a C-terminal helix-loop-helix domain. IKKα and IKKβ preferentially form heterodimers, though they are capable of forming homodimers as well.[92,93,95] Dominant negative versions of IKKα and IKKβ, in which a critical lysine in the ATP-binding domain is replaced with an alanine, were each shown to block induction of an NF-κB reporter gene by TNF-α and IL-1, although dominant negative IKKβ (dnIKKβ) was found to be a more potent inhibitor than dnIKKα.[93-95] Expression of antisense IKKα transcripts in transient transfection experiments blocked these inducers in similar assays, and dominant negative versions of both kinases prevented the nuclear translocation of NF-κB in response to TNF-α induction.[92] Furthermore, immunoprecipitation of the endogenous complex demonstrated that it is activated in response to TNF-α, IL-1, and PMA.[93] Thus, a number of lines of evidence indicate that these kinases are involved in the activation of NF-κB by various inducers. Immunoprecipitations of overexpressed IKKα and IKKβ were shown to phosphorylate both key serines of IκBα and IκBβ.[91-95] The ability of IKKα and IKKβ to directly phosphorylate these residues was subsequently confirmed using purified, recombinant protein kinases.[96,97] Thus, IKKα and IKKβ are bona fide IκB kinases, responsive to known inducers of NF-κB.

Given that IKKα and IKKβ can phosphorylate both critical serines in the target IκB molecules, and that dominant negative versions of either one can inhibit the various inducers of NF-κB, it was originally speculated that the two kinases might play redundant roles. Recent studies of mice in which either gene has been disrupted by homologous recombination indicate that this is not the case.[98-102] The IKKβ knockout mice died before birth, primarily due to severe liver degeneration,[100,102] apparently a result of increased sensitivity to TNF-α-mediated apoptosis of these cells. A similar phenotype had been previously observed in p65 knockout mice,[103] presumably due to a loss of the NF-κB signaling pathway that activates anti-apoptotic programs and thereby protects cells from TNF-α-induced apoptosis. Significantly, double knockout mice in which both the IKKβ and TNF-α Receptor I (TNFRI) genes were disrupted developed normally through birth.[100] The lack of IKKβ also resulted in almost total loss of inducibility by TNF-α or IL-1, directly implicating this IKK in the activation of NF-κB by proinflammatory stimuli. Consistent with this possibility, two groups found no impairment of NF-κB inducibility in response to these cytokines in IKKα-deficient embryonic fibroblasts or thymocytes, indicating that IKKα is not required for these pathways.[98,101] Rather, these mice displayed multiple developmental defects, primarily with limb and skeletal patterning and proliferation, and epidermal keratinocyte differentiation, thereby implicating a role for IKKα in these events. Thus, although IKKα and IKKβ seemed to have redundant physiological functions based on transfection experiments with dominant negative versions of these protein kinases, they appear to actually have distinct roles *in vivo*. Note, however, that a third group analyzing IKKα knockout mice did see diminished induciblity of NF-κB in response to TNF-α and IL-1 treatment.[99] This suggests that although IKKβ clearly is primarily responsible for activation of NF-κB in response to these inducers, IKKα might still contribute to the overall magnitude of this response. Similarly, one of the groups analyzing the IKKβ knockout mice did find some residual, albeit severely impaired, NF-κB response to TNF-α and IL-1 in embryonic fibroblasts.[102] All groups do agree, however, that IKKα and IKKβ have distinct roles, and cannot effectively substitute for each other.

Another component of the IκB kinase complex was identified via genetic complementation of mutant cell lines that were no longer activatable by inducers of NF-κB,[104] and was thus designated the NF-κB essential modulator (NEMO). Using antibodies against this protein, NEMO was shown to be an integral part of the 700 kDa high molecular weight kinase complex containing IKKα and IKKβ. Interestingly, the kinase complex containing IKKα/β isolated from cell lines deficient in NEMO exhibited a smaller size of 300-450 kDa, further demonstrating that this protein is indeed part of the endogenous complex. Additional purification of the high molecular weight IκB kinase complex by other groups also led to the identification of NEMO, alternatively designated IKKγ[105] or IKKAP1.[106] NEMO was independently isolated as adenovirus E3 protein-interacting protein and designated FIP-3.[107] NEMO/IKKγ interacts directly with IKKβ, but not IKKα, and is also capable of self-oligomerization.[104-106] NEMO/IKKγ is not itself a protein kinase, but does possess several protein-protein interaction motifs, suggesting a structural role within the complex, perhaps mediating interactions with upstream activators.

IKKα and IKKβ are protein kinases that themselves are regulated by phosphorylation. This was first indicated by the fact that purified recombinant MEKK1 (mitogen activated protein kinase/extracellular signal regulated protein kinase kinase kinase-1), a member of the mitogen activated protein kinase kinase kinase (MAP3K) family, was able to directly activate the high molecular weight IκB-kinase *in vitro*.[108] This activated complex, in turn, could be inactivated by subsequent treatment with calf intestinal alkaline phosphatase.[108] Subsequent work demonstrated that the activity of the *in vivo*-activated IKK complex isolated from TNF-α-treated cells was diminished when treated *in vitro* with protein phosphatase 2A (PP2A).[91] Once the cDNAs for these kinases were cloned, it was noted that both IKKα and β contain a canonical MAPKK activation loop (SxxxS).[93,109,110] Substitution of these serines with alanine residues inactivated both kinases, while phosphomimetic glutamic acid substitutions at these positions in IKKβ resulted in a constitutively active kinase.[93,109,106,110] Thus, significant evidence supports the idea that phosphorylation of one or both of the serines in the activation loop of IKKα and IKKβ is responsible for their activation.

Two distinct MAP3Ks were initially implicated as upstream activators of the IKKs. As mentioned above, purified recombinant MEKK1 was shown to directly activate the high molecular weight IκB-kinase *in vitro*.[108] Furthermore, MEKK1 induces the site-specific phosphorylation of IκBα in transient transfection experiments, and a dominant negative version of MEKK1 partially suppresses activation of an NF-κB reporter gene by TNF-α.[108,111] Another MAP3K, NF-κB inducing kinase (NIK),[112] has also been implicated in the activation of IKKα and β. NIK was originally identified in a two hybrid screen that employed as a bait TRAF2, a component of the TNF-α receptor complex. A dominant negative version of NIK has been shown to strongly inhibit the induction of an NF-κB reporter gene. Both MEKK1 and NIK can activate NF-κB reporters *in vivo*, as well as increase the kinase activity of cotransfected IKKα and β.[94-96,109,113,114] Both MEKK1 and NIK, when isolated from transfected cells, were shown to phosphorylate one or both of these serines.[109,110,115] Further support was provided by the demonstration that purified recombinant MEKK1 can phosphorylate a synthetic peptide corresponding to the activation loop of IKKβ *in vitro*.[96] Finally, a recent report examined the effect of MEKK1 and NIK on the activity of IKKα and IKKβ *in vivo*, using low expression levels such that both of the IKKs were incorporated into the endogenous high molecular weight complex.[110] Under these conditions, the activity of IKKα and of IKKβ was stimulated with the physiological inducers TNF-α and IL-1. Comparison of wild type versions of these kinases with mutants, in which either one or both serines in the activation loop were replaced with alanines, revealed that only mutations in IKKβ caused a decrease in the inducible kinase activity of the transfected protein, while the corresponding mutations in IKKα had no effect. Thus, *in vivo*, it appears that phosphorylation of the activation loop of IKKβ is the critical, ultimate regulatory event in activating the endogenous IKK complex rather than that of IKKα, in agreement with the results from the knockout mice (see above).

The evidence discussed thus far regarding a specific role of NIK or MEKK1 is insufficient to distinguish which kinase plays an actual role in activating the IKKα/β complex *in vivo*. Several groups have placed more emphasis on NIK as the true upstream kinase, since very low levels of this protein were able to strongly activate an NF-κB reporter gene in transient transfection experiments, and since dominant negative NIK (dnNIK) is a much more potent inhibitor of IKK activity than dnMEKK1 when overexpressed in cells.[109,114,116] Note, however, that although overexpression experiments are useful in determining specific pathways that may be affected by particular inducers, they are certainly not equivalent to physiological conditions. Relative potencies of transfected kinases should therefore not be interpreted as a strict reflection of their relative roles *in vivo*.[96] Hence, the evidence discussed above does not decisively assign a more credible role for NIK over MEKK1 in the activation of the IKKα/β complex.

In support of this view, recent experiments using mice in which the endogenous NIK gene has been ablated demonstrate that fibroblasts isolated from these mice show full NF-κB inducibility in response to TNF-α and IL-1, indicating that NIK is not essential for this response.[117] Rather, these mice have severe defects in the development of lymph nodes, Peyer's patches, spleen, and thymus. This is a similar, though more severe, phenotype as that observed in mice lacking the lymphotoxin β receptor (LtβR). Significantly, activation through the LtβR involves members of the TRAF family of proteins, and interactions with these proteins, in fact, led to the initial identification of NIK. Interestingly, the recent characterization of mice displaying alymphoplasia, an autosomal recessive condition characterized by the systemic absence of lymph nodes and Peyers patches, mapped the cause of this defect to a single point mutation in the noncatalytic domain of NIK.[118] Further experiments demonstrated that transgenic complementation with wild type NIK restored normal structures to the affected organs. Thus, these results imply that NIK is involved in a distinctive but certainly not universal pathway leading to NF-κB activation.

The physiological role of MEKK1 in NF-κB activation is also unclear. Mice in which the MEKK1 gene has been knocked out display no defects in NF-κB inducibility in response to TNF-α.[119,120] One possible explanation for these seemingly contradictory results comes from the demonstration that other MAP3Ks, including MEKK2,[121] MEKK3,[121] Cot/Tpl-2,[122] TAK,[123] and MLK3,[124] are able to activate the IKKs. This indicates possibly redundant roles among these kinases, allowing one to compensate for the loss of others in knockout experiments. Further experiments, including double and triple knockouts, will be required to fully investigate these relationships. Additionally, a number of protein kinases distinct from MAP3Ks have also been implicated in the activation of the IKKα/β complex, including certain isoforms of protein kinase C (PKC)[125] and Akt/protein kinase B(PKB).[126-128]

It should also be noted that the 700-kDa IKK complex can be independently activated by ubiquitination (Fig. 6-3).[49] Like the ubiquitination that targets IκBα for degradation by the proteasome, this ubiquitination requires an E1 and the Ubc4/Ubc5 subfamily of E2s. In marked contrast to the ubiquitination mediated degradation of IκBα, however, the ubiquitination-mediated activation of the 700-kDa complex does not require the proteasome. Ubiquitination therefore plays two distinct roles in the activation of NF-κB. First, it can activate the IKK complex in a proteasome-independent manner. Second, it can induce degradation of phosphorylated IκBα in a proteasome-dependent manner. Although ubiquitination-induced endocytosis of α-factor pheromone receptor Ste2 has been reported in S. cerevesaie,[129] ubiquitin-induced enzyme activation is unprecedented in mammalian cells. It is not known whether an E3 which, if present, presumably would reside in the 700-kDa complex itself, is necessary for this activation. It also is not known what the target of the ubiquitination machinery is in the 700-kDa complex. In any case, this dual mode of activation of the IKK complex suggests that the complex may be an integrator of signals leading to the activation of NF-κB.

The extensive characterization of the IKKα/β/γ complex does not necessarily imply that this complex contains the sole IκB kinases. In fact, a recent report describes the identification of a distinct IκB kinase complex that is responsive to a subset of inducers of NF-κB, namely the phorbol ester PMA and cross-linked T cell receptor.[130] An IKK-related kinase has been characterized from this complex, and designated IKKε. Though recombinant IKKε directly phosphorylates only serine 36 of IκBα, the PMA-activated endogenous IKKε complex phosphorylates both critical serine residues. Remarkably, this activity is due to the presence of a distinct kinase in this complex that is neither IKKα nor IKKβ, and which is able to phosphorylate both serine 32 and 36 when separated from the activated, endogenous complex. This associated kinase may therefore be an as yet unidentified IKK, and the physiological role for IKKε may in fact be to phosphorylate the activation loop of this novel IKK and thus stimulate its activity. Consistent with this possibility, another IKK-related kinase designated TBK1/NAK, which exhibits high sequence similarity with IKKε, has recently been identified.[131,132] Evidence thus far indicates that TBK/NAK phosphorylates and activates the IKKα/β complex.[131,132]

UPSTREAM REGULATORS OF NF-κB

NF-κB can be activated by a large number of stressful stimuli; understanding how these stimuli are transmitted to the cytoplasm is of fundamental interest. The greatest progress that has been made in this area of research is with regard to TNF-α signaling. Although a detailed discussion of TNF-α receptor-mediated signaling of NF-κB (recently reviewed in ref. 133) is beyond the scope of this chapter, a brief mention is merited. There are two forms of the receptor, TNFR1 (p55) and TNFR2 (p75), which are both ubiquitously expressed. Activation of both probably involves trimerization mediated by binding of the trimeric TNF-α protein. When liganded to TNF-α, both receptors can activate NF-κB, although it is believed that TNFR1 plays a more important role in most situations. Both receptors activate NF-κB by recruitment to the receptor of the TRAF2 protein.[134,135] It should be noted that activation of the IL-1 receptor involves a homologous protein, TRAF6.[136] In the case of TNFR2, recruitment is made directly by the cytoplasmic domain of the receptor,[134] whereas in the case of TNFR1, it is indirect and mediated by TRADD, which serves as a bridging protein between the cytoplasmic domain of the receptor and TRAF2.[135] Overexpression of TRAF2 or TRAF6 alone activates NF-κB. TRAF2 or TRAF6 binds NIK,[112] while TRAF2 binds to MEKK1.[137] However, as discussed above, it appears that neither of these MAP3Ks is an essential mediator of TNF-α activation of NF-κB.

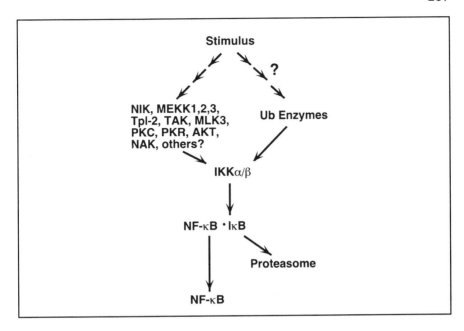

Figure 6-3. Signal-induced phosphorylation of IκBα. After exposure of cells to stressful stimuli such as TNF-α, X-irradiation, or lipopolysaccharide (LPS), selective protein kinases that include NIK, and MEKK1,2,3, among others, are activated. These protein kinases then activate IKKα and IKKβ. IKK activity can be independently activated by ubiquitination and therefore potentially serves as an integrator of phosphorylation- and ubiquitination-dependent signals. Phosphorylation of IκBα targets it for degradation by the ubiquitin-proteasome pathway, thereby allowing the nuclear translocation of NF-κB.

It remains to be seen if simple recruitment of these molecules to the TNF-α receptor complex alone is sufficient to activate the cascade or if other events are necessary. For example, RIP, a component of the TNFR1 complex, has been shown to be necessary for TNF-α-induced NF-κB activation, but its precise role in this activation remains to be defined.[138,139] In addition, it has been shown that dominant negative Rac1, Cdc42, and Rho A can inhibit NF-κB activation in response to TNF-α;[140,141] thus, these molecules may also participate in TNF-α-induced activation of NF-κB. Ceramide has also been proposed as a second messenger in TNFR-mediated activation of NF-κB;[142] its relationship to the above-mentioned molecules remains to be determined.

The HTLV-1 Tax protein activates numerous transcription factors in T cells, including CREB, ATF-2, and NF-κB.[143,144] Tax activation of NF-κB has been shown to be mediated, at least in part, by induced site-specific phosphorylation of IκBα at ser-32 and -36.[61] Two mechanisms have been proposed to account for this. In the first, Tax binds to and stimulates MEKK1, which in turn, activates NF-κB.[145]

In the second, Tax binds to NEMO/IKKγ, and either by recruitment of a yet to be identified factor or by direct stimulation, then activates IKKα/β.[146-148] Interestingly, a mutant form of Tax which is not able to activate NF-κB also exhibited a markedly decreased interaction with NEMO/IKKγ. Tax has also been shown to interact with the proteasome subunits HsN3 and HC9 and promote p105 processing[149] and thus potentially activate NF-κB in a manner distinct from the IκB pathway.

PKR (double-stranded RNA-activated kinase) is an activator of NF-κB. Antisense RNA to PKR blocks NF-κB activation in response to dsRNA but not TNF-α.[150] More tellingly, mouse embryonic fibroblasts containing germline deletions of the PKR gene display defective NF-κB responses to dsRNA but not TNF-α.[151] Thus, PKR plays a critical role in dsRNA- but not TNF-α-induced NF-κB activation. PKR has been proposed to activate NF-κB by phosphorylating IκB and inducing its dissociation from NF-κB.[152] However, because tyrosine phosphorylation is the only type of phosphorylation that induces dissociation of IκB without its degradation *in vivo*, and PKR is a serine/threonine kinase, the significance of the *in vitro* findings remains uncertain. More recent evidence indicates that PKR, in fact, activates NF-κB by activating the IKK complex.[153,154]

Raf-1 activates NF-κB in transient transfection experiments.[155] Raf-1, like PKR, has also been proposed to activate NF-κB by dissociating the NF-κB: IκB complex.[156] However, Raf-1, like PKR, is a serine/threonine kinase and for the same reasons given above is unlikely to act by this mechanism. Intriguingly, Raf-1 is a MAP3K like MEKK1. However, Raf itself does not activate IKKβ directly, but appears to do so through the activation of MEKK1.[157]

Additional studies have also implicated one or both of the IKKs in the induction of NF-κB by LPS, ionizing radiation, CD28 crosslinking, and other stimuli.[158-161] Many stimuli that activate NF-κB also generate reactive oxygen intermediates (ROIs), such as superoxide, within the cell. Indeed, oxidizing agents such as hydrogen peroxide can directly activate NF-κB.[162] Furthermore, antioxidants inhibit NF-κB activation induced not only by oxidants but also by a wide variety of stimuli, such as TNF-α, IL-1, PMA, dsRNA, and lipopolysaccharide.[162] Thus, it has been proposed that ROIs serve as second messengers in the activation of NF-κB (reviewed in ref. 163). The molecular target(s) of these ROIs within the cell has yet to be identified.

SUMMARY AND FUTURE DIRECTIONS

In spite of the intense scrutiny to which the NF-κB system has been subjected, many fundamental questions regarding it remain unanswered. One is the nature of the three-dimensional structure of IκB kinase-α/β, both alone and in complex with IκB. Knowledge of this structure should allow a detailed understanding of how this critical enzyme:substrate interaction occurs. Another is the structure of phosphorylated IκB complexed with β-TrCP, the E3 that specifically recognizes it. Information in this regard will provide insights into the mechanism by which IκB is targeted for degradation by the ubiquitin-proteasome pathway.

The recent major advances that have identified components of the signal transduction cascades leading to the activation of NF-κB raise a number of new issues. One is the assessment of the relative importance of different signaling molecules in response to different stimuli. As has been discussed, experiments based on transient overexpression of such molecules may not necessarily allow definitive conclusions to be drawn. It is anticipated that studies on mice containing germline deletions of single, and possibly multiple, genes encoding these molecules might therefore be necessary. Yet another issue is a detailed understanding of the molecular mechanisms that initiate these signals, many of which probably occur at the cell membrane.

It is anticipated that additional features involved in the processing of p105 and the degradation of IκB by the ubiquitin-proteasome pathway will be identified – for example, specific ubiquitination enzymes such as E3s involved in p105 processing and IKK complex activation. The detailed mechanism by which the IKK complex might serve as a signal integrator, responding to both phosphorylation- and ubiquitination-dependent signals, remains to be determined. Furthermore, the identification of IKK-α and IKK-β raises the question of whether other substrates, potentially outside of the NF-κB pathway, exist. Last but not least, there is the promise of future pharmacologic inhibitors of the pathway, which should both promote mechanistic studies and open the door to novel therapeutic approaches in the treatment of inflammatory and immune diseases.

REFERENCES

1. Thanos D, Maniatis T. NF-κB: a lesson in family values. Cell 1995; 80:529-532.
2. Verma IM, Stevenson JK, Schwarz EM et al. Rel/NF-κB/IκB Family: intimate tales of association and dissociation. Genes Dev 1995; 9:2723-2735.
3. Baldwin AS. The NF-κB and IκB proteins: new discoveries and insights. Annu Rev Immunol 1996; 14:649-681.
4. Baeuerle PA, Baltimore D. NF-κB: ten years after. Cell 1996; 87:13-20.
5. Kieran M, Blank V, Logeat F et al. The DNA binding subunit of NF-κB is identical to factor KBF1 and homologous to the rel oncogene product. Cell 1990; 62:1007-1018.
6. Ghosh S, Gifford AM, Riviere LR et al. Cloning of the p50 DNA binding subunit of NF-κB: homology to rel and dorsal . Cell 1990; 62:1019-1029.
7. Bours V, Villalobos J, Burd PR et al. Cloning of a mitogen-inducible gene encoding a κB DNA-binding protein with homology to the rel oncogene and to cell-cycle motifs. Nature 1990; 348:76-80.

8.   Nolan GP, Ghosh S, Liou H-C et al. DNA binding and IκB inhibition of the cloned p65 subunit of NF-κB, a rel-related polypeptide. Cell 1991; 64:961-969.

9.   Ruben SM, Dillon PJ, Schreck R et al. Isolation of a rel-related human cDNA that potentially encodes the 65-kd subunit of NF-κB. Science 1991; 251:1490-1493.

10.   Schmid RM, Perkins ND, Duckett CS et al. Cloning of an NF-κB subunit which stimulates HIV transcription in synergy with p65. Nature 1991; 352:733-736.

11.   Neri A, Chang C-C, Lombardi L et al. B cell lymphoma-associated chromosomal translocation involves candidate oncogene lyt-10, homologous to NF-κB p50. Cell 1991; 67:1075-1087.

12.   Gilmore TD, Temin HM. v-rel oncoproteins in the nucleus and in the cytoplasm transform chicken spleen cells. J Virol 1988; 62:703-714.

13.   Ghosh G, Van Duyne G, Ghosh S et al. Structure of NF-κB p50 homodimer bound to a κB site. Nature 1995; 373:303-310.

14.   CW, Rey FA, Sodeoka M et al. Structure of the NF-κB p50 homodimer bound to DNA. Nature 1995; 373:311-317.

15.   Fan C-M, Maniatis T. Generation of p50 subunit of NF-κB by processing of p105 through an ATP-dependent pathway. Nature 1991; 354:395-398.

16.   Grimm S, Baeuerle PA. The inducible transcription factor NF-κB: structure-function relationship of its protein subunits. Biochem J 1993; 290:297-308.

17.   Schmitz ML, Baeuerle PA. The p65 subunit is responsible for the strong transcription activating potential of NF-κB. EMBO J 1991; 10:3805-3817.

18.   Ballard DW, Dixon EP, Peffer NJ et al. The 65-kDa subunit of human NF-κB functions as a potent transcriptional activator and a target for v-Rel-mediated repression. Proc Natl Acad Sci U S A 1992; 89:1875-1879.

19.   Perkins ND, Felzien LK, Betts JC et al. Regulation of NF-κB by cyclin-dependent kinases associated with the p300 coactivator. Science 1997; 275:523-527.

20.   Gerritsen M, Williams AJ, Neish AS et al. CREB-binding protein/p300 are transcriptional coactivators of p65. Proc Natl Acad Sci U S A 1997; 94:2927-2932.

21.   Zhong H, Voll RE, Ghosh S. Phosphorylation of NF-κB p65 by PKA stimulates transcriptional activity by promoting a novel bivalent interaction with the coactivator CBP/p300. Mol Cell 1998; 1:661-671.

22.   Ogryzko VV, Schiltz RL, Russanova V et al. The transcriptional coactivators p300 and CBP are histone acetyltransferases. Cell 1996; 87:95-104.

23.   Bannister AJ, Kouzarides N. The CBP co-activator is a histone acetyltransferase. Nature 1996; 384:641-643.

24.   Kee B, Arias J, Montminy MR. Adaptor-mediated recruitment of RNA polymerase II to a signal-dependent activator. J Biol Chem 1996; 271:2373-2375.

25.   Kerr LD, Ransone LJ, Wamsley P et al. Association between proto-oncoprotein Rel and TATA-binding protein mediates transcriptional activation by NF-κB. Nature 1993; 365:412-419.

26.   Xu X, Prorock C, Ishikawa H et al. Functional interaction of the v-Rel and C-Rel oncoproteins with the TATA-binding protein and association with transcription factor IIB. Mol Cell Biol 1993; 13:6733-6741.

27.   Schmitz ML, Stelzer G, Altmann H et al. Interaction of the COOH-terminal transactivation domain of p65 NF-κB with TATA-binding protein, transcription factor IIB, and coactivators. J Biol Chem 1995; 270:7219-7226.

28.   Sha WC, Liou HC, Tuomanen EI et al. Targeted disruption of the p50 subunit of NF-κB leads to multifocal defects in immune responses. Cell 1995; 80:321-330.

29. Haskill S, Beg AA, Tompkins SM et al. Characterization of an immediate-early gene induced in adherent monocytes that encodes IκB-like activity. Cell 1991; 65:1281-1289.

30. Inoue J-I, Kerr LD, Kakizuka A et al. IκBγ, a 70 kd protein identical to the C-terminal half of p110 NF-κB: a new member of the IκB family. Cell 1992; 68:1109-1120.

31. Wulczyn FG, Naumann M, Scheidereit C. Candidate proto-oncogene bcl-3 encodes a subunit-specific inhibitor of transcription factor NF-κB. Nature 1992; 358:597-599.

32. Rice NR, MacKichan ML, Israel A. The precursor of NF-κB p50 has IκB-like functions. Cell 1992; 71:243-253.

33. Thompson JE, Phillips RJ, Erdjument-Bromage H et al. IκB-β regulates the persistent response in a biphasic activation of NF-κB. Cell 1995; 80:573-582.

34. Whiteside ST, Epinat J-C, Rice NR et al. IκBε, a novel member of the IκB family, controls RelA and cRel NF-κB activity. EMBO J 1997; 16;1413-1426.

35. Huxford T, Huang DB, Malek S et al. The crystal structure of the IκBα/NF-κB complex reveals mechanisms of NF-κB inactivation. Cell 1998; 95:759-770.

36. Jacobs MD, Harrison SC. Structure of an IκBα/NF-κB complex. Cell 1998; 95:749-758.

37. Gorina S, Pavletich NP. Structure of the p53 tumor suppressor bound to the ankyrin and SH3 domains of 53BP2. Science 1996; 274:1001-1005.

38. Rogers S, Wells R, Reichsteiner M. Amino acid sequences common to rapidly degraded proteins: the PEST hypothesis. Science 1986; 234:364-368.

39. Franzoso G, Bours V, Park S et al. The candidate oncoprotein Bcl-3 is an antagonist of p50/ NF-κB-mediated inhibition. Nature 1992; 359:339-342.

40. Bours V, Franszoso G, Azarenko V et al. The oncoprotein Bcl-3 directly transactivates through κB motifs via association with DNA-binding p50B homodimers. Cell 1993; 72:729-739.

41. Fujita T, Nolan GP, Liou H-C et al. The candidate proto-oncogene bcl-3 encodes a transcriptional coactivator that activates through NF-κB p50 homodimers. Genes Dev 1993; 7:1354-1363.

42. Phillips RJ, Ghosh S. Regulation of IκBβ in WEHI 231 mature B cells. Mol Cell Biol 1997; 17:4390-4396.

43. Henkel T, Zabel U, van Zee K et al. Intramolecular masking of the nuclear location signal and dimerization domain in the precursor for the p50 NF-κB subunit. Cell 1992; 68:1121-1133.

44. Baeuerle PA, Baltimore D. Activation of DNA-binding activity in an apparently cytoplasmic precursor of the NF-κB transcription factor. Cell 1988; 53:211-217.

45. Palombella VJ, Rando OJ, Goldberg AL et al. The ubiquitin-proteasome pathway is required for processing the NF-κB1 precursor protein and the activation of NF-κB. Cell 1994; 78:773-785.

46. Traenckner EB-M, Wilk S, Baeuerle PA. A proteasome inhibitor prevents activation of NF-κB and stabilizes a newly phosphorylated form of IκB-α that is still bound to NF-κB. EMBO J 1994; 13:5433-5441.

47. Chen ZJ, Hagler J, Palombella V et al. Signal-induced site-specific phosphorylation of IκB-α targets IκB-α to the ubiquitin-proteasome pathway. Genes Dev 1995; 9:1586-1597.

48. Alkalay I, Yaron A, Hatzubai A et al. Stimulation-dependent IκBα phosphorylation marks the NF-κB inhibitor for degradation via the ubiquitin-proteasome pathway. Proc Natl Acad Sci U S A 1995; 92:10599-10603.

49.   Chen ZJ, Parent L, Maniatis T. Site-specific phosphorylation of IκB-α by a novel ubiquitin-dependent protein kinase activity. Cell 1996; 84:853-862.

50.   Scheffner M, Huibregtse JM, Howley PM. Identification of a human ubiquitin-conjugating enzyme that mediates the E6-AP-dependent ubiquitination of p53. Proc Natl Acad Sci U S A 1994; 91:8797-8801.

51.   Rolfe M, Beer-Romero P, Glass S et al. Reconstitution of p53-ubiquitination reactions from purified components: the role of human ubiquitin-conjugating enzyme UBC4 and E6-associated protein (E6AP). Proc Natl Acad Sci U S A 1995; 92:3264-3268.

52.   Orian A, Whiteside S, Israel A et al. Ubiquitin-mediated processing of NF-κB transcriptional activator precursor p105. J Biol Chem 1995; 270:21707-21714.

53.   Yaron, A, Hatzubai, A, Davis, M et al. Identification of the receptor component of the IκBα-ubiquitin ligase. Nature 1998; 396:590-594.

54.   Spencer E, Jiang J, Chen ZJ. Signal-induced ubiquitination of IκBα by the F-box protein Slimb/β-TrCP. Genes Dev 1999; 13:284-294.

55.   Winston JT, Strack P, Beer-Romero P et al. The SCF β-TRCP-ubiquitin ligase complex associates specifically with phosphorylated destruction motifs in IκBα and β-catenin and stimulates IκBα ubiquitination in vitro. Genes Dev 1999; 13:270-283.

56.   Maniatis T. A ubiquitin ligase complex essential for the NF-κB, Wnt/Wingless, and Hedgehog signaling pathways. Genes Dev 1999; 13:505-510.

57.   Hochstrasser M. Ubiquitin, proteasomes, and the regulation of intracellular protein degradation. Curr Opin Cell Biol 1995; 7:215-223.

58.   Lin L, Ghosh S. A glycine rich region in NF-κB p105 functions as a processing signal for the generation of the p50 subunit. Mol Cell Biol 1996; 16:2248-2254.

59.   Thevenin C, Kim SJ, Rieckmann P et al. Induction of nuclear factor-κB and the human immunodeficiency virus long terminal repeat by okadaic acid, a specific inhibitor of phosphatases 1 and 2A. New Biol 1990; 2:793-800.

60.   Brown K, Gerstberger S, Carlson L et al. Control of IκB-α proteolysis by site-specific, signal-induced phosphorylation. Science 1995; 267:1485-1491.

61.   Brockman JA, Scherer DC, Hall SM et al. Coupling of signal-response domain in IκB-α to multiple pathways for NF-κB activation. Mol Cell Biol 1995; 15:2809-2818.

62.   Traenckner EB-M, Pahl HL, Henkel T et al. Phosphorylation of human IκB-α on serines 32 and 36 controls IκB-α proteolysis and NF-κB activation in response to diverse stimuli. EMBO J 1995; 14:2876-2883.

63.   DiDonato J, Mercurio F, Rosette C et al. Mapping of the inducible IκB phosphorylation sites that signal its ubiqutination and degradation. Mol Cell Biol 1996; 16:1295-1304.

64.   Scherer DC, Brockman JA, Chen ZJ et al. Signal-induced degradation of IκBα requires site-specific ubiquination. Proc Natl Acad Sci U S A 1995; 92:11259-11263.

65.   Baldi L, Brown K, Franzoso G et al. Critical role for lysines 21 and 22 in signal-induced, ubiquitin-mediated proteolysis of IκB-α. J Biol Chem 1996; 271:376-379.

66.   McKinsey TA, Brockman JA, Scherer DC et al. Inactivation of IκBβ by the tax protein of human T-cell leukemia virus type 1: a potential mechanism for constitutive activation of NF-κB. Mol Cell Biol 1996; 16:2083-2090.

67.   Weil R, Laurent-Winter C, Israel A. Regulation of IκBβ degradation. J Biol Chem 1997; 272:9942-9949.

68.   Reach M, Galindo RL, Towb P et al. A gradient of Cactus protein degradation

establishes dorsoventral polarity in the Drosophila embryo. Dev Biol 1996; 180:353-364.

69. Bergmann A, Stein D, Geisler R et al. A gradient of cytoplasmic Cactus degradation establishes the nuclear localization gradient of the dorsal morphogen in Drosophila. Mech Dev 1996; 60:109-123.

70. Zhong H, Yang HS, Erdjument-Bromage H et al. The transcriptional activity of NF-κB is regulated by IκB-associated PKA catalytic subunit through a cyclic AMP-independent mechanism. Cell 1997; 89:413-424.

71. Finco TS, Beg AA, Baldwin AS Jr. Inducible phosphorylation of IκBα is not sufficient for its dissociation from NF-κB and is inhibited by protease inhibitors. Proc Natl Acad Sci U S A 1994; 91:11884-11888.

72. Alkalay I, Yaron A, Hatzubai A et al. *In vivo* stimulation of IκB phsophorylation is not sufficient to activate NF-κB. Mol Cell Biol 1995; 15:1294-1301.

73. DiDonato JA, Mercurio F, Karin M. Phosphorylation of IκBα precedes but is not sufficient for its dissociation from NF-κB. Mol Cell Biol 1995; 15:1302-1311.

74. Johnson ES, Gonda DK, Varshavsky A. Cis-trans recognition and subunit-specific degradation of short-lived proteins. Nature 1990; 346:287-291.

75. Sun S-C, Ganchi PA, Ballard D et al. NF-κB controls expression of inhibitor IκB: evidence for an inducible autoregulatory pathway. Science 1993; 259:1912-1915.

76. Chiao PJ, Miyamoto S, Verma IM. Autoregulation of IκBα activity. Proc Natl Acad Sci U S A 1994; 91:22-32.

77. Zabel U, Baeuerle PA. Purified human IκB can rapidly dissociate the complex of the NF-κB transcription factor with its cognate DNA. Cell 1990; 61:255-265.

78. Arenzana-Seisdedos F, Thompson J, Rodriguez MS et al. Inducible nuclear expression of newly synthesized IκBα negatively regulates DNA-binding and transcriptional activities of NF-κB. Mol Cell Biol 1995; 15:2689-2696.

79. Sen R, Baltimore D. Multiple nuclear factors interact with the immunoglobulin enhancer sequences. Cell 1986; 46:705-716.

80. Miyamoto S, Chiao PJ, Verma IM. Enhanced IκBα degradation is responsible for constitutive NF-κB activity in mature murine B-cell lines. Mol Cell Biol 1994; 14:3276-3282.

81. Imbert V, Rupec RA, Livolsi A et al. Tyrosine phosphorylation of IκB-α activates NF-κB without proteolytic degradation of IκB-α. Cell 1996; 86:787-798.

82. Barroga CF, Stevenson JK, Schwarz, EM et al. Constitutive phosphorylation of IκB-α by casein kinase II. Proc Natl Acad Sci U S A 1995; 92:7637-7641.

83. Kuno K, Ishikawa Y, Ernst MK et al. Identification of an IκBα-associated protein kinase in a human monocytic cell line and determination of its phosphorylation sites on IκBα. J Biol Chem 1995; 270:27914-27919.

84. McElhinny JA, Trushin SA, Bren GD et al. Casein kinase II phosphorylates IκBα at S-283, S289, S-293, and T-291 and is required for its degradation. Mol Cell Biol 1996; 16:899-906.

85. Lin R, Beauparlant P, Makris C et al. Phosphorylation of IκBα in the C-terminal PEST domain by casein kinase II affects intrinsic protein stability. Mol Cell Biol 1996; 16:1401-1409.

86. Schwarz EM, Van Antwerp DV, Verma IM. Constitutive phosphorylation of IκBα by casein kinase II occurs preferentially at serine 293: requirement for degradation of free IκBα. Mol Cell Biol 1996; 16:3554-3559.

224

87.   Van Antwerp DJ, Verma IM. Signal-induced degradation of IκBα: association with NF-κB and the PEST sequence in IκBα are not required. Mol Cell Biol 1996; 16:6037-6045.

88.   Aoki T, Sano Y, Ymamoto T et al. The ankyrin repeats but not the PEST-like sequences are required for signal-dependent degradation of IκBα. Oncogene 1996; 12:1159-1164.

89.   Brown K, Franzoso G, Baldi L et al. The signal response of IκBα is regulated by transferable N- and C-terminal domains. Mol Cell Biol 1997; 17:3021-3027.

90.   Bennett BL, Lacson RG, Chen CC et al. Identification of signal-induced IκB-α kinases in human endothelial cells. J Biol Chem 1996; 271:19680-19688.

91.   DiDonato JA, Hayakawa M, Rothwarf DM et al. A cytokine-responsive IκB kinase that activates the transcription factor NF-κB. Nature 1997; 388:548-554.

92.   Zandi E, Rothwarf DM, Delhase M et al. The IκB kinase complex (IKK-) contains two kinase subunits, IKK-α and IKK-β, necessary for IκB phosphorylation and NF-κB activation. Cell 1997; 91:243-252.

93.   Mercurio F, Zhu H, Murray BW et al. IKK--1 and IKK--2: cytokine-activated IκB kinases essential for NF-κB activation. Science 1997; 278:860-866.

94.   Regnier CH, Song HY, Gao X et al. Identification and characterization of an IκB kinase. Cell 1997; 90:373-383.

95.   Woronicz JD, Gao X, Cao Z et al. IκB kinase-β: NF-κB activation and complex formation with IκB kinase-α and NIK. Science 1997; 278:866-869.

96.   Lee FS, Peters RT, Dang LC et al. MEKK1 activates both IκB kinase α and IκB kinase β. Proc Natl Acad Sci U S A 1998; 95:9319-9324.

97.   Zandi E, Chen Y, Karin M. Direct Phosphorylation of IκB by IKKα and IKKβ: Discrimination Between Free and NF-κB-Bound Substrate. Science 1998; 281:1360-1363.

98.   Hu Y, Baud V, Delhase M et al. Abnormal morphogenesis but intact IKK activation in mice lacking the IKKα subunit of IκB kinase. Science 1999; 284:316-320.

99.   Li Q, Lu Q, Hwang, JY et al. IKK1-deficient mice exhibit abnormal development of skin and skeleton. Genes Dev 1999; 13:1322-1328.

100.   Li Q, Van Antwerp D, Mercurio F et al. Severe liver degeneration in mice lacking the IκB kinase 2 gene. Science 1999; 284:321-325.

101.   Takeda K, Takeuchi O, Tsujimura T et al. Limb and skin abnormalities in mice lacking IKKα. Science 1999; 284:313-316.

102.   Tanaka M, Fuentes ME, Yamaguchi K et al. Embryonic lethality, liver degeneration, and impaired NF-κB activation in IKK-β-deficient mice. Immunity 1999; 10:421-429.

103.   Beg AA, Sha WC, Bronson RT et al. Embryonic lethality and liver degeneration in mice lacking the RelA component of NF-κB. Nature 1995; 376:167-70.

104.   Yamaoka S, Courtois G, Bessia C et al. Complementation cloning of NEMO, a component of the IκB kinase complex essential for NF-κB activation. Cell 1998; 93:1231-1240.

105.   Rothwarf DM, Zandi E, Natoli G et al. IKK-γ is an essential regulatory subunit of the IκB kinase complex. Nature 1998; 395:297-300.

106.   Mercurio F, Murray BW, Shevchenko A et al. IκB kinase (IKK)-associated protein 1, a common component of the heterogeneous IKK complex. Mol Cell Biol 1999; 19:1526-1538.

107.   Li Y, Kang J, Horwitz MS. Interaction of an adenovirus E3 14.7-kilodalton protein with a novel tumor necrosis factor α-inducible cellular protein containing leucine zipper domains. Mol Cell Biol 1998; 18:1601-1610.

108. Lee FS, Hagler J, Chen ZJ et al. Activation of the IκBα kinase complex by MEKK1, a kinase of the JNK pathway. Cell 1997; 88:213-222.

109. Ling L, Cao Z, Goeddel, DV. NF-κB-inducing kinase activates IKK-α by phosphorylation of Ser- 176. Proc Natl Acad Sci U S A 1998; 95:3792-3797.

110. Delhase M, Hayakawa M, Chen Y et al. Positive and negative regulation of IκB kinase activity through IKKβ subunit phosphorylation. Science 1999; 284:309-313.

111. Hirano M, Osada S, Aoki T, Hirai S et al. MEK kinase is involved in tumor necrosis factor α-induced NF-κB activation and degradation of IκB-α. J Biol Chem 1996; 271:13234-13238.

112. Malinin NL, Boldin MP, Kovalenko AV et al. MAP3K-related kinase involved in NF–κB induction by TNF, CD95 and IL-1. Nature 1997; 385:540-544.

113. Nakano H, Shindo M, Sakon S et al. Differential regulation of IκB kinase α and β by two upstream kinases, NF-κB-inducing kinase and mitogen-activated protein kinase/ERK kinase kinase-1. Proc Natl Acad Sci U S A 1998; 95:3537-3542.

114. Nemoto S, DiDonato JA, Lin, A. Coordinate regulation of IκB kinases by mitogen-activated protein kinase kinase kinase 1 and NF–κB-inducing kinase. Mol Cell Biol 1998; 18:7336-7343.

115. Lin X, Mu Y, Cunningham ET et al. Molecular determinants of NF-κB-inducing kinase action. Mol Cell Biol 1998; 18:5899-5907.

116. Karin M, Delhase M. JNK or IKK, AP-1 or NF-κB, which are the targets for MEK kinase 1 action? Proc Natl Acad Sci U S A 1998; 95:9067-9069.

117. Karin M, Ben-Neriah Y. Phosphorylation meets ubiquitination: the control of NF-κB activity. Annu Rev Immunol 2000; 18:621-663.

118. Shinkura R, Kitada K, Matsuda F et al. Alymphoplasia is caused by a point mutation in the mouse gene encoding NF-κB-inducing kinase. Nat Genet 1999; 22:74-77.

119. Xia Y, Makris C, Su B et al. MEK kinase 1 is critically required for c-Jun N-terminal kinase activation by proinflammatory stimuli and growth factor-induced cell migration. Proc Natl Acad Sci U S A 2000; 97:5243-5248.

120. Yujiri T, Ware M, Widmann C et al. MEK kinase 1 gene disruption alters cell migration and c-Jun NH2-terminal kinase regulation but does not cause a measurable defect in NF-kappa B activation. Proc Natl Acad Sci U S A 2000; 97:7272-7277.

121. Zhao Q, Lee, FS. Mitogen-activated protein kinase/ERK kinase kinases 2 and 3 activate nuclear factor-κB through IκB kinase-α and IκB kinase-β. J Biol Chem 1999; 274:8355-8358.

122. Lin X, Cunningham ET Jr, Mu Y et al. The proto-oncogene Cot kinase participates in CD3/CD28 induction of NF-κB acting through the NF-κB-inducing kinase and IκB kinases. Immunity 1999; 10:271-280.

123. Ninomiya-Tsuji J, Kishimoto K, Hiyama A et al. The kinase TAK1 can activate the NIK-IκB as well as the MAP kinase cascade in the IL-1 signaling pathway. Nature 1999; 398:252-256.

124. Hehner SP, Hofmann TG, Ushmorov A et al. Mixed-lineage kinase 3 delivers CD3/CD28-derived signals into the IκB kinase complex. Mol Cell Biol 2000; 20:2556-2568.

125. Lallena MJ, Diaz-Meco MT, Bren G et al. Activation of IκB kinase beta by protein kinase C isoforms. Mol Cell Biol 1999; 19:2180-8.

126. Kane LP, Shapiro VS, Stokoe D et al. Induction of NF-κB by the Akt/PKB kinase. Curr Biol 1999; 9:601-604.

127. Ozes ON, Mayo LD, Gustin JA et al. NF-κB activation by tumour necrosis factor requires the Akt serine-threonine kinase. Nature 1999; 401:82-85.

128.   Romashkova JA., Makarov SS. NF-κB is a target of AKT in anti-apoptotic PDGF signaling. Nature 1999; 401:86-90.

129.   Hicke L, Riezman H. Ubiquitination of a yeast plasma membrane receptor signals its ligan-stimulated endocytosis. Cell 1996; 84:277-287.

130.   Peters RT, Liao S-M, Maniatis T. IKKε Is Part of a Novel PMA-Inducible IκB Kinase Complex. Mol Cell 2000; 5:513–522.

131.   Pomerantz JL, Baltimore D. NF-κB activation by a signaling complex containing TRAF2, TANK and TBK1, a novel IKK-related kinase. EMBO J 1999; 18:6694-6704.

132.   Tojima Y, Fujimoto A, Delhase M et al. NAK is an IκB kinase-activating kinase. Nature 2000; 404:778-782.

133.   Nagata S. Apoptosis by death factor. Cell 1997; 88:355-365.

134.   Rothe M, Sarma V, Dixit VM et al. TRAF2-mediated activation of NF-κB by TNF receptor 2 and CD40. Science 1995; 269:1424-1427.

135.   Hsu H, Shu H-B, Pan M-G et al. TRADD-TRAF2 and TRADD-FADD interactions define two distinct TNF receptor 1 signal transduction pathways. Cell 1996; 84:299-308.

136.   Cao Z, Xiong J, Takeuchi M et al. TRAF6 is a signal transducer for interleukin-1. Nature 1996; 383:443-446.

137.   Baud V, Liu Z-G, Bennett B et al. Signaling by proinflammatory cytokines: oligomerization of TRAF2 and TRAF6 is sufficient for JNK and IKK activation and target gene induction via an amino-terminal effector domain. Genes Dev. 1999; 13:1297-1308.

138.   Hsu H, Huang J, Shu H-B et al. TNF-dependent recruitment of the protein kinase RIP to the TNF receptor-1 signaling complex. Immunity 1996; 4:387-396.

139.   Ting AT, Pimentel-Muinos FX, Seed B. RIP mediates tumor necrosis factor receptor 1 activation of NF–κB but not Fas/APO-1-initiated apoptosis. EMBO J 1996; 15:6189-6196.

140.   Sulciner DJ, Irani K, Yu Z-X et al. rac1 regulates a cytokine-stimulated, redox-dependent pathway necessary for NF-κB activation. Mol Cell Biol 1996; 16:7115-7121.

141.   Perona R, Montaner S, Saniger L et al. Activation of the nuclear factor-κB by Rho, CDC42, and Rac-1 proteins. Genes Dev 1997; 11:463-475.

142.   Schutze S, Potthoff K, Machleidt T et al. TNF activates NF-κB by phosphatidylcholine-specific phospholipase C-induced "acidic" sphingomyelin breakdown. Cell 1992; 71:765-776.

143.   Leung K, Nabel GJ. HTLV-1 transactivation induces interleukin-2 receptor expression through an NF-κB-like factor. Nature 1988; 333:776-778.

144.   Ballard DW, Bohnlein E, Lowenthal JW et al. HTLV-I Tax induces cellular proteins that activate the κB element in the IL-2 receptor a gene. Science 1988; 241:1652-1655.

145.   Yin MJ, Christerson LB, Yamamoto Y et al. HTLV-I Tax protein binds to MEKK1 to stimulate IκB kinase activity and NF-κB activation. Cell 1998; 93:875-884.

146.   Harhaj EW, Sun SC. IKKγ serves as a docking subunit of the IκB kinase (IKK) and mediates interaction of IKK with the human T-cell leukemia virus Tax protein. J Biol Chem 1999; 274:22911-22914.

147.   Chu Z-L, Shin Y-A, Yang J-M et al. IKKγ mediates the interaction of cellular IκB kinases with the tax transforming protein of human T cell leukemia virus type 1. J Biol Chem 1999; 274:15297-15300.

148.   Jin D-Y, Giordano V, Kibler KV et al. Role of adapter function in oncoprotein-mediated activation of NF-κB. Human T-cell leukemia virus type I tax interacts directly with IκB kinase. J Biol Chem 1999; 274:17402-17405.

149. Rousset R, Desbois C, Bantignies F et al. Effects of NF-κB1/p105 processing of the interaction between the HTLV-1 transactivator Tax and the proteasome. Nature 1996; 381:328-331.

150. Maran A, Maitra RK, Kumar A et al. Blockage of NF-κB signaling by selective ablation of an mRNA target by 2-5A antisense chimeras. Science 1994; 265:789-792.

151. Yang YL, Reis LFL, Pavlovic J et al. Deficient signaling in mice devoid of double-stranded RNA-dependent protein kinase. EMBO J 1995; 14:6095-6106.

152. Kumar A, Haque J, Lacoste J et al. Double-stranded RNA-dependent protein kinase activates transcription factor NF-κB by phosphorylating IκB. Proc Natl Acad Sci U S A 1994; 91:6288-6292.

153. Zamanian-Daryoush M, Mogensen TH, DiDonato JA et al. NF-κB activation by double-stranded-RNA-activated protein kinase (PKR) is mediated through NF-κB-inducing kinase and IκB kinase. Mol Cell Biol 2000; 20:1278-1290.

154. Chu WM, Ostertag D, Li ZW et al. JNK2 and IKKβ are required for activating the innate response to viral infection. Immunity 1999; 11:721-731.

155. Finco TS, Baldwin AS. κB site-dependent induction of gene expression by diverse inducers of nuclear factor κB requires raf-1. J Biol Chem 1993; 268:17676-17679.

156. Li S, Sedivy JM. Raf-1 protein kinase activates the NF-κB transcription factor by dissociating the cytoplasmic NF-κB--IκB-α complex. Proc Natl Acad Sci U S A 1993; 90:9247-9251.

157. Baumann B, Weber CK, Troppmair J et al. Raf induces NF-κB by membrane shuttle kinase MEKK1, a signaling pathway critical for transformation. Proc Natl Acad Sci USA 2000; 97:in press.

158. Bender K, Gottlicher M, Whiteside S et al. Sequential DNA damage-independent and -dependent activation of NF- κB by UV. EMBO J 1998; 17:5170-5181.

159. Harhaj EW, Sun SC. IκB kinases serve as a target of CD28 signaling. J Biol Chem 1998; 273:25185-90.

160. Li N, Karin M. Ionizing radiation and short wavelength UV activate NF-κB through two distinct mechanisms. Proc Natl Acad Sci U S A 1998; 95:13012-13017.

161. O'Connell MA, Bennett BL, Mercurio F et al. Role of IKK1 and IKK2 in lipopolysaccharide signaling in human monocytic cells. J Biol Chem 1998; 273:30410-30414.

162. Schreck R, Rieber P, Baeuerle PA. Reactive oxygen intermediates as apparently widely used messengers in the activation of the NF-κB transcription factor and HIV-1. EMBO J 1991; 10:2247-2258.

163. Baeuerle PA, Henkel T. Function and activation of NF-κB in the immune system. Annu Rev Immunol 1994; 12:141-179.

Chapter 7

THE BASAL TRANSCRIPTION APPARATUS

Stephen F. Anderson and Jeffrey D. Parvin

Division of Molecular Oncology
Department of Pathology
Brigham and Women's Hospital and Harvard Medical School
75 Francis Street
Boston, MA 02115

INTRODUCTION

Research on gene expression tends to focus on the factors which bind to enhancers and upstream promoter elements to provide specific regulation of a given gene. It is unclear, however, how any of these factors exert their influence to either activate or repress the target gene. The downstream target for all of these specific activators and repressors is the basal transcription apparatus, a set of proteins which is capable of mRNA synthesis from any RNA polymerase II promoter. In this review the functions of the individual basal transcription factors will be described, as well as the way upstream activating factors may influence the basal apparatus. The basal apparatus can be found in a pre-assembled complex, known as the RNA polymerase II holoenzyme, which contains other bridging factors that may link upstream activators to the basal machinery. Chromatin-remodeling proteins, which may be part of a holoenzyme, will be discussed in a holistic model of activator interactions with the basal transcription apparatus on chromatin templates.

REVIEW OF CURRENT RESEARCH

The transcription of any gene can be divided into five steps: 1) recruitment of the polymerase to the promoter to form a pre-initiation complex; 2) formation of an open complex, which contains a region of single-stranded DNA into which the polymerase is inserted to read the template strand; 3) initiation, the synthesis of the first phosphodiester bonds; 4) promoter clearance/elongation, where the polymerase makes a transition from a promoter-bound initiation mode to an unrestricted elongation mode; and 5) reinitiation, or recruitment of additional polymerases to a recently vacated promoter so that thousands of transcripts are made from a single gene. Any of these steps may become rate-limiting for a given promoter under given cellular conditions, although most research into gene regulation currently focuses on the formation or stabilization of the pre-initiation complex.

THE PRE-INITIATION COMPLEX

In order to initiate transcription, RNA polymerase II (Pol II) must be loaded onto the promoter by associating with a number of accessory proteins, known as basal transcription factors. This pre-initiation complex can be assembled on the DNA from free basal factors binding in a defined order.[1,2] Many of the interactions between basal transcription factors and transcriptional activators increase the assembly or stability of the pre-initiation complex in vitro, and thus the ordered assembly model of such a complex will be discussed first. Figure 7-1 shows a complete transcriptional cycle from initiation through reinitiation. In step 1, TFIID, a multisubunit complex containing a TATA-box binding protein (TBP) and TBP-associated factors (TAFs), binds to the TATA box, a conserved DNA sequence just upstream of the mRNA start site (bent arrow). This complex is further stabilized by the binding of TFIIA in step 2 and the binding of TFIIB in step 3. RNA polymerase II and TFIIF bind in step 4. The pre-initiation complex is completed by the binding of TFIIE and TFIIH in step 5. Step 6 depicts the formation of an open complex, a single-stranded region surrounding the start site, which allows the polymerase to access the bases of the coding strand. Open complex formation requires the hydrolysis of ATP. Step 7 shows initiation, promoter clearance, and elongation. Once the polymerase has left the promoter, some basal factors dissociate from the polymerase while others remain bound to the promoter. The promoter can then undergo further rounds of pre-initiation complex assembly, and these reinitiation events may proceed more quickly due to factors which remain bound to the promoter.[3,4]

Much of gene regulation research has historically focused on the recruitment of basal transcription factors, particularly TFIID and TFIIB, to the promoter or stabilization of the pre-initiation complex by protein–protein interactions with bound activators.[5,6] It has recently been established that many if not all pre-initiation complexes are not assembled from free basal factors, but exist in a pre-assembled complex called the holoenzyme.[7–11] In light of this development, much of the basal factor recruitment model of activation can be incorporated into a model in which an activator recruits not just a single basal factor but an entire holoenzyme.[12] Furthermore, another mechanism by which transcription can be regulated has recently been described: several activators, including the HIV Tat protein, apparently modify the

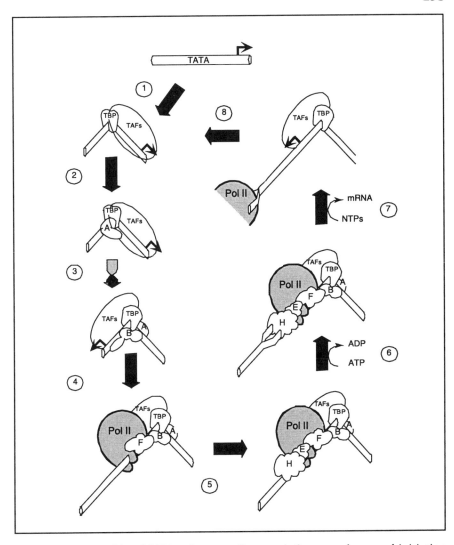

Figure 7-1. Assembly of RNA polymerase II transcription complexes and initiation of transcription. The structures of TBP, TFIIA, TFIIB, Pol II and the bent promoter DNA are based on x-ray or electron diffraction structural data.[36--39,73,98,129,130] **1)** Binding of TFIID, which consists of TATA-box binding protein (TBP) and TBP-associated factors (TAFs) to a TATA box upstream of an mRNA start site (bent arrow). **2)** Binding of TFIIA to the TFIID-DNA complex. **3)** Binding of TFIIB to the DA complex. Note that the complex is rotated by 180° to show the binding of the remaining basal factors. The N-terminal domain of TFIIB, for which no strucural data is available, is shown outlined with a dotted line. **4)** Binding of TFIIF and Pol II to the DAB complex. **5)** Binding of TFIIE and TFIIH to the DABPolF complex. **6)** Hydrolysis of ATP to form an open complex. **7)** Phosphodiester bond formation: initiation and promoter clearance. Additional ATP hydrolysis occurs at this point as the C-terminal domain (CTD) of Pol II is phosphorylated by TFIIH. Basal transcription factors dissociate from the polymerase. **8)** TFIID remains bound to the promoter and can nucleate repeated pre-initiation complex formation.

polymerase to increase processivity[13] through an interaction which appears to involve TFIIH.[14—17] To understand these new models of gene regulation, it is necessary to review the factors involved.

Core RNA polymerase II: RNA polymerase II catalyzes the synthesis of all protein-encoding mRNA in the cell. Pol II exists in two forms, a "core" complex of 12 subunits which is capable of basal transcription,[18,19] as well a second form, discussed below, known as the holoenzyme which contains the core subunits plus many other associated factors.[7-11] The core polymerase is a 500 kd complex of tightly associated proteins. The largest subunit, a 220 kd protein, contains a heptapeptide sequence, YSPTSPS, repeated multiple times at the C-terminus: this seven-residue sequence is present in 27 near-perfect repeats in yeast, while in humans this sequence is repeated 52 times.[20,21] A low-resolution electron diffraction structure determination has mapped this C-terminal domain (CTD) to a point on the polymerase near the binding site of TFIIB, i.e., the farthest point from the mRNA start site.[22] The CTD is unphosphorylated in polymerases entering into the pre-initiation complex[23] and remains so even when the polymerase is paused after 20-30 nt of transcript have been synthesized, but upon transition from initiation to elongation modes it becomes highly phosphorylated.[24] It appears that phosphorylation is not the causative event in the mechanism by which polymerase becomes processive, but rather a signal for productive mRNA synthesis. TFIIH contains a kinase activity specific for the CTD,[25] and phosphorylation of the CTD by TFIIH results in the recruitment of mRNA capping factors to the nascent mRNA.[26] A phosphorylated CTD also binds to mRNA splicing factors.[27] These data suggest that phosphorylation may not affect the processivity of the polymerase, but instead affect the processing of the nascent mRNA.

It is remarkable that very little is known about the other 11 subunits of the core polymerase, even though this is one of the most critical enzymes in the cell. This gap in research interest probably reflects the general impression that the core polymerase is a passive machine subject to regulation by enhancer-binding factors and cofactors. Since promoter-specific RNA-dependent RNA polymerases in bacteriophage can have as little as 99 kd of mass,[28] it is likely that much of the 500 kd of mass in the human polymerase is devoted to mediating regulation. Several of the subunits have been shown to be direct targets of upstream activators: the yeast RPB5 and RPB7 subunits play a role in transcriptional activation;[29] the tumorigenic EWS-Fli factor binds RPB7;[30] and the BRCA1 tumor suppressor makes direct contacts with subunit 2.[31] In addition, the smallest subunit, RPB10α, appears to mediate transcriptional activation *in vitro* from multiple activators.[31]

The phosphorylated polymerase, once engaged in elongation, does not require any basal transcription factors to synthesize an RNA transcript. However, the polymerase cannot initiate transcription on its own: basal factors are needed to recruit the polymerase to a given promoter and regulatory factors are necessary to modulate its initiation and elongation activity.[32]

TBP: Many mRNA promoters contain a consensus DNA sequence 31-25 bp upstream of the start site known as the TATA box. This DNA motif is A-T rich, with the optimal sequence TATAAAA.[33] This sequence is recognized by the TATA-binding protein (TBP) subunit of TFIID. On several promoters lacking a TATA box, other sequences about 30 nt downstream of the start site seem to be specifically recognized by TBP,[34,35] but it is not clear whether these interactions will occur at all promoters. TBP contains a highly conserved DNA-binding domain and an N-terminal domain, which varies somewhat from species to species. The crystal structure of the DNA-binding domain of TBP has been solved alone[36,37] and bound to a TATA box-containing DNA fragment.[38,39] These structures show that TBP is a crescent or saddle-shaped molecule, with a "stirrup" hanging down on each side of the DNA helix. Most surprisingly, the DNA at the TATA box is bent about 80° by the binding of TBP, which splays open the minor groove in an unusual protein-DNA interaction based upon hydrophobic interactions. The reason for this drastic bending is unknown, but such an unusual structural feature may provide a means for other basal transcription factors or holoenzyme complexes to recognize a promoter. Alternatively, severe DNA bending might serve to reorient nucleosomes.

TBP *in vitro*[40] and TFIID *in vivo*[41] have been demonstrated to exist in an equilibrium between an inactive dimer and a DNA-binding monomer. One point at which transcription can be regulated is therefore at the level of availability of monomeric TBP/TFIID. Another point at which transcription can be regulated is at the TATA-binding step. The affinity of TFIID for the TATA box can be strongly affected by the topological state of the DNA template. TFIID binding to the TATA box is strongest on supercoiled DNA[42] and TBP has been shown to bind with higher affinity to pre-bent DNA.[43]

TFIID: The TBP polypeptide is found in multiple complexes, each of which is essential for transcription with either RNA polymerase I, II, or III.[44,45] In Pol II transcription, the essential TBP-containing complex is TFIID. Mammalian TFIID is a complex of at least 13 different polypeptide subunits including TBP and a dozen TBP-associated factors (TAFs).[46] Other nonuniversal TAFs have been reported, although it is unclear whether these proteins are in fact specialized TAF subunits or merely tightly associated coactivators[47,48] (discussed below). The TBP subunit alone in the presence of the other general transcription factors will support basal transcription, but the TAF subunits are required for regulation of transcription.[46] These TAFs can be divided into two groups: subunits involved in DNA binding interactions and subunits involved in interactions with regulatory domains.

The DNA-binding subunits of TFIID include TBP and a number of TAFs with homology to histones. These histone-fold proteins, hTAF$_{II}$80, hTAF$_{II}$31, and hTAF$_{II}$20/15, are capable of interacting with histones, indicating a similar set of protein-interaction motifs, and with each other in relationships analogous to those of H4, H3, and H2B, respectively.[49] The crystal structure of a complex of *Drosophila* dTAF$_{II}$42 and dTAF$_{II}$62 shows that these proteins form a structure resembling the (H3/H4)2 nucleosome core tetramer.[50] Thus, it is likely that the human homologues, hTAF$_{II}$80 and hTAF$_{II}$31, along with hTAF$_{II}$20/15 form a structure analogous to a histone octamer. Footprinting of TFIID on the hsp-70 promoter shows that TFIID minimally protects 17-18 bp of DNA centered over the TATA box, but on the adenovirus type 2 major late promoter TFIID protection extends downstream to +35, spanning about 75 bp of DNA.[51] On a different DNA fragment also containing an adenovirus major late (AdML) promoter, the footprint downstream of the start site contains hypersensitive sites every ten base pairs, consistent with the DNA being bound to a surface with one side exposed to solvent,[52] similar to the patterns seen for nucleosome-bound DNA. A nucleosome core contains about 140 bp of DNA, so a 70 bp footprint suggests that a TAF octamer wraps DNA around itself only once. On the adenovirus E4 promoter, TFIID alone protects only the 20 bp over the TATA box, but in the presence of the activator ATF or GAL4, TFIID again protects a 70 bp region of the promoter.[53,54] Thus, some DNA sequences may interact with a TAF subunit while others may not, and upstream transcriptional activator binding may induce wrapping of the promoter DNA around the histone-like TAF subunits. This wrapping induces negative supercoils in the DNA,[42,55] which enhances the ability of the polymerase to initiate transcription,[56] provided that the polymerase is able to access the promoter. It has been shown that a number of TAFs are potentially close enough to the start site to block polymerase entry, but that addition of TFIIA to a TFIID-DNA complex causes a conformational change which clears most of the TAFs from the start site.[55] Taken together, this data suggests a model for activation involving TFIIA, TAFs and an activation domain (Fig. 7-2). The conformation of TFIID when it first binds to a promoter may not allow the DNA to be wrapped around the histone-like TAFs (Fig. 7-2 A), but on some promoters the presence of a transcriptional activator will induce DNA wrapping (Fig. 7-2 B), perhaps by binding to and constraining TAFs. The DNA becomes supercoiled, but the start site may be inaccessible to the other basal factors until TFIIA binds (Fig. 7-2 C), shifting TAFs away from the start site. The other basal factors and polymerase can then bind to the exposed supercoiled DNA and initiate transcription (Fig. 7-2 D). Thus, wrapping of DNA around the histone-like core of TFIID may both enhance transcription by inducing negatively supercoiled DNA as well as inhibit transcription by steric interference with the start site, an inhibition which can be overcome by the presence of TFIIA.

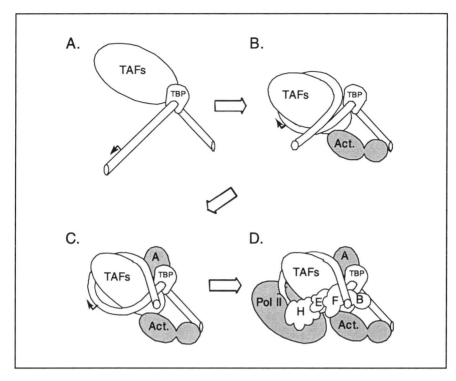

Figure 7-2. A possible model for activation incorporating the histone-like TAF subunits and TFIIA. **A.** The initial conformation of TFIID bound to a TATA box does not encourage DNA wrapping. **B.** Binding of an activator induces wrapping of the promoter DNA around the histone octamer-like TAF complex. Supercoiling of the DNA, which enhances Pol II initiation, occurs but the start site is inaccessible to the polymerase and basal factors. **C.** Binding of TFIIA causes a rearrangement of TAF subunits to expose the start site. **D.** RNA polymerase II and basal factors can bind to the exposed, supercoiled start site and initiate transcription.

The remaining subunits of TFIID are involved in protein-protein interactions with transcriptional activator proteins. One function of multiple TAFs is to allow for synergistic transcription: by increasing the number of target sites for potential activators, several upstream activators can bind at once, each contributing to the enhancement of transcription. For example, the *Drosophila* transcriptional activator proteins Bicoid and Hunchback bind to $dTAF_{II}110$ and $dTAF_{II}60$, respectively. When these two factors bind together upstream of the same promoter, the effects are synergistic, yielding transcription activation greater than each factor will produce individually.[57] Furthermore, Bicoid was found to contain a second activation domain which binds $dTAF_{II}60$, while both of these domains together activate transcription synergistically.[58] Thus, a single transcriptional regulator-TFIID interaction may consist of multiple interactions between regulatory domains and multiple TAFs.

It is unclear how the interaction of a bound transcriptional activator with a TAF subunit leads to activation of transcription. In most systems, TFIID and the other basal transcription factors alone are insufficient for activation of transcription. Thus, the paradigm of activation being a simple recruitment of TFIID to the promoter is not complete. A class of proteins known as coactivators, discussed below, are necessary for activated transcription *in vitro* when highly purified transcription factors are used. Thus, activator-TAF interactions must send some sort of mechanical signal to other parts of the pre-initiation complex through these coactivators.[59]

Interestingly, $TAF_{II}250$ has other functions besides its role as a target for activation domains. It contains a protein kinase activity which phosphorylates TFIIF,[60] which is reported to stimulate transcription[61] although the physiological significance of this mechanism is not yet known. When $TAF_{II}250$ was cloned, it was found to be the same as CCG1,[62,63] a cell cycle gene which when mutated causes cells to arrest in G1.[64,65] Thus, the presence of functional TFIID is a requirement for cells entering S phase.

TATA-less promoters and initiator elements: Some promoters lack a TATA box, yet are recognized efficiently by TFIID. These promoters often have a conserved DNA sequence, YYAN(T/A)YY, overlapping with or near the start site called the initiator element (Inr),[66,67] which is recognized by a 120 kd basal factor, TFII-I.[68] TFIID can recognize a TFII-I bound initiator,[69] and pre-initiation complex assembly presumably proceeds normally once TFIID binds nonspecifically to the DNA around the initiator. It is not yet clear whether pre-initiation complexes containing TFII-I on TATA-less promoters are functionally equivalent in basal and activated transcription to complexes on TATA-box containing promoters with or without TFII-I. Other initiator-binding factors such as YY1 have been found which can replace certain basal factor requirements *in vitro*.[70] It is therefore not yet clear how many modes of promoter recognition exist.

TFIIB: TFIIB is a single polypeptide of 35 kd[71] which stabilizes the binding of the TBP subunit of TFIID to the TATA box.[1,72] The crystal structure of the C-terminal two-thirds of TFIIB bound to a TBP-DNA complex has been solved[73] and shows that the TFIIB C-terminal domain is a bi-lobed structure (denoted by solid lines in step 3 of Fig. 7-1) which binds to the C-terminal "stirrup" of TBP. TFIIB binds within the bend in the DNA created by TBP binding, and is thus able to contact DNA both upstream and downstream of the TATA box. Binding of TFIIB across the bend in the promoter stabilizes the structure (Fig. 7-1). Because the bend in the DNA was initially generated by the binding of TBP, the presence of TFIIB stabilizes the TBP-DNA interaction. Thus, one function of TFIIB is to maintain TBP bound to a promoter.

Mutations in yeast TFIIB alter the position of the transcription start site[74] and mutations in the large subunit of RNA polymerase II have been found which have the same phenotype.[75] TFIIB and Pol II associate *in vitro*,[76] suggesting that this interaction between TFIIB and Pol II defines the mRNA start site. Footprinting studies have shown that the presence of the full TFIIB protein can result in protection of the DNA as far downstream as the start site,[1,77] and a prediction based upon the crystal structure of the C-terminal domain of TFIIB complexed with TBP and DNA[73] indicates that the N-terminal domain would extend downstream, as would be expected if TFIIB in fact defines the start site (Fig. 7-1).

Binding of activators to TFIIB results in a conformational change in TFIIB which exposes binding sites for other basal transcription factors, driving the formation of the pre-initiation complex.[78] TFIIB was first shown to bind directly to the Herpes Simplex Virus activator VP16[6,79] and a large number of other activators have since been shown to bind TFIIB as well. From these interactions with transcription factors, it is clear that another function of TFIIB is to stabilize the pre-initiation complex through contacts with activators.

TFIIF: TFIIF is a 220 kd heterotetramer consisting of two copies each of a 74 kd and a 30 kd subunit.[80-82] Unlike the previous basal factors, TFIIF joins the pre-initiation complex simultaneously with the polymerase,[83] recognizing the N-terminus of TFIIB.[84] It will catalyze the elution of polymerase from nonspecific DNA sequence elements[85] and thus functions to target the polymerase to a promoter. As well as being an essential component of the pre-initiation complex, TFIIF stimulates polymerase elongation of an RNA transcript.[86,87] *In vitro* experiments with purified factors suggest that TFIIF dissociates from the polymerase some time after the first ten nucleotides have been transcribed, although it is capable of reassociation to a stalled polymerase[88] and may enhance elongation by somehow restarting stalled polymerases. Unlike TFIID and TFIIB, only a few transcriptional activation domains have been found to bind TFIIF.[89,90]

TFIIE: TFIIE is a 200 kd heterotetramer composed of two 56 kd and two 34 kd subunits.[91-93] In the presence of TFIIH, TFIIE stabilizes the open complex formed by the action of the TFIIH helicases.[94] TFIIE has been shown to bind to Pol II, TFIIF, TFIID and TFIIH,[95] and may function primarily as a spacer between these other basal transcription factors. Like TFIIF, TFIIE is relatively infrequently bound by activation domains, although several examples have been found.[90,96,97] A low-resolution electron diffraction structure has been determined which places TFIIE at the end of the polymerase closest to the active site and furthest from TFIIB.[98] Thus, TFIIE is bound downstream of the polymerase and is therefore, with the possible exception of TFIIH, the furthest downstream of the basal factors.

TFIIH: TFIIH is a multisubunit protein whose function during transcriptional initiation is to catalyze the unwinding of the promoter DNA and generate an open complex, a region of single-stranded DNA, for the polymerase to enter and initiate transcription. This activity is absolutely required on linear or relaxed DNA templates and enhances transcription from some but not all promoters on supercoiled DNA.[56,99-102] TFIIH contains 9 polypeptide subunits of size 89, 80, 62, 52, 44, 43, 37, 36, and 34 kd.[103] TFIIH complexes vary in size according to the species and the method of preparation, and thus the stoichiometry of these subunits is not certain. The two largest subunits of TFIIH encode DNA helicases and function not only in transcription, but in nucleotide excision repair as well.[104,105] The p89 subunit (ERCC3/XP-B) has a 3'-5' helicase activity,[106] meaning that it moves 3' to 5' on single-stranded DNA. It is likely that TFIIH is pushed along ahead of or dragged alongside the polymerase, unwinding DNA until the polymerase is fully inserted into the open complex. If this is the case, then p89 would travel along the template strand to unwind the DNA ahead of polymerase. The p80 subunit (ERCC2/XP-D) has a 5'-3' helicase activity,[107] and would in this model unwind DNA from the opposite strand. In yeast, the p80 homologue, RAD3, can be mutated such that it no longer possesses helicase activity, and yet the TFIIH is still active in basal transcription *in vitro*.[108] In contrast, the helicase activity of the yeast homologue of p89, RAD25, is essential for transcription activity.[109] It has recently been reported that a naturally occurring mutation in human p89 reduces both helicase and transcription activity,[110] although it has not yet been possible to eliminate the possibility that both helicases function together as a redundant system to ensure that the essential open complex is formed.

In addition to the helicase activities of p80 and p89, TFIIH has a protein kinase activity.[25] This kinase activity requires three subunits: the 43 kd cdk7/MO15 subunit,[111,112] the 37 kd cyclin H subunit,[113,114] and the 36 kd MAT1 subunit.[115] About half of the cdk7 kinase subunits in a cell are found independent of TFIIH[116,117] and are capable of an activating phosphorylation of Cdc2 and Cdk2.[114] Thus, these subunits have cdk-activating kinase (CAK) activity. When the three CAK subunits are in the TFIIH complex, the kinase specificity shifts to include the Pol II CTD as well as cdks.[116] The phosphorylation of the CTD occurs at the same time as the polymerase clears the promoter,[24] although CTD phosphorylation has not yet been shown to be the cause of promoter clearance. Mutation of the kinase activity[116] or inhibition with the ATP analog H-8[118] have no effect on basal or activated transcription *in vitro*. Reconstitution of the nine subunit TFIIH complex has been successful using baculovirus expression system and has revealed that all of the five subunit core of TFIIH (p89, p62, p52, p44, and p34) are required for function, and that the remaining five subunits stimulate the level of basal transcription.[119] Several mutations in the helicase domains of TFIIH have been analyzed and shown to be critical for TFIIH function in basal transcription.[120]

Coactivators

When using highly purified mammalian transcription factors, the binding of upstream activators to a promoter does not result in stimulation of transcription *in vitro*. A crude chromatographic column fraction termed USA (for upstream factor stimulatory activity) was found to be required for activated transcription *in vitro*,[121,122] and this fraction has been further fractionated to purify several stimulatory factors. These factors have very little effect on the basal transcription reaction, but are required for the function of upstream activators and are referred to as coactivators. The mode or modes of operation of these coactivators is largely unknown. In general, these coactivators are promiscuous, supporting activation from multiple activation domains on most if not all of the promoters assayed, and require TFIID and TFIIA for activity.

TFIIA

TFIIA was originally identified independent of the USA fraction as an activity which enhanced Pol II transcription levels.[32] TFIIA is a complex of three polypeptides with molecular weights of 35, 19 and 12 kd[123,124] which, after cloning and purification, has turned out not to be required for basal transcription in highly purified systems *in vitro*.[125-128] TFIIA does, however, stimulate basal transcription in the presence of TFIID but not TBP,[126] and dramatically elevates levels of activated transcription.[125-128] For these reasons, TFIIA has been demoted from a general transcription factor to a coactivator.

The crystal structure of the yeast TFIIA-TBP-DNA complex has been solved.[129,130] From these structures, it is known that TFIIA binds to the N-terminal stirrup of the TBP saddle, that is it binds on the opposite side of the DNA from TFIIB (Fig. 7-1). TFIIA binds on the upstream face of TBP, facing away from the rest of the pre-initiation complex (as shown in Step 2 of Fig. 7-1). One domain of TFIIA extends perpendicular to the plane defined by the bend in the DNA and mutations in this region decrease transcriptional activity *in vitro*.[131] Photocrosslinking studies show that at least three TAFs occupy the same region as TFIIA[55] although it is not yet known which of these TAFs are in physical contact with TFIIA. These interactions between TFIIA and TFIID and the promoter DNA provide the most detailed model currently available for the function of a coactivator (Fig. 7-2). It is not yet known if other coactivators will similarly potentiate activation by rearrangement of TAF orientations to allow access to the start site, or if these other coactivators each potentiate activation by their own mechanisms.

Topoisomerase I

DNA topoisomerase I (topo I) has been found to both repress basal transcription and stimulate activated transcription.[132] Neither activity requires topo I DNA relaxation activity, and topo I from vaccinia virus, *E. coli,* or yeast cannot be substituted for the human protein. Thus, this coactivator/repressor functions by species-specific protein-protein interactions. The inhibitory effect on basal transcription can be overcome by TFIIA, suggesting that topo I competes for the same TFIID binding site or sites as TFIIA. This suggests that one role of TFIIA is to prevent the formation of inhibitory complexes.

HMG-1/2

These nonhistone chromosomal proteins allow activation of transcription from some but not all promoters *in vitro* in the presence of TFIIA.[133] Unlike topo I, HMG-2 does not increase activation levels by repressing basal transcription, suggesting that not all coactivators share the same mechanism. Titration of HMG-2 indicates that only a few copies of the protein per promoter are required for optimal activation, after which further addition of protein leads to decreased activation. HMG proteins are known to bind DNA nonspecifically and induce severe DNA bending,[134] and might be expected to have a high affinity for the bent DNA around the promoter. The DNA-binding domain of HMG-2 is necessary and sufficient to support activation, although footprinting shows no evidence that this protein actually binds the promoter DNA.[133] Thus, it is unclear whether HMG-1 and HMG-2 function by binding to the promoter DNA and thus stabilize some other interaction, or through protein-protein interactions.

PC4

The USA coactivator preparation was fractionated to isolate multiple positively acting and negatively acting factors, including a 15 kd protein, called positive component 4 (PC4), which potentiates the activation of transcription.[59,135] PC4 binds to the activation domains of VP16 and other activators, as well as to TFIIA.[135] This protein can be inactivated by phosphorylation and contains a cryptic DNA binding domain although this domain is uncovered only when the N-terminus of the protein is deleted.[59]

From the common elements of the coactivators discussed above, a general model for coactivator potentiation of activation can be constructed (Fig. 7-3). Figure 7-3 A depicts a pre-initiation complex bound to the promoter, incorrectly positioned in the absence of TFIIA, perhaps due to the presence of an inhibitor at the TFIIA binding site. The binding of TFIIA and an activator to the pre-initiation complex in Figure 7-3 B constrain the TAF, but the machinery by which activation signals are transmitted to the polymerase is still out of the reach of the TAF. In Figure 7-3 C, a coactivator recognizes elements of the constrained pre-initiation complex, which might include the positions of the activation domain, various TAFs, TFIIA, and/or the conformation of the DNA. The bound coactivator makes protein-protein contacts with some element of the pre-initiation complex, transmits a mechanical signal to the complex, and by some as yet unknown mechanism, transcription is activated.

It should be noted that the search for coactivators has also turned up negative cofactors such as NC1[121] and NC2,[122,136] a subunit which has also been characterized as Dr1.[137] These inhibitors tend to interfere with binding of TFIIA and/or B to TFIID, inhibiting pre-initiation complex formation.[122,135,139] Additionally, the above coactivators will, at high enough concentrations, themselves inhibit basal transcription,[132,135,138,140] perhaps by binding to the pre-initiation complex or to the promoter DNA. When associated with an upstream activator, TFIIA and other basal factors and coactivators can often diminish this nonspecific repression, resulting in a modest apparent stimulation of transcription.[125,126,141] This stimulation is not strictly activation but is more accurately referred to as antirepression. This phenomenon is not restricted to coactivators and describes the dramatic increase in transcription observed when nucleosomes are removed from a DNA template by competition with activators.[142] The model for activators interacting with a histone octamer-like TAF complex put forward in Figure 7-2 might predict that, as with DNA-histone interactions, the initial TAF-DNA conformation may be inhibitory and that some part of the TFIIA- or activator-induced rearrangement can be considered as antirepression. This has not, however, been confirmed experimentally. Thus, many situations in which transcription is stimulated by activators and coactivators will consist at least in part of antirepression.

RNA Polymerase II Holoenzyme

Pol II exists in eukaryotic cells in at least two forms: as a core enzyme of 12 subunits and a total size of 500 kd and as a holoenzyme containing a large number of subunits and a total mass of 60 S as measured by sucrose sedimentation,[10] putting it in the same size range as a ribosome. The yeast holoenzyme is responsible for all mRNA transcription,[143] and in yeast cells the recruitment of a holoenzyme to a promoter is sufficient to activate transcription.[12] The mammalian Pol II holoenzyme has been purified,[8-11,144] although the exact composition remains murky. The size of the holoenzyme complex renders it unstable to the hydrostatic and osmotic pressures of biochemical purification. Thus, the method of purification determines which subunits are present in the complex. An added difficulty in characterizing this complex is that it does not appear to have a single composition, but rather is represented by multiple populations of holoenzyme complexes.[145] Subpopulations of Pol II holoenzyme complexes may be specifically primed for activation of specific pathways.

242

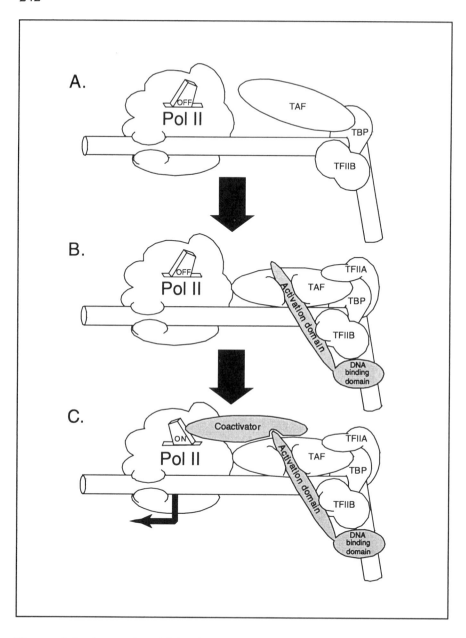

Figure 7-3. A general model for the mechanism by which coactivators transmit an activation signal from a transcriptional activator. **A.** In the absence of activators, basal factors may be incorrectly positioned. **B.** TFIIA and transcriptional activators constrain the basal transcription apparatus but basal factors alone are incapable of making the contacts necessary to activate the polymerase. **C.** A coactivator recognizes the basal factor-activator conformation and is able to mechanically transmit the signal for activation to the polymerase, initiating transcription from the promoter (arrow).

The Pol II holoenzyme is best defined in yeast, where it has been defined using both genetic and biochemical means. The polymerase in a yeast holoenzyme is associated with nine polypeptides known as SRB factors.[146,147] The SRB (suppressors of R\NA polymerase B mutations) proteins were selected based on their ability to associate tightly with the CTD domain of the core Pol II. These factors include SRB10 and SRB11, a kinase/cyclin pair[148] also involved in phosphorylation of the polymerase CTD and whose human homologues, cycC and Cdk8,[149] have been found in the mammalian Pol II holoenzyme.[10,144,150] The human homologue of yeast SRB7 has also been found in the holoenzyme[9] suggesting that many, if not all, of the yeast SRB proteins will be found to have a homologue in the mammalian holoenzyme. Other similar factors, known collectively as "mediator," have been identified in the yeast and mammalian Pol II holoenzymes.[151,152] It has been proposed that the mediator complex associates with Pol II in the holoenzyme at the initiation step, and with phosphorylation of the CTD, the mediator falls off and subunits which are important in the elongation phase of transcription then associate to make an elongator holoenzyme complex.[153] Perhaps the elongator complex contains capping and mRNA processing factors, as discussed in the polymerase section of this review. As will be discussed below, the mediator components (including SRB polypeptides) are also present in a separate complex which functions as a coactivator.

In addition to the SRBs, most of the basal transcription factors are present in the holoenzyme. In yeast it has been shown that all of the basal transcription factors except TFIID and TFIIE are found in the Pol II holoenzyme.[7] For the mammalian holoenzyme, there are conflicting reports. In some examples, all of the factors required for basal transcription were found in the holoenzyme[8,154] while in other preparations only TFIIF, TFIIE, and TFIIH were associated with the holoenzyme.[9,144,155]

The yeast holoenzyme has been shown to contain transcriptional coactivators,[7] and the mammalian counterparts similarly have transcriptional coactivators. Semantically, a coactivator must be separable from a basal factor, thus it has been suggested that a holoenzyme-bound coactivator should be termed a "regulatory target."[11] The CREB binding protein (CBP) and the related p300 protein have both been shown as a Pol II holoenzyme components.[156,157] Activation of transcription in vitro by phosphorylated CREB protein has been shown to be mediated by two essential pathways. The first is via a glutamine-rich domain of CREB binding to $TAF_{II}130$ of TFIID, and the second is a phosphorylation-specific interaction of CREB to CBP bound in the Pol II holoenzyme. Together these create a stimulatory signal for transcription.[156] These results suggests that CBP and p300 bridge the basal machinery of the Pol II holoenzyme to the upstream activators. CBP and p300 mediate activation signals from many transcriptional regulators, and multiple activators binding to highly structured promoter elements, termed enhanceosomes, create optimal binding surfaces for CBP, which stimulates transcription in vitro.[158] As will be discussed below, CBP and p300 function in other coactivator complexes as well as regulatory targets in the holoenzyme complex, making these factors central to transcriptional regulation by many activators.

Chromatin

In this review, transcription has thus far been treated for the most part as if it occurs on naked DNA, whereas in reality, the pre-initiation complex must contend with chromatin.[159] It has long been recognized that the binding of nucleosomes to an RNA polymerase II promoter results in a dramatic inhibition of transcription.[160,161] Activators can stimulate transcription from nucleosome-bound templates, although not to the same levels as on naked DNA.[162] However, as basal transcription is virtually nonexistent in the presence of nucleosomes, the ratio of activated transcription to basal transcription in chromatin is higher. Thus, chromatin amplifies the modest physical advantages of an activated pre-initiation complex into a much larger differential between transcription of activated and unactivated genes. Nucleosome formation at a promoter can be prevented by the binding of a pre-initiation complex,[163,164] TFIID,[165] or even TBP,[166] and thus the regulation of transcription depends entirely on the ability of a pre-initiation complex to nucleate on a chromatin-free promoter. As DNA is typically nucleosome-bound except during the momentary passage of a replication fork, TFIID or Pol II holoenzyme binding must typically first involve the removal, or remodeling, of nucleosomes from a promoter.

It has been shown in model systems *in vitro* that the binding of transcriptional activators can allow access to the DNA of a nucleosome.[142] This may provide a mechanism for the transcription of genes whose activators bind close to the promoter and can thus allow the pre-initiation complex access, but activator binding alone is unlikely to free a nucleosome-bound promoter when the activator binding sites are kilobases up- or downstream of the promoter. In such cases, transcriptional activation must rely on other mechanisms to remove nucleosomes from a promoter.

A generally occurring process likely involves enhancer-binding factors which can bind to their enhancer sequences even when present in chromatin. Such a factor then starts a chain of events which leads to the modification of chromatin, which in turn facilitates other enhancer-binding factors to bind. This leads eventually to binding to the promoter of the transcriptional machinery.[167]

THE SWI/SNF COMPLEX

A complex has been found in yeast which is capable of altering the wrapping of DNA on nucleosome cores in an ATP-dependent manner.[168,169] Several of the genes encoding proteins in this complex had previously been implicated in the switch of sexual mating types or sucrose metabolism (sucrose nonfermenters), and thus the complex is named SWI/SNF. This complex consists of at least 10 proteins in yeast, and stimulates the binding of transcriptional activators to nucleosomal DNA in an ATP-dependent manner.[170] The SWI/SNF complex is recruited to promoters by enhancer binding regulators, allowing chromatin remodeling to occur at the promoter;[167] thus SWI/SNF is an early acting complex in gene activation. This complex has been reported to be associated with the yeast holoenzyme,[171] which would suggest that histone remodeling may also be concurrent with pre-initiation complex recruitment. Mammalian homologues of some SWI/SNF genes have been identified,[172-174] and a partially purified complex containing one of these homologues was found to have similar activities to the yeast complex.[175] Biochemical data indicate that the SWI/SNF complex also functions during the elongation phase of transcription.[176]

Histone Acetyl Transferases

The affinity of nucleosome cores for DNA depends on the acetylation state of the amino-terminal "tails" of the histone subunits. The acetylation of lysine residues in histones results in a loosening of the nucleosome which may make the DNA more accessible for transcription.[177,178] A protein complex binds to p300 and CBP, which has a histone acetyl transferase (HAT) activity.[179] This p300/CBP-associated factor (PCAF) is in a protein complex[180] which is homologous to the yeast SAGA complex.[181] The PCAF complex appears to be a counterpart of TFIID, as it contains multiple TAF subunits without TBP,[180] and in some unknown way regulates the activation of transcription on chromatin templates via acetylation.

The converse of HAT activity is the deacetylation of chromatin by complexes containing transcriptional repressors.[182] Transcriptional repressors, such as the retinoblastoma tumor suppressor protein, recruit the deacetylation complex to promoters to suppress transcription.[183] Methylation of DNA also recruits the deacetylation protein complex in order to render the template silent.[184]

Mediator and Similar Coactivator Complexes

The SRB and mediator complexes have been found in other assemblies besides the Pol II holoenzyme. In yeast this complex has been called the mediator, but in mammalian cells it has been given multiple acronymic names including ARC, CRSP, SMCC, TRAP, NAT, DRIP, and Mediator.[185] According to one model, this complex binds to core Pol II in order to constitute the holoenzyme.[186] Data from our laboratory indicate that this mediator complex functions separately from the Pol II holoenzyme and can function to recruit holoenzyme to promoter elements.[145] Data suggest that the effects of this mediator complex are most pronounced when using chromatin templates.[187,188]

SRC/p160 Coactivator Complexes

Yet another coactivator complex also contains CBP along with the p160 class of coactivator proteins, including SRC1, AIB1, and GRIP1.[189] These factors regulate the functions of nuclear hormone receptors, and possibly NF-κB-dependent gene expression (Chapter 2). Because these coactivator complexes contain CBP, the chromatin around the promoter can become acetylated by these factors, and thus stimulate the mRNA synthesis.

Activation via Initiation Versus Elongation

With all the interactions which have been documented between activation domains and basal transcription factors, it is easy to assume that all activation occurs by a single mechanism, the recruitment of basal factors or a holoenzyme to a promoter by bound activators (depicted in Fig. 7-4 A). However, it is apparent that some activators have nothing to do with recruitment of basal factors to a promoter, instead affecting the ability of the polymerase to elongate the RNA transcript.[13] Thus, this mechanism of activation occurs after the pre-initiation complex has been assembled, and may involve one of the enzymatic activities of TFIIH or other kinase (Fig. 7-4 B).

Polymerases may pause at sites throughout a gene, but pause sites are often located at the 5' end of the gene, often just downstream of the promoter. There appears to be no single consensus sequence which initiates a pause, although some sequences such as HIV TAR form an RNA secondary structure which interacts with the polymerase.[190,191] Studies of the *Drosophila* hsp70 promoter show that the polymerase remains bound at the pause site 20-30 nt downstream of the start site until the CTD is hyperphosphorylated, at which point the polymerase proceeds to transcribe the remainder of the gene.[24] It has been shown that Pol II requires ATP β-γ bond hydrolysis to transcribe more than 30 nucleotides,[192] suggesting that CTD phosphorylation and promoter clearance are linked and not merely temporally related events. It is not yet known whether a paused polymerase remains physically tethered to the promoter, and thus Figure 7-5 depicts two models of promoter clearance. Figure 7-5 details a mechanism in which the polymerase is tethered to the promoter by the unphosphorylated CTD. It has been demonstrated that yeast SRB2 binds to TBP,[193] tethering the polymerase to the promoter via the CTD, the SRBs, and TFIID. Upon initiation, the polymerase may be prevented from moving more than a short distance downstream by the tethered CTD. When an activator binds, an interaction with the TFIIH kinase activity may stimulate phosphorylation of the CTD. This phosphorylation would disrupt the CTD-SRB interactions, freeing the polymerase to extend the nascent RNA transcript.

In the case of HIV transcription, elongation by Pol II is regulated by the Tat protein binding to the TAR element of the nascent transcript. Tat binds to transcriptional elongation factor b (P-TEFb), a kinase shown to regulate transcriptional elongation *in vitro*.[194] The factor P-TEFb contains the polypeptides, CycT and cdk9.[195,196] These factors function in the pathway outlined in Figure 7-5 to stimulate transcriptional elongation through pause sites via phosphorylation of the CTD of Pol II.

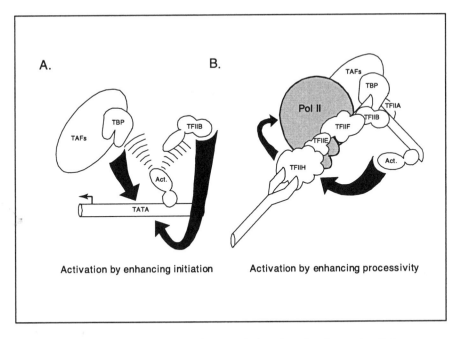

Figure 7-4. Two mechanisms of transcriptional activation. **A.** Enhancement of initiation by recruitment of basal factors, either free or assembled into a holoenzyme complex, to a promoter by bound activators. **B.** Enhancement of processivity by activator contact with an assembled pre-initiation complex. This contact sends a signal, most likely through TFIIH, to modify RNA polymerase II in a manner which enhances elongation.

248

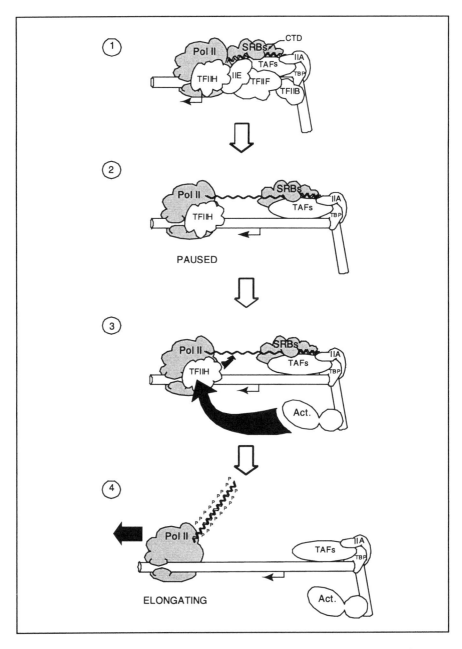

Figure 7-5. Model for promoter clearance involving phosphorylation of the RNA polymerase II C-terminal domain. 1) RNA polymerase II initiates transcription at a promoter (small arrow). The polymerase transcribes 20–30 nucleotides and pauses because it is tethered to the promoter by the CTD (2). Binding of an activator induces TFIIH to phosphorylate the CTD (3). Phosphorylation releases the CTD from the promoter.[4]

SUMMARY AND FUTURE DIRECTIONS

With each passing year, the mechanisms of transcription seem to become more complex rather than simpler. Where it was once thought that transcription consisted of a straightforward binding of a handful of basal factors to a promoter, we now discover that a bewildering array of activators, repressors, coactivators, negative cofactors, chromatin-modifying factors, and chromosomal proteins engage in a delicate balancing act. The balance between stimulatory and inhibitory signals, both specific and nonspecific, makes these myriad pathways essential in the cell for even simple operations.

The relatively new technology of gene arrays to analyze transcriptional profiles in cells will allow a more complete understanding of how these factors impact actual genes. This has begun in yeast using temperature sensitive alleles of basal transcription factors in order to identify genetic pathways sensitive to the loss of one factor or another.[197] Similar analyses which disrupt basal factor or coactivator function will reveal which genes are affected, and these assays can then be repeated in multiple cell types which have genetic modifications. Also important in defining how this complex process is actually mediated will be chromatin immunoprecipitation experiments which identify the promoter DNA to which a given factor or coactivator is bound. Together, these new technologies will likely lead to a new understanding of the complex process of gene activation.

REFERENCES

1. Buratowski S, Hahn S, Guarente L, et al. Five intermediate complexes in transcription initiation by RNA polymerase II. Cell 1989; 56:549-561.

2. Zawel L, Reinberg D. Common themes in assembly and function of eukaryotic transcription complexes. Annu Rev Biochem 1995; 64:533-561.

3. Van Dyke MW, Sawadogo M, Roeder RG. Stability of transcription complexes on class II genes. Mol Cell Biol 1989; 9:342-344.

4. Jiang Y, Gralla JD. Uncoupling of initiation and reinitiation rates during HeLa RNA polymerase II transcription *in vitro*. Mol Cell Biol 1993; 13:4572-4577.

5. Stringer KF, Ingles CJ, Greenblatt J. Direct and selective binding of an acidic transcriptional activation domain to the TATA-box factor TFIID. Nature 1990; 345:783-786.

6. Lin YS, Green MR. Mechanism of action of an acidic transcriptional activator *in vitro*. Cell 1991; 64:971-981.

7. Koleske AM, Young RA. An RNA polymerase II holoenzyme responsive to activators. Nature 1994; 368:466-469.

8. Ossipow V, Tassan JP, Nigg EA, et al. A mammalian RNA polymerase II holoenzyme containing all components required for promoter-specific transcription initiation. Cell 1995; 83:137-146.

9. Chao DM, Gadbois EL, Murray PJ, et al. A mammalian SRB protein associated with an RNA polymerase holoenzyme. Nature 1996; 380:82-85.

10. Maldonado E, Shiekhattar R, Sheldon M, et al. A human RNA polymerase II complex associated with SRB and DNA-repair proteins. Nature 1996; 381:86-89.

11. Parvin, J. D., Young, R.A. Regulatory targets in the RNA polymerase II holoenzyme. Curr. Opin. Genet. Dev. 1998; 8:565-570.

12. Barberis A, Pearlberg J, Simkovich N, et al. Contact with a component of the polymerase II holoenzyme suffices for gene activation. Cell 1995; 81:359-368.

13. Marciniak RA, Sharp PA. HIV-1 Tat protein promotes formation of more-processive elongation complexes. EMBO J 1991; 10:4189-4196.

14. Xiao H, Pearson A, Coulombe B, et al. Binding of basal transcription factor TFIIH to the acidic activation domains of VP16 and p53. Mol Cell Biol 1994; 14:7013-7024.

15. Tong X, Drapkin R, Reinberg D, et al. The 62- and 80-kDa subunits of transcription factor IIH mediate the interaction with Epstein-Barr virus nuclear protein 2. Proc Natl Acad Sci U S A 1995; 92:3259-3263.

16. Blau J, Xiao H, McCracken S, et al. Three functional classes of transcriptional activation domains. Mol Cell Biol 1996; 16:2044-2055.

17. Yankulov KY, Pandes M, McCracken S, et al. TFIIH functions in regulating transcriptional elongation by RNA polymerase II in *Xenopus* oocytes. Mol Cell Biol 1996; 16:3291-3299.

18. Sawadogo M, Sentenac A. RNA polymerase B (II) and general transcription factors. Annu Rev Biochem 1990; 59:711-754.

19. Acker, J., de Graaff, M., Cheynel, I., Khazak, V., Kedinger, C., Vigneron, M. Interactions between the human RNA polymerase II subunits. J Biol Chem. 1997; 272:16815-16821.

20. Nonet M, Sweetser D, Young RA. Functional redundancy and structural polymorphism in the large subunit of RNA polymerase II. Cell 1987; 50:909-915.

21. Corden JL. Tails of RNA polymerase II. Trends Biochem Sci 1990; 15:383-387.

22. Meredith GD, Chang W-H, Li Y, et al. The C-terminal domain revealed in the structure of RNA polymerase II. J Mol Biol 1996; 258:413-419.

23. Lu H, Flores O, Weinmann R, et al. The nonphosphorylated form of RNA polymerase II preferentially associates with the initiation complex. Proc Natl Acad Sci U S A 1991; 88:10004-10008.

24. O'Brien T, Hardin S, Greenleaf A, et al. Phosphorylation of RNA polymerase II C-terminal domain and transcriptional elongation. Nature 1994; 370:75-77.

25. Lu H, Zawel L, Fisher L et al. Human general transcription factor IIH phosphorylates the C-terminal domain of RNA polymerase II. Nature 1992; 358:641-645.

26. Cho, E. J., Rodriguez, C.R., Takagi, T., Buratowski, S. Allosteric interactions between capping enzyme subunits and the RNA polymerase II carboxy-terminal domain. Genes Dev. 1998; 12:3482-3487.

27. Mortillaro, M. J., Blencowe, B.J., Wei, X., Nakayasu, H., Du, L., Warren, S.L., Sharp, P.A., Berezney, R. A hyperphosphorylated form of the large subunit of RNA polymerase II is associated with splicing complexes and the nuclear matrix. Proc Natl Acad Sci U S A 1996; 93:8253-8257.

28. McAllister, W. T. Structure and function of the bacteriophage T7 RNA polymerase (or, the virtues of simplicity). Cell Mol Biol Res. 1993; 39:385-391.

29. Miyao, T., Woychik, N.A. RNA polymerase subunit RPB5 plays a role in transcriptional activation. Proc Natl Acad Sci U S A 1998; 95:15281-15286.

30. Petermann, R., Mossier, B.M., Aryee, D.N., Khazak, V., Golemis, E.A., Kovar, H. Oncogenic EWS-Fli1 interacts with hsRPB7, a subunit of human RNA polymerase II. Oncogene 1998; 17:603-610.

31. Schlegel, B. P., Green, V.J., Ladias, J.A., Parvin, J.D. BRCA1 interaction with RNA polymerase II reveals a role for hRPB2 and hRPB10α in activated transcription. Proc Natl Acad Sci U S A 2000; 97:3148-3153.

32. Matsui T, Segall J, Weil A, et al. Multiple factors required for accurate initiation of transcription by purified RNA polymerase II. J Biol Chem 1980; 255:11992-11996.

33. Bucher P. Weight matrix descriptions of four eukaryotic RNA polymerase II promoter elements derived from 502 unrelated promoter sequences. J Mol Biol 1990; 212:563-578.

34. Purnell BA, Emanuel PA, Gilmour DS. TFIID sequence recognition of the initiator and sequences farther downstream in *Drosophila* class II genes. Genes Dev 1994; 8:830-842.

35. Burke TW, Kadonaga JT. *Drosophila* TFIID binds to a conserved downstream basal promoter element that is present in many TATA-box-deficient promoters. Genes Dev 1996; 10:711-724.

36. Nikolov DB, Hu SH, Lin J, et al. Crystal structure of TFIID TATA-box binding protein. Nature 1992; 360:40-46.

37. Chasman DI, Flaherty KM, Sharp PA et al. Crystal structure of yeast TATA-binding protein and model for interaction with DNA. Proc Natl Acad Sci U S A 1993; 90:8174-8178.

38. Kim JL, Nikolov DB, Burley SK. Co-crystal structure of TBP recognizing the minor groove of a TATA element. Nature 1993; 365:520-527.

39. Kim Y, Geiger JH, Hahn S et al. Crystal structure of a yeast TBP/TATA-box complex. Nature 1993; 365:512-520.

40. Coleman RA, Taggart AK, Benjamin LR, et al. Dimerization of the TATA binding protein. J Biol Chem 1995; 270:13842-13849.

41. Taggart AK, Pugh BF. Dimerization of TFIID when not bound to DNA. Science 1996; 272:1331-1333.

42. Tabuchi H, Handa H, Hirose S. Underwinding of DNA on binding of yeast TFIID to the TATA element. Biochem Biophys Res Commun 1993; 192:1432-1438.

43. Parvin JD, McCormick RJ, Sharp PA, et al. Pre-bending of a promoter sequence enhances affinity for the TATA-binding factor. Nature 1995; 373:724-727.

44. Comai L, Tanese N, Tjian R. The TATA-binding protein and associated factors are integral components of the RNA polymerase I transcription factor, SL1. Cell 1992; 68:965-976.

45. Sharp PA. TATA-binding protein is a classless factor. Cell 1992; 68:819-821.

46. Dynlacht BD, Hoey T, Tjian R. Isolation of coactivators associated with the TATA-binding protein that mediate transcriptional activation. Cell 1991; 66:563-576.

47. Brou C, Chaudhary S, Davidson I, et al. Distinct TFIID complexes mediate the effect of different transcriptional activators. EMBO J 1993; 12:489-499.

48. Jacq X, Brou C, Lutz Y et al. Human TAFII30 is present in a distinct TFIID complex and is required for transcriptional activation by the estrogen receptor.Cell 1994; 79:107-117.

49. Hoffman A, Chiang C-M, Oelgeschläger T, et al. A histone octamer-like structure within TFIID. Nature 1996; 380:356-359.

50. Xie X, Kobuko T, Cohen S, et al. Structural similarity between TAFs and the heterotetrameric core of the histone octamer. Nature 1996; 380:316-322.

51. Nakajima N, Horikoshi M, Roeder RG. Factors involved in specific transcription by mammalian RNA polymerase II: purification, genetic specificity, and TATA box-promoter interactions of TFIID. Mol Cell Biol 1988; 8:4028-4040.

52. Sawadogo M, Roeder RG. Interaction of a gene-specific transcription factor with the adenovirus major late promoter upstream of the TATA box region. Cell 1985; 43:165-175.

53. Horikoshi M, Carey MF, Kakidani H, et al. Mechanism of action of a yeast activator: direct effect of GAL4 derivatives on mammalian TFIID-promoter interactions. Cell 1988; 54:665-669.

54. Horikoshi M, Hai T, Lin YS, et al. Transcription factor ATF interacts with the TATA factor to facilitate establishment of a preinitiation complex. Cell 1988; 54:1033-1042.

55. Oelgeschläger T, Chiang C-M, Roeder RG. Topology and reorganization of a human TFIID-promoter complex. Nature 1996; 382:735-738.

56. Parvin JD, Sharp PA. DNA topology and a minimal set of basal factors for transcription by RNA polymerase II. Cell 1993; 73:533-540.

57. Sauer F, Hansen SK, Tjian R. Multiple TAFIIs directing synergistic activation of transcription. Science 1995; 270:1783-1788.

58. Sauer F, Hansen SK, Tjian R. DNA template and activator-coactivator requirements for transcriptional synergism by *Drosophila* Bicoid. Science 1995; 270:1825-1828.

59. Kretzschmar M, Kaiser K, Lottspeich F, et al. A novel mediator of class II gene transcription with homology to viral immediate-early transcriptional regulators. Cell 1994; 78:525-534.

60. Dikstein R, Ruppert S, Tjian R. TAFII250 is a bipartite protein kinase that phosphorylates the basal transcription factor RAP74. Cell 1996; 84:781-790.

61. Kitajima S, Chibazakura T, Yonaha M, et al. Regulation of the human general transcription initiation factor TFIIF by phosphorylation. J Biol Chem 1994; 269:29970-29977.

62. Ruppert S, Wang EH, Tjian R. Cloning and expression of human TAFII250: a TBP-associated factor implicated in cell-cycle regulation. Nature 1993; 362:175-179.

63. Hisatake K, Hasegawa S, Takada R, et al. The p250 subunit of native TATA box-binding factor TFIID is the cell-cycle regulatory protein CCG1. Nature 1993; 362:179-181.

64. Sekiguchi T, Miyata T, Nishimoto T. Molecular cloning of the cDNA of human X chromosomal gene (CCG1) which complements the temperature-sensitive G1 mutants, tsBN462 and ts13, of the BHK cell line. EMBO J 1988; 7:1683-1687.

65. Sekiguchi T, Nohiro Y, Nakamura Y, et al. The human CCG1 gene, essential for progression of the G1 phase, encodes a 210-kilodalton nuclear DNA-binding protein. Mol Cell Biol 1991; 11:3317-3325.

66. Smale ST, Baltimore D. The "initiator" as a transcription control element. Cell 1989; 57:103-113.

67. Javahery R, Khachi A, Lo K, et al. DNA sequence requirements for transcriptional initiator activity in mammalian cells. Mol Cell Biol 1994; 14:116-127.

68. Roy AL, Meisterernst M, Pognonec P, et al. Cooperative interaction of an initiator-binding transcription initiation factor and the helix-loop-helix activator USF. Nature 1991; 354:245-248.

69. Roy AL, Malik S, Meisterernst M, et al. An alternative pathway for transcriptional initiation involving TFII-I. Nature 1993; 365:355-359.

70. Usheva A, Shenk T. TATA-binding protein-independent initiation: YY1, TFIIB, and RNA polymerase II direct basal transcription on supercoiled template DNA. Cell 1994; 76:1115-1121.

71. Ha I, Lane W, Reinberg D. Cloning of a human gene encoding the general transcription initiation factor IIB. Nature 1991; 352:689-695.

72. Imbalzano AN, Zaret KS, Kingston RE. Transcription factor (TF) IIB and TFIIA can independently increase the affinity of the TATA-binding protein for DNA. J Biol Chem 1994; 269:8280-8286.

73. Nikolov DB, Chen H, Halay ED, et al. Crystal structure of a TFIIB-TBP-TATA-element ternary complex. Nature 1995; 377:119-128.

74. Pinto I, Ware DE, Hampsey M. The yeast SUA7 gene encodes a homolog of human transcription factor TFIIB and is required for normal start site selection *in vivo*. Cell 1992; 68:977-988.

75. Berroteran RW, Ware DE, Hampsey M. The sua8 suppressors of *Saccharomyces cerevisiae* encode replacements of conserved residues within the largest subunit of RNA polymerase II and affect transcription start site selection similarly to sua7 (TFIIB) mutations. Mol Cell Biol 1994; 14:226-237.

76. Tschochner H, Sayre MH, Flanagan PM, et al. Yeast RNA polymerase II initiation factor e: isolationand identification as the functional counterpart of human transcription factor IIB. Proc Natl Acad Sci U S A 1992; 89:11292-11296.

77. Maldonado E, Ha I, Cortes P, et al. Factors involved in specific transcription by mammalian RNA polymerase II: role of transcription factors IIA, IID, and IIB during formation of a transcription-competent complex. Mol Cell Biol 1990; 10:6335-6347.

78. Roberts SG, Green MR. Activator-induced conformational change in general transcription factor TFIIB. Nature 1994; 371:717-720.

79. Lin YS, Ha I, Maldonado E, et al. Binding of general transcription factor TFIIB to an acidic activating region. Nature 1991; 353:569-571.

80. Sopta M, Burton ZF, Greenblatt J. Structure and associated DNA-helicase activity of a general transcription initiation factor that binds to RNA polymerase II. Nature 1989; 341:410-414.

81. Aso T, Vasavada HA, Kawaguchi T, et al. Characterization of cDNA for the large subunit of the transcription initiation factor TFIIF. Nature 1992; 355:461-464.

82. Finkelstein A, Kostrub CF, Li J, et al. A cDNA encoding RAP74, a general initiation factor for transcription by RNA polymerase II. Nature 1992; 355:464-467.

83. Flores O, Lu H, Killeen M, et al. The small subunit of transcription factor IIF recruits RNA polymerase II into the preinitiation complex. Proc Natl Acad Sci U S A 1991; 88:9999-10003.

84. Ha I, Roberts S, Maldonado E, et al. Multiple functional domains of human transcription factor IIB: distinct interactions with two general transcription factors and RNA polymerase II. Genes Dev 1993; 7:1021-1032.

85. Conaway JW, Conaway RC. An RNA polymerase II transcription factor shares functional properties with *Escherichia coli* sigma 70. Science 1990; 248:1550-1553.

86. Flores O, Maldonado E, Reinberg D. Factors involved in specific transcription by mammalian RNA polymerase II: factors IIE and IIF independently interact with RNA polymerase II. J Biol Chem 1989; 264:8913-8921.

87. Price DH, Sluder AE, Greenleaf AL. Dynamic interaction between a *Drosophila* transcription factor and RNA polymerase II. Mol Cell Biol 1989; 9:1465-1475.

88. Zawel L, Kumar KP, Reinberg D. Recycling of the general transcription factors during RNA polymerase II transcription. Genes Dev 1995; 9:1479-1490.

89. Joliot V, Demma M, Prywes R. Interaction with RAP74 subunit of TFIIF is required for transcriptional activation by serum response factor. Nature 1995; 373:632-635.

90. Martin ML, Lieberman PM, Curran T. Fos-Jun dimerization promotes interaction of the basic region with TFIIE-34 and TFIIF. Mol Cell Biol 1996; 16:2110-2118.

91. Peterson MG, Inostroza J, Maxon ME, et al. Structure and function of the recombinant subunits of human TFIIE. Nature 1991; 354:369-373.

92. Ohkuma Y, Sumimoto H, Hoffman A, et al. Structural motifs and potential s holomogies in the large subunit of human general transcription factor TFIIE.

93. Sumimoto H, Ohkuma Y, Sinn E, et al. Conserved sequence motifs in the small subunit of human general transcription factor TFIIE. Nature 1991; 354:401-404.

94. Holstege FCP, Tantin D, Carey M, et al. The requirement for the basal transcription factor IIE is determined by the helical stability of promoter DNA. EMBO J 1995; 14:810-819.

95. Maxon ME, Goodrich JA, Tjian R. Transcription factor IIE binds preferentially to RNA polymerase IIa and recruits TFIIH: a model for promoter clearance. Genes Dev 1994; 8:515-524.

96. Tong X, Wang F, Thut CJ, et al. The Epstein-Barr virus nuclear protein 2 acidic domain can interact with TFIIB, TAF40, and RPA70 but not with TATA-binding protein. J Virol 1995; 69:585-588.

97. Sauer F, Fondell JD, Ohkuma Y, et al. Control of transcription by Kruppel through interactions with TFIIB and TFIIE beta. Nature 1995; 375:162-164.

98. Leuther KK, Bushnell DA, Kornberg RD. Two-dimensional crystallography of TFIIB- and IIE-RNA polymerase II complexes: implications for start site selection and initiation complex formation. Cell 1996; 85:773-779.

99. Parvin JD, Shykind BM, Meyers RE et al. Multiple sets of basal factors initiate transcription by RNA polymerase II. J Biol Chem 1994; 269:18414-18421.

100. Timmers HTM. Transcription initiation by RNA polymerase II does not require hydrolysis of the beta-gamma phosphoanhydride bond of ATP. EMBO J 1994; 13:391-399.

101. Goodrich JA, Tjian R. Transcription factors IIE and IIH and ATP hydrolysis direct promoter clearance by RNA polymerase II. Cell 1994; 77:145-156.

102. Pan G, Greenblatt J. Initiation of transcription by RNA polymerase II is limited by melting of the promoter DNA in the region immediately upstream of the initiation site. J Biol Chem 1994; 269:30101-30104.

103. Drapkin R, Reinberg D. The multifunctional TFIIH complex and transcriptional control. Trends Biochem Sci 1994; 19:504-508.

104. Schaeffer L, Roy R, Humbert S, et al. DNA repair helicase: a component of BTF2 (TFIIH) basic transcription factor. Science 1993; 260:58-63.

105. Drapkin R, Reardon JT, Anseri A, et al. Dual role of TFIIH in DNA excision repair and in transcription by RNA polymerase II. Nature 1994; 368:769-772.

106. Ma L, Siemssen ED, Noteborn MHM, et al. The xeroderma pigmentosum group B protein ERCC3 produced in the baculovirus system exhibits DNA helicase activity. Nucl Acids Res 1994; 22:4095-4102.

107. Sung P, Bailly V, Weber C, et al. Human xeroderma pigmentosum group D gene encodes a DNA helicase. Nature 1993; 365:852-855.

108. Feaver WJ, Svjstrup JQ, Bardwell AJ, et al. Dual roles of a multiprotein complex from S. cerevisiae in transcription and DNA repair. Cell 1993; 75:1379-1387.

109. Park E, Guzder SN, Kokken MHM, et al. RAD25 (SSL2), the yeast homolog of the human xeroderma pigmentosum group B DNA repair gene, is essential for viability. Proc Natl Acad Sci U S A 1992; 89:11416-11420.

110. Hwang JR, Moncollin V, Vermeulen W, et al. A 3'-5' XPB helicase defect in repair/transcription factor TFIIH of xeroderma pigmentosum group B affects both DNA repair and transcription. J Biol Chem 1996; 271:15898-15904.

111. Shuttleworth J, Godfrey R, Colman A. p40MO15, a cdc2-related protein kinase involved in negative regulation of meiotic maturation of Xenopus oocytes. EMBO J 1990; 9:3233-3240.

112. Roy R, Adamczewski JP, Seroz T, et al. The MO15 cell cycle kinase is associated with the TFIIH transcription-DNA repair factor. Cell 1994; 79:1093-1101.

113. Serizawa H, Mäkelä TP, Conaway JW, et al. Association of Cdk-activating kinase subunits with transcription factor TFIIH. Nature 1995; 374:280-283.

256

114. Shiekhattar R, Mermelstein F, Fisher RP, et al. Cdk-activating kinase complex is a component of human transcription factor TFIIH. Nature 1995; 374:283-287.

115. Fisher RP, Jin P, Chamberlin HM, et al. Alternative mechanisms of CAK assembly require an assembly factor or an activating kinase. Cell 1995; 83:47-57.

116. Mäkelä TP, Parvin JD, Kim J, et al. A kinase-deficient transcription factor IIH is functional in basal and activated transcription. Proc Natl Acad Sci U S A 1995; 92:5174-5178.

117. Drapkin R, Le Roy G, Cho H, et al. Human cyclin-dependent kinase-activating kinase exists in three distinct complexes. Proc Natl Acad Sci U S A 1996; 93:6488-6493.

118. Serizawa H, Conaway JW, Conaway RC. Phosphorylation of C-terminal domain of RNA polymerase II is not required in basal transcription. Nature 1993; 363:371-374.

119. Tirode, F., Busso, D., Coin, F., Egly, J.M. Reconstitution of the transcription factor TFIIH: assignment of functions for the three enzymatic subunits, XPB, XPD, and cdk7. Mol Cell 1999; 3:87-95.

120. Moreland, R. J., Tirode, F., Yan, Q., Conaway, J.W., Egly, J.M., Conaway, R.C. A role for the TFIIH XPB DNA helicase in promoter escape by RNA polymerase II. J Biol Chem. 1999; 274:22127-22130

121. Meisterernst M, Roy AL, Lieu HM, et al. Activation of class II gene transcription by regulatory factors is potentiated by a novel activity. Cell 1991; 66:981-993.

122. Meisterernst M, Roeder RG. Family of proteins that interact with TFIID and regulate promoter activity. Cell 1991; 67:557-567.

123. Cortes P, Flores O, Reinberg D. Factors involved in specific transcription by mammalian RNA polymerase II: purification and analysis of transcription factor IIA and identification of transcription factor IIJ. Mol Cell Biol 1992; 12:413-421.

124. Coulombe B, Killeen M, Liljelund P, et al. Identification of three mammalian proteins that bind to the yeast TATA box protein TFIID. Gene Expr 1992; 2:99-110.

125. DeJong J, Roeder RG. A single cDNA, hTFIIA/a, encodes both the p35 and p19 subunits of human TFIIA. Genes Dev 1993; 7:2220-2234.

126. Ma D, Watanabe H, Mermelstein F et al. Isolation of a cDNA encoding the largest subunit of TFIIA reveals function important for activated transcription. Genes Dev 1993; 7:2246-2257.

127. Sun X, Ma D, Sheldon, et al. Reconstitution of human TFIIA activity from recombinant polypeptides: a role in TFIID-mediated transcription. Genes Dev 1994; 8:2336-2348.

128. DeJong J, Bernstein R, Roeder RG. Human general transcription factor TFIIA: characterization of a cDNA encoding the small subunit and requirement for basal and activated transcription. Proc Natl Acad Sci U S A 1995; 92:3313-3317.

129. Tan S, Hunziker Y, Sargent DF, et al. Crystal structure of a yeast TFIIA/TBP/DNA complex. Nature 1996; 381:127-134.

130. Geiger JH, Hahn S, Lee S, et al. Crystal structure of the yeast TFIIA/TBP/DNA complex. Science 1996; 272:830-836.

131. Kang JJ, Auble DT, Ranish JA, et al. Analysis of the yeast transcription factor TFIIA: distinct functional regions and a polymerase II-specific role in basal and activated transcription. Mol Cell Biol 1995; 15:1234-1243.

132. Merino A, Madden KR, Lane WS, et al. DNA topoisomerase I is involved in both repression and activation of transcription. Nature 1993; 365:227-232.

133. Shykind BM, Kim J, Sharp PA. Activation of the TFIID-TFIIA complex with HMG-2. Genes Dev 1995; 9:1354-1365.

134. Grosschedl R, Giese K, Pagel J. HMG domain proteins: architectural elements in the assembly of nucleoprotein structures. Trends Genet 1994; 10:94-100.

135. Ge H, Roeder RG. Purification, cloning, and characterization of a human coactivator, PC4, that mediates transcriptional activation of class II genes. Cell 1994; 78:513-523.

136. Goppelt A, Stelzer G, Lottspeich F, et al. A mechanism for repression of class II gene transcription through specific binding of NC2 to TBP-promoter complexes via heterodimeric histone fold domains. EMBO J 1996; 15:3105-3116.

137. Inostroza JA, Mermelstein FH, Ha I, et al. Dr1, a TATA-binding protein-associated phosphoprotein and inhibitor of class II gene expression. Cell 1992; 70:477-489.

138. Ge H, Roeder RG. The high mobility group protein HMG1 can reversibly inhibit class II gene transcription by interaction with the TATA-binding protein. J Biol Chem 1994; 269:17136-17140.

139. Kim TK, Zhao Y, Ge H, et al. TATA-binding protein residues implicated in a functional interplay between negative cofactor NC2 (Dr1) and general factors TFIIA and TFIIB. J Biol Chem 1995; 270:10976-10981.

140. Stelzer G, Goppelt A, Lottspeich F, et al. Repression of basal transcription by HMG2 is counteracted by TFIIH-associated factors in an ATP-dependent process. Mol Cell Biol 1994; 14:4712-4721.

141. Ma D, Olave I, Merino A, et al. Separation of the transcriptional coactivator and antirepression functions of transcription factor IIA. Proc Natl Acad Sci U S A 1996; 93:6583-6588.

142. Workman JL, Kingston RE. Nucleosome core displacement *in vitro* via a metastable transcription factor-nucleosome complex. Science 1992; 258:1780-1784.

143. Thompson CM, Young RA. General requirement for RNA polymerase II holoenzymes *in vivo*. Proc Natl Acad Sci U S A 1995; 92:4587-4590.

144. Scully, R., Anderson, S.F., Chao, D.M., Wei, W., Ye, L., Young, R.A., Livingston, D.M., Parvin, J.D. BRCA1 is a component of the RNA polymerase II holoenzyme. Proc Natl Acad Sci U S A 1997; 94:5605-5610

145. Chiba, N., Suldan, Z., Freedman, L.P., Parvin, J.D. Binding of liganded vitamin D receptor to the vitamin D receptor interacting protein coactivator complex induces interaction with RNA polymerase II holoenzyme. J. Biol Chem 2000; 275:10719-10722

146. Thompson CM, Koleske AJ, Chao DM, et al. A multisubunit complex associated with the RNA polymerase II CTD and TATA-binding protein in yeast. Cell 1993; 73:1361-1375.

147. Hengartner CJ, Thompson CM, Zhang J, et al. Association of an activator with an RNA polymerase II holoenzyme. Genes Dev 1995; 9:897-910.

148. Liao SM, Zhang J, Jeffery DA et al. A kinase-cyclin pair in the RNA polymerase II holoenzyme. Nature 1995; 374:193-196.

149. Tassan J-P, Jaquenoud M, Léopold P, et al. Identification of human cyclin-dependent kinase 8, a putative protein kinase partner for cyclin C. Proc Natl Acad Sci U S A 1995; 92:8871-8875.

150. Rickert P, Seghezzi W, Shanahan F, et al. Cyclin C/CDK8 is a novel CTD kinase associated with RNA polymerase II. Oncogene 1996; 12:2631-2640.

151. Kim YJ, Bjorklund S, Li Y, et al. A multiprotein mediator of transcriptional activation and its interaction with the C-terminal repeat domain of RNA polymerase II. Cell 1994; 77:599-608.

152. Myers, L. C., Gustafsson, C.M., Bushnell, D.A., Lui, M., Erdjument-Bromage, H., Tempst, P., Kornberg, R.D. The Med proteins of yeast and their function through the RNA polymerase II carboxy-terminal domain. Genes Dev 1998; 12:45-54

153. Svejstrup, J. Q., Li, Y., Fellows, J., Gnatt, A., Bjorklund, S., Kornberg, R.D. Evidence for a mediator cycle at the initiation of transcription. Proc Natl Acad Sci U S A 1997; 94:6075-6078

154. Pan, G., Aso, T., Greenblatt, J. Interaction of elongation factors TFIIS and elongin A with a human RNA polymerase II holoenzyme capable of promoter-specific initiation and responsive to transcriptional activators. J Biol Chem 1997; 272:24563-24571

155. Cho, H., Maldonado, E., Reinberg, D. Affinity purification of a human RNA polymerase II complex using monoclonal antibodies against transcription factor IIF. J Biol Chem 1997; 272:11495-11502

156. Nakajima, T., Uchida, C., Anderson, S.F., Parvin, J.D., Montminy, M. Analysis of a cAMP-responsive activator reveals a two-component mechanism for transcriptional induction via signal-dependent factors. Genes Dev 1997; 11:738-747

157. Neish, A. S., Anderson, S.F., Schlegel, B.P., Wei, W., Parvin, JD. Factors associated with the mammalian RNA polymerase II holoenzyme. Nuc. Acids Res. 1998; 26:847-853.

158. Kim, T. K., Kim, T.H., Maniatis, T. Efficient recruitment of TFIIB and CBP-RNA polymerase II holoenzyme by an interferon-beta enhanceosome in vitro. Proc Natl Acad Sci U S A 1998; 95:12191-12196

159. Owen-Hughes T, Workman JL. Experimental analysis of chromatin function in transcription control. Crit Rev Eukaryot Gene Expr 1994; 4:403-441.

160. Knezetic JA, Luse DS. The presence of nucleosomes on a DNA template prevents initiation by RNA polymerase II in vitro. Cell 1986; 45:95-104.

161. Lorch Y, LaPointe JW, Kornberg RD. Nucleosomes inhibit the initiation of transcription but allow chain elongation with the displacement of histones. Cell 1987; 49:203-210.

162. Workman JL, Roeder RG, Kingston RE. An upstream transcription factor, USF (MLTF), facilitates the formation of preinitation complexes during an in vitro chromatin assembly. EMBO J 1990; 9:1299-1308.

163. Matsui T. Transcription of adenovirus type 2 major late and peptide IX genes under conditions of in vitro nucleosome assembly. Mol Cell Biol 1987; 7:1401-1408.

164. Knezetic JA, Jacob GA, Luse DS. Assembly of RNA polymerase II preinitiation complexes before assembly of nucleosomes allows efficient initation by RNA polymerase II *in vitro*. Cell 1986; 45:95-104.

165. Workman JL, Roeder RG. Binding of transcription factor TFIID to the major late promoter during *in vitro* nucleosome assembly potentiates subsequent initiation by RNA polymerase II. Cell 1987; 51:613-622.

166. Meisterernst M, Horikoshi M, Roeder RG. Recombinant yeast TFIID, a general transcription factor, mediates activation by the gene-specific factor USF in a chromatin assembly assay. Proc Natl Acad Sci U S A 1990; 87:9153-9157.

167. Cosma, M. P., Tanaka, T., Nasmyth, K. Ordered recruitment of transcription and chromatin remodeling factors to a cell cycle- and developmentally regulated promoter. Cell 1999; 97:299-311

168. Cairns BR, Kim YJ, Sayre MH, et al. A multisubunit complex containing the SWI1/ADR6, SWI2/SNF2, SWI3, SNF5, and SNF6 gene products isolated from yeast. Proc Natl Acad Sci U S A 1994; 91:1950-1954.

169. Peterson CL, Dingwall A, Scott MP. Five SWI/SNF gene products are components of a large multisubunit complex required for transcriptional enhancement. Proc Natl Acad Sci U S A 1994; 91:2905-2908.

170. Côté J, Quinn J, Workman JL, et al. Stimulation of GAL4 derivative binding to nucleosomal DNA by the yeast SWI/SNF complex. Science 1994; 265:53-60.

171. Wilson CJ, Chao DM, Imbalzano AN, et al. RNA polymerase II holoenzyme contains SWI/SNF regulators involved in chromatin remodeling. Cell 1996; 84:235-244.

172. Khavari PA, Peterson CL, Tamkun JW, et al. BRG1 contains a conserved domain of the SWI2/SNF2 family necessary for normal mitotic growth and transcription. Nature 1993; 366:170-174.

173. Muchardt C, Yaniv M. A human homologue of *Saccharomyces cerevisiae* SNF2/SWI2 and *Drosophila* brm genes potentiates transcriptional activation by the glucocorticoid receptor. EMBO J 1993; 12:4279-4290.

174. Muchardt C, Sardet C, Bourachot B et al. A human protein with homology to *Saccharomyces cerevisiae* SNF5 interacts with the potential helicase hbrm. Nucl Acids Res 1995; 23:1127-1132.

175. Kwon H, Imbalzano AN, Khavari PA, et al. Nucleosome disruption and enhancement of activator binding by a human SWI/SNF complex. Nature 1994; 370:477-481.

176. Brown, S. A., Imbalzano, A.N., Kingston, R.E. Activator-dependent regulation of transcriptional pausing on nucleosomal templates. Genes Dev. 1996; 10:1479-1490

177. Lee DY, Hayes JJ, Pruss D et al. A positive role for histone acetylation in transcription factor access to nucleosomal DNA. Cell 1993; 72:73-84.

178. Wolffe AP, Pruss D. Targeting chromatin disruption: transcription regulators that acetylate histones. Cell 1996; 84:817-819.

179. Yang X-J, Ogryzko VV, Nishikawa J-I, et al. A p300/CBP-associated factor that competes with the adenoviral oncoprotein E1A. Nature 1996; 382:319-324.

180. Ogryzko, V. V., Kotani, T., Zhang, X., Schlitz, R.L., Howard, T., Yang, X.J., Howard, B.H., Qin, J., Nakatani, Y. Histone-like TAFs within the PCAF histone acetylase complex. Cell 1998; 94:35-44.

181. Grant, P. A., Duggan, L., Cote, J., Roberts, S.M., Brownell, J.E., Candau, R., Ohba, R., Owen-Hughes, T., Allis, C.D., Winston, F., Berger, S.L., Workman, J.L. Yeast Gcn5 functions in two multisubunit complexes to acetylate nucleosomal histones: characterization of an Ada complex and the SAGA (Spt/Ada) complex. Genes Dev. 1997; 11:1640-1650.

182. Hassig, C. A., Fleischer, T.C., Billin, A.N., Schreiber, S.L., Ayer, D.E. Histone deacetylase activity is required for full transcriptional repression by mSin3A. Cell 1997; 89:341-347.

183. Brehm, A., Miska, E.A., McCance, D.J., Reid, J.L., Bannister, A.J., Kouzarides, T. Retinoblastoma protein recruits histone deacetylase to repress transcription. Nature 1998; 391:597-601.

184. Kass, S. U., Pruss, D., Wolffe, A.P. How does DNA methylation repress transcription? Trends Genet. 1997; 13:444-449

185. Kingston, R. E. A shared but complex bridge. Nature 1999; 399:199-200

186. Asturias, F. J., Jiang, Y.W., Myers, L.C., Gustafsson, C.M., Kornberg, R.D. Conserved structures of mediator and RNA polymerase II holoenzyme. Science 1999; 283:985-987

187. Naar, A. M., Beaurang, P.A., Zhou, S., Abraham, S., Solomon, W., Tjian, R. Composite co-activator ARC mediates chromatin-directed transcriptional activation. Nature 1999; 398:828-832

188. Rachez, C., Lemon, B.D., Suldan, Z., Bromleigh, V., Gamble, M., Naar, A.M., Erdjument-Bromage, H., Tempst, P., Freedman, L.P. Ligand-dependent transcription activation by nuclear receptors requires the DRIP complex. Nature 1999; 398:824-828

189. Freedman, L. P. Increasing the complexity of coactivation in nuclear receptor signaling. Cell 1999; 97:5-8

190. Wu-Baer F, Sigman D, Gaynor RB. Specific binding of RNA polymerase II to the human immunodeficiency virus trans-activating region RNA is regulated by cellular cofactors and Tat. Proc Natl Acad Sci U S A 1995; 92:7153-7157.

191. Keen NJ, Gait MJ, Karn J. Human immunodeficiency virus type-1 Tat is an integral component of the activated transcription-elongation complex. Proc Natl Acad Sci U S A 1996; 93:2505-2510.

192. Jiang Y, Yan M, Gralla JD. A three-step pathway of transcription initiation leading to promoter clearance at an activated RNA polymerase II promoter. Mol Cell Biol 1996; 16:1614-1621.

193. Koleske AJ, Buratowski S, Nonet M, et al. A novel transcription factor reveals a functional link between the RNA polymerase II CTD and TFIID. Cell 1992; 69:883-894.

194. Zhu, Y., Pe'ery, T., Peng, J., Ramanathan, Y., Marshall, N., Marshall, T., Amendt, B., Mathews, M.B., Price, D.H. Transcription elongation factor P-TEFb is required for HIV-1 tat transactivation *in vitro*. Genes Dev. 1997; 11:2622-2632

195. Wei, P., Garber, M.E., Fang, S.M., Fischer, W.H., Jones, K.A. A novel CDK9-associated C-type cyclin interacts directly with HIV-1 Tat and mediates its high-affinity, loop-specific binding to TAR RNA. Cell 1998; 92:451-462

196.   Chen, D., Fong, Y., Zhou, Q. Specific interaction of Tat with the human but not rodent P-TEFb complex mediates the species-specific Tat activation of HIV-1 transcription. Proc Natl Acad Sci U S A 1999; 96:2728-2733

197.   Holstege, F. C., Jennings, E.G., Wyrick, J.J., Lee, T.I., Hengartner, C.J., Green, M.R., Golub, T.R., Lander, E.S., Young, R.A. Dissecting the regulatory circuitry of a eukaryotic genome. Cell 1998; 95:717-728

Chapter 8

THERAPEUTIC REGULATION OF LEUKOCYTE ADHESION MOLECULE EXPRESSION

Mary E. Gerritsen

Department of Cardiovascular Research
Genentech
MS 42, 1 DNA Way
South San Francisco, CA 94080

INTRODUCTION

Leukocyte adhesion to the vascular endothelium is the initial event which is required for and precedes leukocyte extravasation in tissue, and thus is a critical step in the inflammatory response. The use of monoclonal antibodies, soluble ligands, anti-sense agents and targeted gene disruption has clearly demonstrated that interfering with leukocyte-endothelial ligand receptor interaction and/or the upregulated expression/activation of leukocyte adhesion molecules reduces the inflammatory response and provides therapeutic benefit in a variety of acute and chronic models of inflammation.[1-9] Recent reports suggest that these approaches may have therapeutic value in the clinic as well.[10, 11]

There are numerous strategies for interfering with leukocyte trafficking, and many of these are currently being tested in the clinic. However, the long term success of the current approaches is uncertain – the use of monoclonal antibodies, anti-sense oligonucleotides, and peptide antagonists, while potentially useful for acute or short term indications such as shock, myocardial infarction, or stroke, may not be cost-effective, tolerated, or useful in more chronic indications such as rheumatoid arthritis, multiple sclerosis, and inflammatory bowel disease. Considerable effort has therefore been directed towards the identification and development of small, orally active molecules that would interfere with one or more aspects of leukocyte trafficking. Many groups initially screened for inhibitors of the leukocyte receptors, i.e., inhibitors of ligand-receptor interaction. However, the molecular interaction of the leukocyte adhesion receptors with their cognate ligands has been problematic. Difficulties in the development of high throughput, non-cell based assays, with the multiple sites of interaction between the adhesion receptors and their ligands, with overlapping and potentially redundant ligand-receptor interactions, as well as with the relatively low affinity of these interactions, probably explain the limited number of molecules that have been identified to date. However, small molecule antagonists of ICAM/LFA1 and VCAM/VLA4 have been reported including both peptidomimetics[12,13] and clerodane diterpenes.[14] A number of oligosaccacharide-based antagonists have been described for the selectins.[15-23] However, the utility of many of these compounds may be limited by their lack of oral activity, their low potency, and/or their relatively high cost of production. Molecular modeling studies may provide future advances in the development of new small molecule inhibitors of leukocyte adhesion receptor-ligand binding.

Resting or non-activated endothelial cells present a non-adhesive substrate to leukocytes. However, upon stimulation with cytokines such as tumor necrosis factor-α (TNF-α) and interleukin-1 (IL-1), leukocyte adhesion to human endothelial cells is markedly increased. This increase in adhesion occurs via the upregulated surface expression of the leukocyte adhesion molecules, ICAM-1 (CD54), VCAM-1 (CD106), and E-selectin (CD62E), a consequence of transcriptional activation of these genes. P-selectin mRNA is not upregulated by these cytokines, but can be upregulated by IL-4 and oncostatin M.[24]

Although the majority of adhesion molecule expression-inhibiting compounds reported to date were identified in cell-based assays, there are numerous molecular targets at which these compounds could conceivably interfere. The transmission of the signal from the membrane receptor to the target gene in the nucleus involves a complex cascade of events including protein-protein interactions, protein kinases, and protein phosphatases. Transcriptional regulation of all three vascular endothelial cell adhesion proteins, i.e., ICAM-1, VCAM-1, and E-selectin, is regulated by nuclear factor-κB (NF-κB). NF-κB is a family of dimeric transcription factor complexes of which the p50/p65 heterodimer is the predominant endothelial species (reviewed in 25). Cytokine responsiveness of E-selectin, VCAM-1, and ICAM-1 requires the presence of conserved NF-κB sites in the promoters of all three genes, and p50/p65 has been shown to bind the NF-κB sites in ICAM, E-selectin, and VCAM-1.[25] The activation of NF-κB requires phosphorylation and subsequent degradation of a cytoplasmic inhibitor, IκBα (reviewed in 26-28). The transcriptional activity of NF-κB is also controlled by phosphorylation of the dimer subunits, p50 and p65, and phosphorylation of p65 has been shown to positively regulate its transcriptional activity.[29-34]

Two kinases have been identified that will phosphorylate IκB at serines 32 and 36, namely IκB kinases α and β (IKKα and IKKβ).[35-39] Based on genetic deletion studies in the mouse, IKKβ appears to be the more important controller of NF-κB activity in response to cytokines.[40-44] The IKK are thought to be present in a high molecular weight complex which contains scaffolding proteins (IKKγ and IKAP) as well as NF-κB, IκB, and other proteins.[45-47] In turn, the activities of IKK are regulated by NF-κB-interacting kinase (NIK) and the MAP/ERK kinases 1-3 (MEKK1-3).[48-54]

The Rho family of small molecules, Rac1, Cdc42hs, and RhoA, may also regulate NF-κB activation.[55-59] Dominant negative forms of Rac1 and RhoA block NF-κB activation by IL1β and TNF-α. Rac1 and Cdc42hs share a number of common effector proteins, including the p21 activated kinases (PAKs),[60] of which four have been identified (PAK1-4).[60-63] Activation of PAKs can in turn activate the ERK, JNK, and p38 MAPK cascades.[60-63] PAK1 has been shown to mediate NF-κB activation by Ras, Raf1, and Rac1, and dominant active PAK1 can stimulate the nuclear translocation of the p65 subunit of NF-κB in the absence of IKK activation.[64]

Activation of the mitogen-activated protein kinase (MAPK) family of serine/threonine protein kinases may also play an important role in cytokine activation of endothelial cells and in the expression of at least one adhesion molecule, E-selectin. As reviewed elsewhere in this volume, the promoter for the E-selectin gene contains four positive regulatory domains (PDs) necessary for cytokine responsiveness. PDI, III, and IV contain NF-κB recognition sequences, while the fourth element, PDII, contains the sequence TGACATCA. The E-selectin PDII site is recognized by several members of the CREB/ATF family of transcription factors including activating transcription factor-2 (ATF-2), ATF-α, and c-Jun.[65,66] c-Jun is phosphorylated by the c-Jun terminal kinases (JNK)[67] and ATF-2 is phosphorylated by both the JNK and p38 subgroups of MAP kinases.[68,69] Additionally, it has been shown that the phosphorylation of c-Jun and ATF-2 in their N-terminal activation domains regulates the transcriptional activity of these transcription factors.[70,71] TNF activation of the "stress kinase cascade" leads to activation of JNK and p38 kinases; thus this parallel pathway may converge with NF-κB to result in maximal expression of E-selectin. The effects of dominant interfering molecules which block JNK-1/p38 signaling demonstrate the importance of these kinases for cytokine-induced E-selectin expression.[72] The inability of mice in which the gene for ATF-2 has been knocked out to express E-selectin in response to LPS provides *in vivo* support for this hypothesis.[73]

An additional kinase pathway has been implicated in the regulation of ICAM-1 expression. In contrast to E-selectin and VCAM-1, the immune interferon, interferon γ (IFNγ), selectively upregulates ICAM-1, and can synergistically, or at least additively, augment ICAM-1 expression induced by cytokines such as TNF-α and IL-1. Most but not all of IFNγ's biological effects on cells are mediated via the JAK (Janus kinase)-STAT (Signal Transduction and Activator or Transcription) signaling pathway. Treatment of cells with IFNγ induces a DNA-binding activity (GAF or gamma interferon activator factor; composed of STAT1α (p91) dimers) as a consequence of tyrosine phosphorylation of STAT1 by members of the JAK family of receptor-associated protein tyrosine kinases. The general scheme of this pathway involves aggregation of ligand-occupied receptors, resulting in the formation of multimeric complexes with JAKS. The JAKS catalyze the phosphorylation of the receptors, themselves, and the STAT proteins. Tyrosine phosphorylated and dimerized STATS translocate to the nucleus and bind to DNA regulatory elements. Members of the JAK-STAT signaling cascade have been shown to be activated in response to a variety of cytokines and growth factors in addition to the interferons.[74-77] An interferon response element (pIRE) has been identified at -76 to -65 of the ICAM-1 promoter, and IFNγ induces binding of a transcription factor, probably related to STAT-1α,[78] to this site. In addition, IRF-1 cooperates with NF-κB to activate VCAM-1 gene transcription.[79]

The development of cell-based assays for evaluating drug effects on adhesion protein expression or transcription factor activation has revealed new mechanisms for a number of well-known anti-inflammatory agents including glucocorticoids, nitric oxide, aspirin, ibuprofen, calcium channel blockers, and antioxidants. Many of these drugs block NF-κB signaling. A variety of natural products, isolated from extracts of various plants and fungi known to have anti-inflammatory properties, have also been reported to inhibit adhesion molecule expression, and the molecular targets for some of these products identified. Additionally, a limited number of novel compounds have been reported. These will be discussed below.

Known Anti-Inflammatory and Immunosuppressive Agents

A remarkable number of well-known anti-inflammatory drugs appear to inhibit endothelial cell adhesion molecule expression, a "side-effect" of the drugs not previously recognized. However, it is important to interpret these data with a note of caution. The "inhibitory" effects of many of the drugs occur at suprapharmacological concentrations, many well beyond what it is reasonable to expect circulating blood levels to be in patients taking such drugs. These instances are noted in the discussion of the various drugs below.

Aspirin and Salicylates: High concentrations of aspirin and sodium salicylate (i.e., >1mM) inhibit activation of NF-κB by stabilizing IκBα.[80] In endothelial cells, aspirin and sodium salicylate inhibit IκBα phosphorylation and subsequent degradation.[81] Aspirin and salicylate block NF-κB activation and suppress cytokine-stimulated adhesion molecule expression as well as leukocyte adhesion.[81,82] Sodium salicylate also suppresses leukocyte transmigration *in vitro* under defined laminar flow[81] and inhibits oxygen radical-mediated leukocyte emigration in postcapillary venules *in vivo*.[83] These effects are independent of the inhibitory effects of these agents on cyclooxygenase,[81,82] adding to a growing list of prostaglandin-independent effects of salicylates (reviewed in 84). Salicylates can act as free radical scavengers, selectively inactivating hydroxyl anions,[85] suggesting that salicylates may act as antioxidants.[86]

Cyclosporin (CsA) and FK506: CsA and FK506 are efficient inhibitors of the Ca^{2+}/calmodulin-dependent phosphatase, calcineurin (reviewed in 87), and both drugs inhibit the catalytic activity of calcineurin. In addition to inhibiting the activity of the NF-AT transcription factors, CsA and FK506 are weak inhibitors of NF-κB activation, an effect apparently mediated by inhibition of the proteosomal cleavage of IκBα and IκBβ.[88,89] However, although these immunosuppressants inhibit lymphocyte adhesion to endothelial cells *in vitro*, neither FK-506[90] nor cyclosporin (M.E. Gerritsen, unpublished observations) appears to inhibit cytokine-induced endothelial adhesion protein expression at reasonable concentrations. High concentrations of CsA (1.5μg/ml) have been reported to exert a modest inhibition of LPS-induced E-selectin expression in porcine aortic endothelial cells.[91]

3-Deazaadenosine: 3-deazaadenosine (c^3ADO) is a structural analog of adenosine that has both immunomodulatory and anti-inflammatory activities *in vitro* and *in vivo*.[92-95] c^3ADO co-treatment of TNF-α stimulated HUVEC monolayers reduces neutrophil adhesion, coincident with a reduction in cytokine-stimulated ICAM-1 protein and mRNA expression. c^3ADO also inhibits IL-1-induced lymphocyte and neutrophil adhesion. Although many of the biological activities of c^3ADO have been attributed to inhibition of critical methylation reactions by S-adenosyl homocysteine (AdoHcy) or S-3-deazaadenosinylhomocysteine, the effects on leukocyte adhesion molecule expression appear to be independent of this mechanism, since periodate-oxidized adenosine, which also increases AdoHcy levels, was not inhibitory.[96]

Deoxyspergualin: Deoxyspergualin, an immunosuppressive agent currently in clinical trials for the treatment of transplant rejection and autoimmune disease progression, inhibits LPS-induced nuclear translocation of NF-κB.[97] Although the effects of this drug on endothelial adhesion protein expression have not been reported to date, the inhibitory activities of deoxyspergualin on NF-κB activation may explain, at least in part, the ability of this drug to suppress humoral immune responses.

Dimethylfumarate: Dimethylfumarate is an active component of Fumaderm, a marketed drug for the treatment of psoriasis. Dimethylfumarate has a number of activities that relate to its anti-inflammatory properties, including anti-proliferative and anti-differentiative effects on keratinocytes.[98] Dimethylfumarate also inhibits TNF-α-induced VCAM-1 and ICAM-1 protein and mRNA expression with an IC_{50} of 50 μM.[99] The inhibitory effects of dimethylfumarate were not due to non-specific cytotoxicity, and were also seen with other inducing agents (IL-4, LPS, PMA). The mechanism of difumarate inhibition is not clear, but may relate to inhibition of NF-κB activation. Gel mobility shift studies have shown that difumarate abrogates arsenite-induced NF-κB translocation in aortic endothelial cells.[10] It has also been suggested that difumarate may induce enzymes responsible for the synthesis and maintenance of cellular glutathione pools.[100]

Glucocorticoids: Glucocorticoids are known to modulate the expression of a wide variety of genes, including cell adhesion molecules and cytokines. This class of steroid has been shown to block basal and phorbol ester-induced ICAM-1 expression in monocytic cells and synovial fibroblasts.[101-106] Nanomolar concentrations of the glucocorticoid dexamethasone have been reported to inhibit endotoxin-induced expression of E-selectin and ICAM-1 in endothelial cells.[106] However, other groups have reported that glucocorticoids do not inhibit TNF-α- or IL-1-induced expression of ICAM-1 in endothelial cells, suggesting that the effect of glucocorticoids may be both cell- and stimulus-specific.[107-109] Glucocorticoids bind to the cytoplasmic glucocorticoid receptor (GR), forming a complex which translocates to the nucleus and modulates the expression of various genes by binding a specific DNA sequence known as the glucocorticoid response element (GRE). However, many of the genes which may be regulated by glucocorticoids, including the adhesion molecule E-selectin, do not contain a GRE. It now appears, in certain circumstances, that the effects of glucocorticoids may be mediated indirectly. Several recent reports suggest that the glucocorticoid receptor complex can physically interact with the Rel A (p65) subunit of NF-κB, preventing the binding of

NF-κB to DNA and blocking the ability of NF-κB to activate transcription.[110-114] Caldenhoven et al[111] found that the inhibitory effect of glucocorticoids on ICAM-1 expression was dependent upon the presence of Rel A in the transactivating NF-κB complex. Glucocorticoids may also alter NF-κB-dependent gene expression in some cell types by increasing the transcription of the IκBα gene.[113,115] However, a recent study reported no significant induction of IκBα nor inhibition of NF-κB translocation upon LPS or TNF-α stimulation in endothelial cells treated with glucocorticoids.[116] The p65 component of NF-κB and the GR have been shown to mutually express each other's ability to activate transcription, and both of these transcriptional activators have been shown to depend upon the coactivators CREB-binding protein (CBP) and steroid receptor coactivator-1 (SRC-1) for maximal activity. Recently it was suggested that part of the cross-talk between the p65 component of NF-κB and GR is due in part to nuclear competition for limiting amounts of the coactivators CBP and SRC-1.[117]

Gold sodium thiomalate: Gold sodium thiomalate is a compound widely used in the treatment of rheumatoid arthritis, although the mechanism of its therapeutic action remains poorly understood. Recently Newman and co-workers evaluated the effects of this drug on endothelial cell adhesion molecule expression. Gold sodium thiomalate inhibited TNF-α-stimulated increases in VCAM and E-selectin, but not ICAM-1. Further, the effects of gold sodium thiomalate could be mimicked by thiomalate, but not gold thioglucose.[118]

Ibuprofen: Ibuprofen, at millimolar concentrations, inhibits TNF-α- and IL-1-induced surface expression of VCAM-1 (and ICAM-1, although to a lesser extent) on HUVEC, and also suppresses endothelial adhesiveness to lymphocytes, basophils, and mast cells, but not to neutrophils. The concentrations of ibuprofen are higher than those required to inhibit prostaglandin synthesis, suggestive of a mechanism other than cyclooxygenase blockade. Moreover, indomethacin, a potent cyclooxygenase inhibitor, failed to mimic the effects of ibuprofen.[119]

Pentoxifylline: Pentoxifylline (1-(5'-oxyhexyl)3,7-dimethylxanthine), a methylxanthine phosphodiesterase inhibitor, is a weak inhibitor of synthesis of proinflammatory cytokines, and inhibits T-cell binding to TNF-α-activated endothelial cells[120] and cytokine-stimulated neutrophil transmigration.[121,122] In pulmonary epithelial cells, pentoxyfylline inhibited cytokine-induced expression of ICAM-1 and the production of IL-8 and MCP-1.[123] Although the mechanism of action of pentoxyfylline is unclear, this drug has been shown to inhibit protein kinase C- and protein kinase A-catalyzed NF-κB activation in cytoplasmic extracts.[124]

Retinoids: Retinoids comprise a family of retinol metabolites and synthetic derivatives used frequently for topical and systemic dermatological disorders such as psoriasis and acne vulgaris. Pretreatment (4-24 h) of dermal microvascular endothelial cells with high concentrations of all trans-retinoic acid (tRA) ($IC_{50} \sim 40\mu M$) selectively inhibited TNF-α-induced VCAM-1, but not ICAM-1 or E-selectin expression.[125] This differential modulation of TNF-α-induced cell adhesion molecule expression was also reflected at steady state mRNA levels, and correlated with a selective inhibition of NF-κB-dependent binding to the VCAM-1 and not the ICAM-1 promoter. In contrast, TNF-α-induced VCAM-1 expression in large vessel endothelial cells (HUVEC) is apparently not inhibited by tRA.[125]

Tepoxalin: Tepoxalin (5-(4-chlorophenyl)-N-hydroxy-(4-methoxyphenyl)-N-methyl-1H-pyrazole-3-propanamide) is a dual inhibitor of 5-lipoxygenase and cyclooxygenase activity with demonstrated efficacy as an anti-inflammatory agent in a number of animal models.[126] Follow-up studies of the anti-inflammatory actions of this compound revealed a unique ability to inhibit cytokine production, and to suppress IL-2 transcription in human peripheral blood lymphocytes. Tepoxalin can inhibit NF-κB,[127] apparently by stabilizing IκB; these effects occur without inhibition of AP-1 activation. The inhibitory effects of tepoxalin are not related to its effects on prostaglandin or leukotriene production, since potent inhibitors of cyclooxygenase (e.g., naproxen) and 5-lipoxygenase (zileuton) do not interfere with NF-κB activation. Tepoxalin inhibits PMA and cytokine activation of NF-κB, although this drug is more effective on PMA-induced NF-κB activation than on that induced by cytokines. Tepoxalin blocks neutrophil migration into cutaneous inflammatory sites induced by either lipopolysaccharide or TNF in the mouse. This effect correlated with reduced expression of E-selectin and ICAM-1 mRNA.[128]

Natural Products

Natural products often provide important lead structures or highly active compounds which can serve as a chemical beginning to a synthetic program to develop potent and selective inhibitors. Herbal remedies with anti-inflammatory properties exist in virtually all cultures, and it is therefore not surprising that compounds from these various extracts have been identified with inhibitory effects on adhesion molecule expression.

Caffeic acid phenyl ester (CAPE): CAPE, structurally related to the flavonoids (see below), is a natural product derived from propolis of honeybee hives. CAPE has been shown to have antiviral, anti-inflammatory and immunomodulatory properties.[129] A recent study demonstrated that CAPE inhibits NF-κB activation induced by cytokines, phorbol esters, and peroxides.[130] CAPE prevents the translocation of the p65 subunit of NF-κB to the nucleus but does not inhibit IκB degradation.[130]

Curcumin: Curcumin (diferuloylmethane) is the major active component of turmeric (*Curcuma longa*), best known for its distinctive flavor and yellow color. Extracts of turmeric have been used for centuries as an anti-inflammatory remedy in Asian medicine, and purified curcumin has recently been confirmed to have both anti-tumor and anti-inflammatory activity.[131-136] Studies by two groups[132,137] demonstrated that treatment of cells with 2-50 μM curcumin inhibits NF-κB activation stimulated by TNF-α, LPS, IL-1, phorbol esters, and hydrogen peroxide. At these concentrations, curcumin is not cytotoxic and the effects of curcumin on NF-κB activation appear to be due to an inhibition of IκBα phosphorylation and degradation by inhibiting IKKs. The effects of curcumin are not NF-κB selective; previous studies have shown that curcumin also down-modulated the DNA binding of AP-1 transcription factors by both reducing the binding and level of c-Jun/AP-1.[138-140]

Flavonoids: Plant extracts containing flavonoids have been and continue to be widely used in many countries as herbal remedies for a variety of disorders including allergy, rheumatic diseases, and cancer. Purified flavonoids exhibit a wide spectrum of biological activities, including inhibition of serine and threonine protein kinases, phospholipases, lipoxygenases, and cyclooxygenase.[141-143] In addition, the flavonoids exhibit a number of potent anti-inflammatory effects *in vivo*, including inhibition of leukocyte migration/accumulation, edema, and granuloma formation.[144-146] Cotreatment of large and small vessel-derived human endothelial cells (human umbilical vein and human synovial microvessel) with certain hydroxyflavones (e.g., apigenin, baiculein) and flavonols (e.g., quercetin) inhibits cytokine-induced ICAM-1, VCAM-1, and E-selectin expression at the transcriptional level.[147-149] The precise mechanism of action of the flavonoids is unclear at present; apigenin does not inhibit activation of NF-κB, but does inhibit TNF-α-induced β-galactosidase activity in SW 480 cells stably transfected with a β-galactosidase reporter construct driven by four NF-κB elements, suggesting an action on NF-κB transcriptional activation.[148] The effects of quercetin were attributed to inhibition of JNK activation.[147] However, flavonoids exhibit a wide spectrum of pharmacological activities, including non-specific antioxidant activity and inhibition of protein kinase activity (including MAPK); thus it cannot be excluded that the effects of apigenin are due to one or more of these other activities.[150,151] Apigenin demonstrates potent anti-inflammatory activity in several animal models, including carrageenan-induced rat paw edema, rat adjuvant arthritis, and delayed type hypersensitivity in the mouse,[146,148] and has been shown to inhibit TNF-α-induced ICAM-1 upregulated *in vivo* in various organs of the rat.[152] Myricetin, a related flavonoid, was shown to inhibit IKKs in a bladder carcinoma cell line, ECV304.[153] A synthetic flavonoid, 2-(3-amino-phenyl)-8-methoxy-chromene-4-one (PD 098063) selectively inhibited TNF induced VCAM expression, but had no effect on ICAM-1 expression in human aortic endothelial cells by a mechanism apparently independent of NF-κB activation.[154]

Gallates: The phenolic compounds plentiful in red wine include galloyl compounds as well as flavonoids. Gallates have been reported to have anti-inflammatory, anti-mutagenic and anti-microbial, as well as radical scavenger activities.[155-158] Recently, certain gallates have been shown to be inhibitors of NF-κB. Ethyl gallate inhibits activation of NF-κB without altering the degradation of IκB.[159] Other gallate derivatives, such as theaflavin-3,3'-digallate, have been shown to inhibit the IKKs.[160]

Gliotoxin: Gliotoxin is a toxic epipolythiodioxopiperazine metabolite produced by opportunistic fungi such as *Aspergillis fumigatus* which exhibits profound immunosuppressive activity *in vivo*.[161,162] A study indicated that nanomolar concentrations of gliotoxin inhibited the activation of NF-κB in response to a variety of stimuli in T and B cells. The toxin appears to prevent degradation of IκBα,[163] and was recently shown to be an effective, non-competitive inhibitor of the 20S proteasome. Gliotoxin has also been shown to inhibit NF-κB activation and ICAM-1 expression in HUVEC.[164]

Panepoxydone: Isolated from the fermentation of basidiomycete *Lentinus critinus*, panepoxydone has been shown to inhibit PMA- or TNF-α-induced activation of a NF-κB reporter construct in COS-7 cells, independent of any effects on cell replication, transcription, or translation. Gel shift studies demonstrated that panepoxydone inhibited NF-κB activation, an effect that appeared to be selective as AP-1 activation was not inhibited. Panepoxydone inhibited IκB degradation by inhibiting its phosphorylation.[165]

Sesquiterpene Lactone Parthenolide: Sesquiterpene lactones (SLs) isolated from extracts of Mexican Indian medicine plants are traditionally used as anti-inflammatory substances. The SL parthenolide prevents the degradation of IκBα and IκBβ induced by a variety of stimuli. However, Hehner and colleagues[166,167] suggest that SL parthenolide may target the multisubunit IκB kinase complex, preventing NIK- and MEKK1-induced IKK activation.

Tanshinone II-A sulfonate (TAN): TAN is a natural product, isolated from the roots of *Salvia milliorrhiza*, with various known anti-inflammatory activities including suppression of platelet adhesion and aggregation, neutrophil activation, and chemokinesis. TAN also inhibits TNF-α-induced ICAM-1, although the effects are rather weak compared to those of other natural products (<50% inhibition at 200 μM). Effects of TAN on other adhesion molecules (VCAM, E-selectin) have not been described, and little is known about the mechanism of TAN's inhibitory activities.[168,169]

Tetrandine: Tetrandrine is a bisbenzoylisoquinolone alkaloid isolated from a natural Chinese herbal medicine with anti-inflammatory and anti-fibrotic activities. In macrophages, tetrandine inhibits NF-κB activation and suppresses signal-induced degradation of IκBα.[170] However, Chang and co-workers[171] reported that tetrandrine did not inhibit cytokine-induced adhesion molecule expression in endothelial cells.

Miscellaneous Inhibitors

Over the last decade, various groups have tested other classes of pharmacological compounds on endothelial cell adhesion molecule expression. Interestingly, the inhibitory compounds identified come from diverse groups of inhibitors, ranging from antioxidants to certain calcium channel blockers and novel compounds. These are discussed below.

Antioxidants: Reactive oxygen intermediates are thought be important intracellular second messengers in the activation of the transcription factor NF-κB,[172] and in the induced expression of endothelial adhesion molecules.[173,174] Antioxidants have been shown to inhibit nuclear translocation of NF-κB, phosphorylation of IκBα, and NF-κB-dependent gene transcription.[28,175-178] Treatment of endothelial cells with cytokines increases superoxide anion production[175,178-182] and oxidants such as oxidized lipids and H_2O_2 have been shown to increase endothelial adhesion protein expression.[179] A number of antioxidants have been shown to inhibit endothelial NF-κB activation in response to various stimuli, and to inhibit VCAM-1 expression. However, ICAM-1 and E-selectin expression are less sensitive to anti-antioxidants.[174,179] In addition to inhibiting endothelial adhesion molecule expression, antioxidants inhibit monocyte adhesion *in vitro*.[179]

The inhibition of cytokine-induced adhesion molecules by antioxidants is not as clear as it might first appear. The literature is based primarily on the use of two reagents, pyrrolidine dithiocarbamate (PDTC) and N-acteyl cystein (NAC). In certain physiological systems, both NAC and PDTC can act as pro-oxidants,[183-185] and thus inhibitory effects of these redox-active compounds can be due to reductive as well as oxidative processes. These mechanisms may differ depending upon the step in the signal transduction cascade with which the redox-active compounds interfere. Thiol modifying agents have also been shown to inhibit cytokine-mediated E-selectin expression in HUVEC, while NAC (up to 30 mM), phenylarsine oxide, and N-ethylmaleimide have little or no effect.[186] These observations may suggest a role for critical protein sulfhydryls in TNF-α and IL-1 signal transduction leading to adhesion molecule expression.[186]

Bay 11-7082 and 11-7085: BAY 11-7082 ((E)3-[(4-methylphenyl)-sulfonyl]-2-propenenitrile) and BAY 11-7085 ((E)3-[4-t-butylphenyl)-sulfonyl]-2-propenenitrile) were identified in cell-based screening assays of inhibitors of TNF-α-induced ICAM-1 expression in HUVEC. These compounds, with IC_{50} values in the low μM range, were found to irreversibly inhibit cytokine-induced expression of ICAM, VCAM, and E-selectin by a mechanism involving inhibition of IκB phosphorylation. The compounds also demonstrated *in vivo* anti-inflammatory activity, inhibiting both adjuvant arthritis and carageenan paw edema in the rat.[187]

2-Carboxamide Derivatives: Investigators from several pharmaceutical companies have described potent inhibitory effects of a novel series of compounds. A series of papers from Parke-Davis/Warner Lambert described the activities of a compound called PD 144795 (5-methoxy-ethoxy)benzo[b]thiophene-2-carboxamide-1-oxide).[188-190] PD 144795 inhibited TNF-induced VCAM and ICAM protein and mRNA, independent of NF-κB activation. PD 144795 was also shown to be orally active in several models of inflammation, and both the *in vitro* and *in vivo* activities of the molecule were shown to reside in the S-enantiomer. No effect is seen when these compounds are added 4 hours post-activation, suggesting that the compounds influence early events in TNF-α activation.[188,189] These compounds also inhibit LPS- and IL-1-induced endothelial adhesion protein expression.[188,189] Recently, another active carboxamide has been described.[190]

ET-18-OCH3: ET-18-OCH3(1-O-Octadecyl-2-O-methyl-rac-glycero-3-phosphocholine), a synthetic diether phospholipid, competitively inhibits phosphatidylserine binding to the regulatory domain of protein kinase C. ET-18-OCH3 inhibits NF-κB activation induced by phorbol esters (but not by cytokines).[191] This drug has been used as an anti-tumor agent; however, the activity of ET-18-OCH3 on endothelial cell adhesion molecule expression has not been reported.

E3330: E3330 ((2E)-3-[5-(2,3-dimethoxy-6-methyl-1,4-benzoquinoyl)]-2-nonyl-2-propenoic acid) is an agent identified on the basis of its hepatoprotective activity. E3330 inhibits LPS-induced TNF-α gene expression in monocytes. E3330 does not inhibit NF-κB translocation to the nucleus, degradation of IκBα, or post-translational modification of p65 (i.e., phosphorylation). Recent data suggest that E3330 inhibition of NF-κB-mediated gene expression may be due to inhibition of NF-κB DNA binding activity.[192,193]

Inhibitors of receptor-mediated endocytosis: Inhibitors of receptor-mediated endocytosis of TNF-α (hypertonicity, cytoplasmic acidification, phenylarsine oxide, putrescine, dansylcadaverine) reduce TNF-α-induced surface expression of E-selectin. Due to the toxic effects of many of these reagents, longer term studies of TNF-α induction of ICAM-1 and VCAM-1 have not been feasible. However, the less toxic primary amines such as putrescine (at 10 mM) and dansylcadaverine (at 100 μM) do inhibit TNF-α-induced ICAM-1 and VCAM-1 expression.[194]

Interleukin-10: Interleukin-10 (IL-10) is a pleiotropic mediator which inhibits production of cytokines in a variety of cell types. In monocyte/macrophages IL-10 suppresses the production of IL-1, IL-6, IL-8, and TNF-α.[195-199] IL-10 also inhibits leukocyte adhesion to IL-1-activated human endothelial cells,[200] an effect that

276

correlated with a decrease in endothelial ICAM-1 and VCAM-1 expression. In primary rat microglia cells, IL-10 inhibits the upregulation of ICAM-1 expression in response to IFNγ and LPS + IFNγ, an effect that appeared to be due to some effect on post-translational processing.[201] Similarly, IL-10 was found to inhibit IFNγ-stimulated surface ICAM-1 expression on monocytes.[202] In monocytes, IL-10 selectively inhibits NF-κB activation; whereas several other transcription factors including NF-IL-6, AP-1, AP-2, GR, CREB, Oct-1, and Sp1 are not affected by IL-10.[203]

Nitric Oxide: Earlier work by many groups has established that nitric oxide (NO) possesses many anti-atherogenic/anti-inflammatory activities including inhibitory effects on vascular smooth muscle proliferation,[204] platelet aggregation,[205] and monocyte adhesion and chemotaxis.[206] NO reduces leukocyte adhesion to endothelial cells by decreasing cytokine-induced VCAM-1, E-selectin, and, to a lesser extent, ICAM-1.[207] The inhibitory effects of NO on leukocyte adhesion are not mediated by cGMP-dependent pathways but instead mediated through inhibition of NF-κB activation.[207] The inhibitory effects of NO on adhesion protein expression are due to both stabilization of IκB (preventing its degradation) and increasing the mRNA expression of IκBα.[208] NO does not alter mRNA levels of the NF-κB subunits p65 or p50.[208]

Phospholipase A$_2$ Inhibitors: TNF-α signal transduction has been associated with the activation of a number of membrane lipid metabolic enzymes, including phosphatidylcholine-specific phospholipase C, sphingomyelinase, and phospholipase A$_2$ (PLA$_2$). PLA$_2$s catalyze the release of unsaturated fatty acids from the sn-2 position of phospholipids. Multiple PLA$_2$s have been identified, including the Ca^{2+}-dependent (group IV) PLA$_2$ (cPLA$_2$), secretory (type II) PLA$_2$ (sPLA$_2$), and the Ca^{2+}-independent iPLA$_2$. The cPLA$_2$ is an 85 kD protein, widely and constitutively expressed, highly selective for arachidonic acid at the sn-2 position, and is activated by Ca^{2+}-dependent phosphorylation. In contrast, the sPLA$_2$ are 14 kD enzymes, active extracellularly at millimolar Ca^{2+}. TNF-α stimulation of keratinocytes in the presence of selective cPLA$_2$ inhibitors (e.g., trifluoromethyll arachidonyl ketone (AACOCF3) and methyl arachidonyl fluorophosphate (MAFP) strongly reduced NF-κB activation.[209] In addition, sPLA$_2$ inhibitors (12-epi-scalardial and LY311727) also suppressed TNF-α-induced NF-κB activation and ICAM-1 expression in human keratinocytes.[209] The PLA$_2$ inhibitor bromophenacylbromide inhibits VCAM-1 expression in endothelial cells.[210] PLA$_2$ activation is a prerequisite for both cyclooxygenase activity (arachidonic acid must be released from membrane phospholipids in order to be metabolized by either cyclooxygense I or II) and thromboxane A_2 synthesis (TxA_2 is an arachidonic acid metabolite synthesized by the sequential actions of cyclooxygenase and thromboxane synthase). Taken together, the inhibitory actions of certain cyclooxygenase inhibitors, PLA_2 inhibitors, and TxA_2 inhibitors (see below) on NF-κB activation and endothelial adhesion molecule expression suggest an important role of the arachidonic acid cascade in modulation of the cytokine response of activated endothelial cells.

Peroxisome Proliferator Activated Receptors (PPAR): The PPAR receptors are nuclear receptors that are highly expressed in adipose tissue, monocytes, and smooth muscle cells. In atherosclerotic plaques, PPARγ is expressed by macrophages and foam cells, and PPARγ ligands can inhibit macrophage activation. PPARγ is also expressed in endothelial cells.[211] The PPARγ ligands 15-deoxy-$\Delta^{12,14}$-prostaglandin J_2 (15d-PGJ_2) and troglitazone markedly attenuate TNF-α-induced VCAM-1 and ICAM-1 (with little or no effect on E-selectin) in HUVECs and HAECs, and reduce monocyte binding to activated endothelial cells. *In vivo*, PPARγ treatment significantly reduced monocyte/macrophage homing to atherosclerotic plaques in ApoE deficient mice.[212,213] The naturally occurring ligands for the PPAR receptors remain relatively poorly defined. However, there are several studies that suggest that various hydroxy and hydroperoxyderivatives of arachidonic acid may be PPARγ ligands. This is of interest since there are a number of reports describing inhibitory effects of arachidonic and related fatty acids on cytokine-induced gene expression in endothelial cells, and it is quite reasonable to postulate that many of these lipids are acting through activation of the PPARγ-receptor pathway.[214-222]

Poly (ADP ribose) synthetase inhibitors: Nicotinamide and 3-aminobenzadine, inhibitors of poly (ADP ribose) synthetase, dose-dependently inhibit surface expression of ICAM-1 induced by IFNγ or phytohaemagglutinin on thyroid cells[223] and by IFNγ on endothelial cells.[224]

Prostacyclin and other agents that increase cAMP: Treatment of endothelial cells with agents that elevate cAMP levels, including dibutryl-cAMP, forskolin, and isobutyl-methylxanthine, inhibits the expression of E-selectin and VCAM-1 in response to TNF-α or IL-1.[225-228] In the case of E-selectin, this effect is due to the inhibition of mRNA synthesis, but is distal to the activation of NF-κB, which is unaffected by elevations in cAMP.[229] Prostacyclin also reduces E-selectin and

VCAM-1 expression,[226] an effect which may be mediated by the increase in intracellular cAMP levels elicited by this autocoid.

Proteasome Inhibitors: The proteasome is a 26S complex which is found in both the nucleus and cytoplasm (reviewed in 230-232). The proteasome represents a non-lysosomal pathway of proteolytic degradation which is primarily responsible for the turnover of either abnormal or rapidly degraded biologically active proteins in intact cells.[232] Drugs that selectively inhibit proteasome activity have recently been demonstrated to inhibit processing of the p105 precursor of NF-κB into the active p50 subunit.[233] Selective inhibitors of the proteasome pathway have been shown to inhibit IκB degradation as well as to inhibit NF-κB activation in a variety of cell types including endothelial cells.[234-235] Recently inhibitors of the proteasome were shown to inhibit cytokine-induced cell surface expression of E-selectin, VCAM-1, and ICAM-1 in endothelial cells,[234] and to exhibit functional activity by blocking adhesion of leukocytes to cytokine-activated endothelial cells under both static and defined flow conditions.[234] Some proteasome inhibitors may block VCAM-1 and ICAM-1 expression by a mechanism that does not decrease nuclear translocation of NF-κB.[235] Initially, proteasome inhibitors were synthetic analogs of ALLN (N-acetyl-Leu-Leu-norleucinal). However, structurally distinct proteasome inhibitors have now been identified, including the natural product lactacystin and CVT-634 (5-methoxy-1-indanone-3-acetyl-leu-D-leu-1-indanylamide).[236] Lactacystin, a natural microbial metabolite,[237] exerts its inhibitory effects through acylation and inhibition of the proteasome.[238] *In vitro*, lactacystin does not react with the proteasome; rather, it undergoes a spontaneous conversion (lactonization) to the active proteasome inhibitor, clasto-lactacystin-β-lactone. When the β-lactone is added to mammalian cells in culture, it rapidly enters the cells and reacts with the sulfhydryl of glutathione to form a thioester adduct that is both structurally and functionally analogous to lactacystin. This adduct, lactathione, like lactacystin, does not react with the proteasome, but can undergo lactonization to yield back the active β-lactone. The β-lactone (not lactacystin) can enter cells, suggesting that the formation of lactathione serves to concentrate the inhibitor inside cells, providing a reservoir for prolonged release of the active β-lactone.[239-244] Recently, a macrophage-derived peptide, PR39, was reported to inhibit the ubiquitin-proteasome-dependent degradation of hypoxia-inducible factor-1α protein.[245] Although the activity of PR39 on adhesion molecule expression has not been described, this agent may provide another useful experimental tool to probe the role of the proteasome in endothelial cell activation.

Other Protease Inhibitors: Certain serine protease inhibitors (e.g., tosyl-lysine-chloromethyl ketone (TLCK), tosyl-phenylalanine-chloromethyl ketone (TPCK) and 3,4-dichloroisocoumarin (DCI)) prevent TNF-α-induced NF-κB activation,[246] an effect that appears to be mediated by inhibitory effects on IκB phosphorylation.[247] These inhibitors do not inhibit NF-κB binding to its cognate recognition sequences or DNA-binding activities of other transcription factors (AP1, Oct-1). The inhibitory effects of DCI and TLCK on adhesion molecule expression appear to be due to acylating and alkylating properties of these drugs, since other non-alklyating or non-acylating inhibitors of cysteinyl (E64), aspartyl (pepstatin), or metallo-(phosphoramidon) proteases do not inhibit the surface expression of E-selectin, VCAM-1, or ICAM-1. At the concentrations that inhibited cytokine-induced adhesion protein expression, TLCK, TPCK, and DCI do not inhibit cellular metabolism, RNA or protein synthesis, suggesting that the observed inhibitory effects of these inhibitors are not the consequence of a non-specific cellular toxicity.[247]

RNA/DNA directed reagents: Anti-sense oligonucleotides have been successfully employed in the suppression of gene expression of various proteins in different cells, obverting or preventing the effects otherwise exerted by these proteins. Several groups have demonstrated that antisense oligodeoxyribonucleotides (PS-ODN), phosphorothioate forms complementary to the mRNAs of ICAM-1, E-selectin, and VCAM-1, inhibit the gene expression of these adhesion molecules and reduce adhesion of leukocytes to endothelial cells *in vitro* and *in vivo*.[248-261] In addition, double stranded phosphorothionate oligonucleotides have been used as "decoys" for NF-κB subunits, although high concentrations of oligonucleotides were required to inhibit NF-κB-dependent gene transcription.[262]

Adenovirus-mediated gene transfer has been successfully used to overexpress regulator proteins in many cell types. A modified IκBα was linked to a nuclear localization sequence and overexpressed in endothelial cells resulting in a specific decrease in NF-κB but not AP-1 binding activity.[263] Overexpression of nuclear IκBα blocked cytokine-induced VCAM-1 transcription and strongly inhibited leukocyte adhesion.[263] This type of therapeutic intervention might theoretically be helpful in suppressing adhesion molecule expression in a variety of inflammatory diseases; however at this juncture it is not clear whether or not adenovirus-mediated gene transfer will become a practical clinical tool.

Topoisomerase II inhibitors: Topoisomerase II selective inhibitors (novobiocin, nalidixic acid, and etoposide) selectively inhibit cytokine-induced VCAM-1 mRNA and surface expression, but not E-selectin or ICAM-1.[264] The effects of the topoisomerase inhibitors were not due to non-specific toxicity, as indexed by assessing inhibition of protein synthesis and β-actin mRNA levels. These observations may suggest a selective role for topoisomerase activity in the regulation of VCAM-1 expression.

Tyrosine Kinase Inhibitors: TNF-α-induced upregulation of VCAM-1 and E-selectin surface expression in HUVEc is dose-dependently inhibited by pretreatment with the protein tyrosine kinase inhibitors herbimycin A (IC_{50} 300 nM) and genistein (IC_{50} 30 μM). Additionally, both inhibitors reduce monocyte adhesion to TNF-α-stimulated HUVEc, and herbimycin A was shown to inhibit the induction of VCAM-1 mRNA by TNF-α, as well as the TNF-α-induced activation of NF-κB. In contrast, the tyrosine kinase inhibitors had little or no effect on ICAM-1 upregulation.[265] However, the effects of herbimycin could be independent of effects on tyrosine kinase. Herbimycin A is thiol-reactive and when added directly to nuclear extracts, can inhibit p50 binding to DNA, an effect reversed by 2-mercaptoethanol. Recent data suggest that herbimycin directly reacts with Cys 62 on p50.[266] In C6 glioma cells, IFNγ activation of the tyrosine protein kinase, JAK2, is also blocked by herbimycin, and this effect correlated with an inhibition of nitric oxide synthase induction by IFNγ.[267]

Protein tyrosine phosphatase inhibitors: In contrast to kinases, little is known about the role of protein phosphatases (PTPases) in cytokine signaling. The PTPase inhibitors phenylarsine oxide, diamine, and pervanadate inhibit TNF-α-induced ICAM-1, VCAM-1, and E-selectin expression in HUVEC, coincident with inhibition of TNF-α-dependent NF-κB activation.[268] This inhibition also correlated with the suppression of monocyte adhesion to cytokine-activated HMVEC. The identity of the phosphatase substrate(s) has not been identified.[268] However, pervanadate mimics interferon gamma induction of ICAM.[269]

Phospholipase C Inhibitors: D609, a phosphatidylcholine-specific phospholipase C inhibitor, reduces VCAM-1 surface expression and VCAM-1 promoter activity in human endothelial cells in a dose-dependent manner. D609 does not inhibit translocation of NF-κB.[270] The molecular mechanism of the effects of D609 on adhesion protein expression has not been elucidated to date.

Thromboxane A_2 synthase inhibitors: Thromboxane A_2 (TxA_2) is a potent eicosanoid metabolite released by activated platelets, macrophages, and leukocytes. There are also reports that vascular endothelial cells may produce TxA_2 in response to TNF-α or platelet activating factor (PAF). Recently the endothelial TxA_2 receptor was cloned and shown to be similar to the previously cloned TxA$_2$ receptor, although there was an alternatively spliced cytoplasmic tail.[271-273] The TxA$_2$ receptor is a member of the G-protein-coupled seven membrane spanning family of receptors. Namakura's laboratory first reported that DP-1904 ([±]-6-[1-imidazoylmethyl]-5,6,7,8-tetrahydronaphthalene-2-carboxylic acid hydrochloride hemihydrate), a TxA$_2$ synthesis inhibitor, suppressed ICAM-1 expression on HUVEC.[274] A later study by the same group provided additional evidence for a regulatory role of TxA$_2$ in ICAM-1 expression. TxA$_2$ receptor blockers suppressed TNF-α, PAF, and U46619 (9,11-dideoxy-9α 11α-epoxymethane prostaglandin F$_{2α}$, a TxA$_2$ receptor agonist) upregulation of ICAM-1.[275] In addition to upregulating ICAM-1, U44619 also increased VCAM-1 and E-selectin expression in HUVEC. SQ 29,548 (a TxA$_2$ receptor antagonist) diminished U46619 upregulation of all three adhesion molecules.[276]

Verapamil: Verapamil is a calcium channel blocker that has been reported to have immunosuppressive activities including inhibition of lymphocyte proliferation, IL-2 production, and IL-2 responsiveness, as well as inhibiting lymphocyte adhesion and migration. At high concentrations (IC$_{50}$~50 μM), verapamil reduced TNF-α-induced VCAM-1 (but not E-selectin or ICAM-1) expression.[277] Verapamil also inhibited IL-4 induction of VCAM.[277] The inhibitory effects of verapamil on VCAM-1 expression also correlated with a reduction in monocytic cell adhesion.[277] The effects of verapamil are unlikely to be due to inhibition of voltage-sensitive calcium channels, since these have never been definitively identified in endothelial cells. Interestingly, high concentrations of verapamil have been reported to inhibit prostaglandin release from endothelial cells; thus the inhibitory effects of verapamil may relate to effects on arachidonic acid metabolism.[278] This remains to be determined.

SUMMARY AND FUTURE DIRECTIONS

Targeting gene expression as a modality for therapeutic intervention is a relatively new concept. Although recent studies have revealed that a number of existing anti-inflammatory drugs can modify gene expression, the criteria for compound selection did not include this accidental, although often beneficial, "side effect." Consequently there remain a number of unresolved issues regarding feasibility, selectivity, safety, and efficacy. Potent inhibitors of ubiquitous transcription factors such as NF-κB, which may be effective anti-inflammatory agents, may have unforeseen mechanism-based toxicities or untoward activities. However, other transcription factors may be cell- and stimulus-specific, offering opportunities for specific intervention. Optimizing compound activities based on inhibition of protein-protein interactions or protein-DNA interactions poses new problems in drug design. The identification of multiple molecular targets offers numerous opportunities for compound identification through high throughput screening. Moreover, the observation that various inhibitors can selectively regulate certain adhesion molecules and not others, and that inhibitory effects of drugs may be cell- or tissue-specific, offers significant opportunities for therapeutic development.

REFERENCES

1.   Issekutz T, and Issekutz A. T lymphocyte migration to arthritic joints and dermal inflammation in the rat: differing migration patterns and the involvement of VLA_4. Clinical Clin Immunol Immunopathol 1991;61: 436-47.

2.   Issekutz A, and Issekutz T. A major portion of polymorphonuclear leukocyte and T lymphocyte migration to arthritic joints in the rat is via LFA-1/Mac-1 independent mechanisms. Clinical Clin Immunol Immunopathol 1993;67: 257-63.

3.   Jasin H, Lightfoot E, Davis L, Rothlein R, Faanes R, and Lipsky P. Amelioration of antigen-induced arthritis in rabbits treated with monoclonal antibodies to leukocyte adhesion molecules. Arthritis Rheum 1992;35: 541-49.

4.   Iigo U, Tkashi T, Tamatani T, and others. ICAM-1-dependent pathway is critically involved in the pathogenesis of adjuvant arthritis in rats. J Immunol 1991;147: 4167-71.

5.   Nolte D, Hecht R, Schmid P, Botzlar A, Menger M, Neumueller C, Sinowatz F, Vestweber D, and Messmer K. Role of Mac-1 and ICAM-1 in ischemia-reperfusion injury in a microcirculation model of Balb/c mice. Am J Physiol 1994;267: H1320-28.

6.   Lo S, Everitt J, and Malik A. Tumor necrosis factor mediates experimental pulmonary edema by ICAM-1 and CD-18 dependent mechanisms. J Clin Invest 1992;89: 981-88.

7.   Isobe M, Yagita H, Okumura K, and Ihara A. Specific acceptance of cardiac allograft after treatment with antibodies to ICAM-1 and LFA-1. Science 1992;255: 1125-26.

8.   Buerke M, Weyrich A, Zheng Z, Gaeta F, Forrest M, and Lefer A. Sialyl Lewis^x-containing oligosaccharide attenuates myocardial reperfusion injury in cats. J Clin Invest 1994;93: 1140-48.

9.   Mulligan M, Watson S, Fennie C, and Ward P. Protective effects of selectin chimeras in neutrophil mediated lung injury. J Immunol 1993;151: 6410-17.

10.   Kavanaugh A, Davis L, Nichols L, Norris S, Rothlein R, Scharschmidt L, and Lipsky P. Treatment of refractory rheumatoid arthritis with a monoclonal antibody to intercellular adhesion molecules-1. Arthritis Rheum 1994;37: 992-99.

11.   Haug G, Covin B, Delmonico F, Auchincloss HJ, Tolkoff-Rubin N, Preffer F, Rothlein R, Norris S, Scharschmidt L, and Cosimi A. A phase I trial of

immunosuppression with anti-ICAM-1 (CD54) mAb in renal allograft recipients. Transplantation 1993;55: 766-73.

12. Welply JK, Steininger CN, Caparon M, Michener ML, Howard SC, Pegg LE, Meyer DM, De Ciechi PA, Devine CS, and Casperson GF. A peptide isolated by phage display binds to ICAM-1 and inhibits binding to LFA-1. Proteins 1996;26: 262-70.

13. McIntyre BW, Woodside DG, Caruso DA, Wooten DK, Simon SI, Neelamegham S, Revelle JK, and Vanderslice P. Regulation of human T lymphocyte coactivation with an alpha4 integrin antagonist peptide. J Immunol 1997;158: 4180-6.

14. Hunter MS, Corley DG, Carron CP, Rowold E, Kilpatrick BF, and Durley RC. Four new clerodane diterpenes from the leaves of Casearia guianensis which inhibit the interaction of leukocyte function antigen 1 with intercellular adhesion molecule 1. J Nat Prod 1997;60: 894-9.

15. Parekh R, and Patel T. Carbohydrate ligands of the LECAM family as candidates for the development of anti-inflammatory compounds. J Pharm Pharmacol 1992;44 Suppl 1: 168-71.

16. Thoma G, Kinzy W, Bruns C, Patton JT, Magnani JL, and Banteli R. Synthesis and biological evaluation of a potent E-selectin antagonist. J Med Chem 1999;42: 4909-13.

17. Park IY, Lee DS, Song MH, Kim W, and Won JM. Cylexin: a P-selectin inhibitor prolongs heart allograft survival in hypersensitized rat recipients. Transplant Proc 1998;30: 2927-8.

18. Davenpeck KL, Berens KL, Dixon RA, Dupre B, and Bochner BS. Inhibition of adhesion of human neutrophils and eosinophils to P-selectin by the sialyl Lewis antagonist TBC1269: preferential activity against neutrophil adhesion in vitro. J Allergy Clin Immunol 2000;105: 769-75.

19. Jain RK, Piskorz CF, Huang BG, Locke RD, Han HL, Koenig A, Varki A, and Matta KL. Inhibition of L- and P-selectin by a rationally synthesized novel core 2-like branched structure containing GalNAc-Lewisx and Neu5Acalpha2-3Galbeta1-3GalNAc sequences. Glycobiology 1998;8: 707-17.

20. Hiramatsu Y, Moriyama H, Kiyoi T, Tsukida T, Inoue Y, and Kondo H. Studies on selectin blockers. 6. Discovery of homologous fucose sugar unit necessary for E-selectin binding. J Med Chem 1998;41: 2302-7.

21. Hiramatsu Y, Tsujishita H, and Kondo H. Studies on selectin blocker. 3. Investigation of the carbohydrate ligand sialyl Lewis X recognition site of P-selectin. J Med Chem 1996;39: 4547-53.

22. Hiramatsu Y, Tsukida T, Nakai Y, Inoue Y, and Kondo H. Study on selectin blocker. 8. Lead discovery of a non-sugar antagonist using a 3D-pharmacophore model. J Med Chem 2000;43: 1476-83.

23. Norman KE, Anderson GP, Kolb HC, Ley K, and Ernst B. Sialyl Lewis(x) (sLe(x)) and an sLe(x) mimetic, CGP69669A, disrupt E-selectin-dependent leukocyte rolling in vivo. Blood 1998;91: 475-83.

24. Xia L, Pan J, Yao L, and McEver RP. A proteasome inhibitor, an antioxidant, or a salicylate, but not a glucocorticoid, blocks constitutive and cytokine-inducible expression of P-selectin in human endothelial cells. Blood 1998;91: 1625-32.

25.   Collins T, Read M, Neish A, Whitley M, Thanos D, and Maniatis T. Transcriptional regulation of endothelial cell adhesion molecules: NF-κB and cytokine-inducible enhancers. FASEB 1995;9: 899-909.

26.   Beg A, and Baldwin A. The IkB proteins: multifunctional regulators of Rel/NF-κB transcription factors. Genes Dev 1993;7: 2064-70.

27.   Verma I, Stevenson J, Schwarz E, Van Antwerp D, and Miyamoto S. Rel/NF-κB/IκB family: intimate tales of association and dissociation. Genes Dev 1995;9: 2723-35.

28.   Baeuerle P, and Henkel T. Function and activation of NF-κB in the immune system. Annual Rev Immunol 1994;12: 141-79.

29.   Li CC, Dai RM, Chen E, and Longo DL. Phosphorylation of NF-KB1-p50 is involved in NF-kappa B activation and stable DNA binding. J Biol Chem 1994;269: 30089-92.

30.   Kushner DB, and Ricciardi RP. Reduced phosphorylation of p50 is responsible for diminished NF-kappaB binding to the major histocompatibility complex class I enhancer in adenovirus type 12-transformed cells. Mol Cell Biol 1999;19: 2169-79.

31.   Zhong H, SuYang H, Erdjument-Bromage H, Tempst P, and Ghosh S. The transcriptional activity of NF-kappaB is regulated by the IkappaB-associated PKAc subunit through a cyclic AMP-independent mechanism. Cell 1997;89: 413-24.

32.   Zhong H, Voll RE, and Ghosh S. Phosphorylation of NF-kappa B p65 by PKA stimulates transcriptional activity by promoting a novel bivalent interaction with the coactivator CBP/p300. Mol Cell 1998;1: 661-71.

33.   Bird TA, Schooley K, Dower SK, Hagen H, and Virca GD. Activation of nuclear transcription factor NF-kappaB by interleukin-1 is accompanied by casein kinase II-mediated phosphorylation of the p65 subunit. J Biol Chem 1997;272: 32606-12.

34.   Wang D, and Baldwin AS, Jr. Activation of nuclear factor-kappaB-dependent transcription by tumor necrosis factor-alpha is mediated through phosphorylation of RelA/p65 on serine 529. J Biol Chem 1998;273: 29411-6.

35.   DiDonato JA, Hayakawa M, Rothwarf DM, Zandi E, and Karin M. A cytokine-responsive IkappaB kinase that activates the transcription factor NF-kappaB. Nature 1997;388: 548-54.

36.   Zandi E, Rothwarf DM, Delhase M, Hayakawa M, and Karin M. The IkappaB kinase complex (IKK) contains two kinase subunits, IKKalpha and IKKbeta, necessary for IkappaB phosphorylation and NF-kappaB activation. Cell 1997;91: 243-52.

37.   Mercurio F, Zhu H, Murray BW, Shevchenko A, Bennett BL, Li J, Young DB, Barbosa M, Mann M, Manning A, and Rao A. IKK-1 and IKK-2: cytokine-activated IkappaB kinases essential for NF-kappaB activation. Science 1997;278: 860-6.

38.   Regnier CH, Song HY, Gao X, Goeddel DV, Cao Z, and Rothe M. Identification and characterization of an IkappaB kinase. Cell 1997;90: 373-83.

39. Woronicz JD, Gao X, Cao Z, Rothe M, and Goeddel DV. IkappaB kinase-beta: NF-kappaB activation and complex formation with IkappaB kinase-alpha and NIK. Science 1997;278: 866-9.

40. Takeda K, Takeuchi O, Tsujimura T, Itami S, Adachi O, Kawai T, Sanjo H, Yoshikawa K, Terada N, and Akira S. Limb and skin abnormalities in mice lacking IKKalpha. Science 1999;284: 313-6.

41. Hu Y, Baud V, Delhase M, Zhang P, Deerinck T, Ellisman M, Johnson R, and Karin M. Abnormal morphogenesis but intact IKK activation in mice lacking the IKKalpha subunit of IkappaB kinase. Science 1999;284: 316-20.

42. Li Q, Lu Q, Hwang JY, Buscher D, Lee KF, Izpisua-Belmonte JC, and Verma IM. IKK1-deficient mice exhibit abnormal development of skin and skeleton. Genes Dev 1999;13: 1322-8.

43. Li Q, Van Antwerp D, Mercurio F, Lee KF, and Verma IM. Severe liver degeneration in mice lacking the IkappaB kinase 2 gene. Science 1999;284: 321-5.

44. Tanaka M, Fuentes ME, Yamaguchi K, Durnin MH, Dalrymple SA, Hardy KL, and Goeddel DV. Embryonic lethality, liver degeneration, and impaired NF-kappa B activation in IKK-beta-deficient mice. Immunity 1999;10: 421-9.

45. Yamaoka S, Courtois G, Bessia C, Whiteside ST, Weil R, Agou F, Kirk HE, Kay RJ, and Israel A. Complementation cloning of NEMO, a component of the IkappaB kinase complex essential for NF-kappaB activation. Cell 1998;93: 1231-40.

46. Rothwarf DM, Zandi E, Natoli G, and Karin M. IKK-gamma is an essential regulatory subunit of the IkappaB kinase complex. Nature 1998;395: 297-300.

47. Cohen L, Henzel WJ, and Baeuerle PA. IKAP is a scaffold protein of the IkappaB kinase complex . Nature 1998;395: 292-6.

48. Malinin NL, Boldin MP, Kovalenko AV, and Wallach D. MAP3K-related kinase involved in NF-kappaB induction by TNF, CD95 and IL-1. Nature 1997;385: 540-4.

49. Ling L, Cao Z, and Goeddel DV. NF-kappaB-inducing kinase activates IKK-alpha by phosphorylation of Ser-176. Proc Natl Acad Sci U S A 1998;95: 3792-7.

50. Nakano H, Shindo M, Sakon S, Nishinaka S, Mihara M, Yagita H, and Okumura K. Differential regulation of IkappaB kinase alpha and beta by two upstream kinases, NF-kappaB-inducing kinase and mitogen-activated protein kinase/ERK kinase kinase-1. Proc Natl Acad Sci U S A 1998;95: 3537-42.

51. Nemoto S, DiDonato JA, and Lin A. Coordinate regulation of IkappaB kinases by mitogen-activated protein kinase kinase kinase 1 and NF-kappaB-inducing kinase. Mol Cell Biol 1998;18: 7336-43.

52. Zhao Q, and Lee FS. Mitogen-activated protein kinase/ERK kinase kinases 2 and 3 activate nuclear factor-kappaB through IkappaB kinase-alpha and IkappaB kinase-beta. J Biol Chem 1999;274: 8355-8.

53. Ozes ON, Mayo LD, Gustin JA, Pfeffer SR, Pfeffer LM, and Donner DB. NF-kappaB activation by tumour necrosis factor requires the Akt serine-threonine kinase. Nature 1999;401: 82-5.

54. Romashkova JA, and Makarov SS. NF-kappaB is a target of AKT in anti-apoptotic PDGF signalling. Nature 1999;401: 86-90.

55. Sulciner DJ, Irani K, Yu ZX, Ferrans VJ, Goldschmidt-Clermont P, and Finkel T. rac1 regulates a cytokine-stimulated, redox-dependent pathway necessary for NF-kappaB activation. Mol Cell Biol 1996;16: 7115-21.

56. Montaner S, Perona R, Saniger L, and Lacal JC. Activation of serum response factor by RhoA is mediated by the nuclear factor-kappaB and C/EBP transcription factors. J Biol Chem 1999;274: 8506-15.

57. Montaner S, Perona R, Saniger L, and Lacal JC. Multiple signalling pathways lead to the activation of the nuclear factor kappaB by the Rho family of GTPases. J Biol Chem 1998;273: 12779-85.

58. Perona R, Montaner S, Saniger L, Sanchez-Perez I, Bravo R, and Lacal JC. Activation of the nuclear factor-kappaB by Rho, CDC42, and Rac-1 proteins. Genes Dev 1997;11: 463-75.

59. Pan ZK, Ye RD, Christiansen SC, Jagels MA, Bokoch GM, and Zuraw BL. Role of the Rho GTPase in bradykinin-stimulated nuclear factor-kappaB activation and IL-1beta gene expression in cultured human epithelial cells. J Immunol 1998;160: 3038-45.

60. Lim L, Manser E, Leung T, and Hall C. Regulation of phosphorylation pathways by p21 GTPases. The p21 Ras-related Rho subfamily and its role in phosphorylation signalling pathways. Eur J Biochem 1996;242: 171-85.

61. Abo A, Qu J, Cammarano MS, Dan C, Fritsch A, Baud V, Belisle B, and Minden A. PAK4, a novel effector for Cdc42Hs, is implicated in the reorganization of the actin cytoskeleton and in the formation of filopodia. EMBO J 1998;17: 6527-40.

62. Zhao ZS, Manser E, Chen XQ, Chong C, Leung T, and Lim L. A conserved negative regulatory region in alphaPAK: inhibition of PAK kinases reveals their morphological roles downstream of Cdc42 and Rac1. Mol Cell Biol 1998;18: 2153-63.

63. Frost JA, Khokhlatchev A, Stippec S, White MA, and Cobb MH. Differential effects of PAK1-activating mutations reveal activity-dependent and -independent effects on cytoskeletal regulation. J Biol Chem 1998;273: 28191-8.

64. Frost JA, Swantek JL, Stippec S, Yin MJ, Gaynor R, and Cobb MH. Stimulation of NF{kappa}B activity by multiple signaling pathways requires PAK1. J Biol Chem 2000.

65. DeLuca L, Johnson D, Whitley M, Collins T, and Pober J. cAMP and tumor necrosis factor competitively regulate transcriptional activation through nuclear factor binding to the cAMP-responsive element/activating transcription factor element of the Endothelial Leukocyte Adhesion Molecule-1 (E-selectin) promoter. J Biol Chem 1994;269: 19193-96.

66. Kaszubska W, Hooft van Huijsduijnen R, Ghersa P, DeRaemy-Schenk A, Chen B, Hai T, DeLamarter J, and Whelan J. Cyclic-AMP independent ATF family members interact with NF-kB and function in the activation of E-selectin promoter in response to cytokines. Mol Cell Biol 1993;13: 7180-90.

67. Pulverer B, Kyriakis J, Avruch J, Nikolakaki E, and Woodgett J. Phosphorylation of c-Jun mediated by MAP kinases. Nature 1991;353: 6760-64.

68. Gupta S, Campbell D, Derijard B, and Davis R. Transcription factor ATF-2. Regulation by the JNK signal transduction pathway. Science 1995;267: 389-93.

69. Lin A, Minden A, Martinetto H, Claret F-X, Lange-Carter C, Mercurio F, Johnson G, and Karin M. Identification of a dual specificity kinase that activates jun kinases and p38-Mpk2. Science 1995;268: 286-90.

70. Smeal T, Binetruy B, Mercola D, Birrer M, and Karin M. Oncogenic and transcriptional cooperation with Ha-Ras requires phosphorylation of c-Jun on serines 63 and 73. Nautre 1991;354: 494-96.

71. Hibi M, Lin A, Smeal T, Minden A, and Karin M. Identification of an oncoprotein-and UV-responsive protein kinase that binds and potentiates the c-JUN activation domain. Genes Dev 1993;7: 2135-48.

72. Read M, Whitley M, Gupta S, Pierce J, Best J, Davis R, and Collins T. TNFa induced E-selectin expression is activated by the NF-κB and JNK/p38 MAP kinase pathways. J Biol Chem 1997; 272:2753-2761.

73. Reimold A, Grusby M, Kosaras B, Fries J, Mori R, Maniwa S, Clauss I, Collins T, Sidman R, Glimcher M, and Glimcher L. Chondrodysplasia neurological abnormalities in ATF-2 deficient mice. Nature 1996;379: 262-65.

74. Bovolenta C, Gasperini S, and Cassatella M. Granulocyte colony-stimulating factor induces the binding of STAT1 and STAT3 to the IFNgamma response region within the promoter of the Fa(gamma)RI/CD64 gene in human neutrophils. FEBS-Lett 1996;386: 239-42.

75. Tsukada J, Waterman W, Koyama Y, Webb A, and Auron P. A novel STAT-like factor mediates lipopolysaccharide, interleukin 1 (IL-1), and IL-6 signaling and recognizes a gamma interferon activation-like element in the IL1β gene. Mol Cell Biochem 1996;16: 2183-94.

76. Novak U, Harpur A, Paradiso L, Kanagasundaram V, Jaworowski A, Wilks A, and Hamilton J. Colony stimulating factor 1-induced STAT1 and STAT3 activation is accompanied by phosphorylation of Tyk2 in macrophages and Tyk2 and JAK1 in fibroblasts. Blood 1995;86: 2948-56.

77. Tweardy D, Wright T, Ziegler S, Baumann H, Chakraborty A, White S, Dyer K, and Rubin K. Granulocyte colony-stimulating factor rapidly activates a distinct STAT-like protein in normal myeloid cells. Blood 1995;86: 4409-16.

78. Look D, Pelletier M, and Holtzman M. Selective interaction of a subset of interferon-γ response element-binding proteins with the intercellular adhesion molecule-1 (ICAM-1) gene promoter controls the pattern of expression on epithelial cells. J Biol Chem 1994;269: 8952-58.

79. Neish A, Read M, Thanos D, Pine R, Maniatis T, and Collins t. Endothelial interferon regulatory factor 1 cooperates with NF-κB as a transcriptional activator of vascular cell adhesion molecule 1. Mol Cell Biol 1995;15: 2558-69.

80. Kopp E, and Ghosh S. Inhibition of NF-κB by sodium salicylate and aspirin. Science 1994;265: 956-59.

81. Pierce J, Read M, Ding H, Luscinskas F, and Collins T. Salicylates inhibit IκBα phosphorylation, endothelial leukocyte adhesion molecule expresson and neutrophil transmigration. J Immunol 1996;156: 3961-69.

82.   Weber C, Erl W, Pietsch A, and Weber P. Aspirin inhibits nuclear factor-κB mobilization and monocyte adhesion in stimulated human endothelial cells. Circulation 1995;91: 1914-17.

83.   Asako H, Kubes P, Wallace J, Wolf R, and Granger D. Modulation of leukocytes adhesion in rat mesenteric venules by aspirin and salicylate. Gastroenterology 1992;103: 146-52.

84.   Weissmann G. Aspirin. Scientific American 1991;264: 84-90.

85.   Sagone A, and Husney R. Oxidation of salicylates by stimulated granulocytes: evidence that these drugs act as free radical scavengers. J Immunol 1987;138.

86.   Haynes D, Wright P, Gadd S, Whitehouse M, and Vernon-Roberts B. Is aspirin a prodrug for antioxidant and cytokine-modulating oxymetabolites? Agents and Actions 1993;29: 49-58.

87.   Hemenway CS, and Heitman J. Calcineurin. Structure, function, and inhibition. Cell Biochem Biophys 1999;30: 115-51.

88.   Frantz B, Nordby EC, Bren G, Steffan N, Paya CV, Kincaid RL, Tocci MJ, O'Keefe SJ, and O'Neill EA. Calcineurin acts in synergy with PMA to inactivate I kappa B/MAD3, an inhibitor of NF-kappa B. EMBO J 1994;13: 861-70.

89.   Marienfeld R, Neumann M, Chuvpilo S, Escher C, Kneitz B, Avots A, Schimpl A, and Serfling E. Cyclosporin A interferes with the inducible degradation of NF-kappa B inhibitors, but not with the processing of p105/NF-kappa B1 in T cells. Eur J Immunol 1997;27: 1601-9.

90.   Karlsson H, and Nassberger L. FK506 suppresses the mitogen-induced increase in lymphocyte adhesiveness to endothelial cells, but does not affect endothelial cell activation in response to inflammatory stimuli. Transplantation 1997;64: 1217-20.

91.   Charreau B, Coupel S, Goret F, Pourcel C, and Soulillou JP. Association of glucocorticoids and cyclosporin A or rapamycin prevents E-selectin and IL-8 expression during LPS- and TNFalpha-mediated endothelial cell activation. Transplantation 2000;69: 945-53.

92.   Leonard EJ, Skeel A, Chiang PK, and Cantoni GL. The action of the adenosylhomocysteine hydrolase inhibitor, 3-deazaadenosine, on phagocytic function of mouse macrophages and human monocytes. Biochem Biophys Res Commun 1978;84: 102-9.

93.   Jurgensen CH, Wolberg G, and Zimmerman TP. Inhibition of neutrophil adherence to endothelial cells by 3-deazaadenosine. Agents Actions 1989;27: 398-400.

94.   Zimmerman TP, Wolberg G, and Duncan GS. Inhibition of lymphocyte-mediated cytolysis by 3-deazaadenosine: evidence for a methylation reaction essential to cytolysis. Proc Natl Acad Sci U S A 1978;75: 6220-4.

95.   Medzihradsky JL. Regulatory role for the immune complex in modulation of phagocytosis by 3-deazaadenosine. J Immunol 1984;133: 946-9.

96.   Jurgensen CH, Huber BE, Zimmerman TP, and Wolberg G. 3-deazaadenosine inhibits leukocyte adhesion and ICAM-1 biosynthesis in tumor necrosis factor-stimulated human endothelial cells. J Immunol 1990;144: 653-61.

97.   Tepper M, Nadler S, Esselstyn J, and Sterbenz K. Deoxyspergualin inhibits kappa light chain expression in 70Z/3 pre-B cells by blocking

lipolysaccharide-induced NF-kappa B activation. J Immunol 1995;155: 2427-36.

98. Thio HB, Zomerdijk TP, Oudshoorn C, Kempenaar J, Nibbering PH, van der Schroeff JG, and Ponec M. Fumaric acid derivatives evoke a transient increase in intracellular free calcium concentration and inhibit the proliferation of human keratinocytes. Br J Dermatol 1994;131: 856-61.

99. Vandermeeren M, Janssens S, Borgers M, and Geysen J. Dimethylfumarate is an inhibitor of cytokine-induced E-selectin, VCAM-1, and ICAM-1 expression in human endothelial cells. Biochem Biophys Res Commun 1997;234: 19-23.

100. Barchowsky A, Dudek EJ, Treadwell MD, and Wetterhahn KE. Arsenic induces oxidant stress and NF-kappa B activation in cultured aortic endothelial cells. Free Radic Biol Med 1996;21: 783-90.

101. van de Stolpe A, Caldenhoven E, Raaijmakers J, van der Saag P, and Koenderman L. Glucocorticoid-mediated repression of intercellular adhesion molecule-1 expression in human monocytic and bronchial epithelial cell lines. Am J Respir Cell Mol Biol 1993;8: 340-47.

102. van de Stolpe A, Caldnhoven E, Stade B, Koenderman L, Raaijmakers J, Johnson J, and van der Saag P. 12-O-tetradecanoyl-13-acetate- and tumor necrosis factor mediated induction of intercellular adhesion molecule-1 is inhibited by dexamethasone. Functional analysis of the human intercellular adhesion molecule-1 promoter. J Biol Chem 1994;269: 6185-92.

103. Perretti M, Wheller S, Harris J, and Flower R. Modulation of ICAM-1 levels on U-937 cells and mouse macrophages by interleukin-1 beta and dexamethasone. Biochem Biophys Res Comm 1996;223: 112-17.

104. Tessier P, Cattaruzzi P, and Mccoll S. Inhibition of lymphocyte adhesion to cytokine-activated synovial fibroblasts by glucocorticoids involves the attentuation of vascular cell adhesion molecule 1 and intercellular adhesion molecule 1 expression. Arthritis Rheum 1996;39: 226-34.

105. Lenardo M, and Baltimore D. NF-kB: a pleiotrophic mediator of inducible and tissue-specific gene control. Cell 1989;58: 227-.

106. Cronstein B, Kimmel S, Levin R, Martiniuk F, and Weissman G. A mechanism for the anti-inflammatory effects of corticosteroids: the glucocorticoid receptor regulates leukocyte adhesion to endothelial cells and expression of endothelial-leukocyte adhesion molecule 1 and intercellular adhesion molecule 1. Proc Natl Acad Sci USA 1992;89: 9991-95.

107. Rothlein R, Czajkoski M, O'Neill M, Marlin S, Mainolfi E, and Merluzzi V. Induction of intercellular adhesion molecule-1 on primary and continuous cell lines by proinflammatory cytokines. Regulation by pharmacologic agents and neutralizing antibodies. J Immunol 1988;141: 1665-69.

108. Swerlick R, Garcia-Gonzalez Z, Kubota Y, et al. Studies of the modulation of MHC antigen and cell adhesion molecule expression on human dermal microvascular endothelial cells. J Invest Dermatol 1991;97: 190-96.

109. Kaiser J, Bickel C, Bochner B, and Schleimer R. The effects of the potent glucocorticoid budesonide on adhesion of eosinophils to human vascular endothelial cells and on endothelial expression of adhesion molecules. J Pharmacol Exper Ther 1993;267: 245-49.

110. Mukaida N, Morita M, Ishikawa N, Rice N, Okamoto S, Kasahara T, and Matsushima K. Novel mechanism of glucocorticoid-mediated gene repression: nuclear factor-κB is a target for glucorticoid-mediated interleukin 8 gene repression. J Biol Chem 1994;269: 13289-95.

111. Caldenhoven E, Liden J, Wissink S, Van de Stolpe A, Raaijmakers J, Koenderman L, Okret S, Gustafsson J, and Van der Saag P. Negative cross-talk between Rel A and the glucocorticoid receptor: a possible mechanism for the anti-inflammatory action of glucocorticoids. Mol Endocrinol 1995;9: 401-12.

112. Ray A, and Prefontaine K. Physical association and functional antagonism between the p65 subunit of transcription factor NF-kB and the glucocorticoid receptor. Proc Natl Acad Sci USA 1994;91: 752-56.

113. Scheinman R, Cogswell P, Lofqvist A, et al. Role of transcriptional activation of IκBα in mediation of immunosuppression by glucocorticoids. Science 1995;270: 283-86.

114. Scheinman R, Gualberta A, Jewell C, Cidlowski J, and Baldwin A. Characterization of mechanisms involved in transrepression of NF-κB by activated glucocorticoid receptors. Mol Cell Biol 1995;15: 943-53.

115. Auphan N, DiDonato C, Rosette A, Helmberg A, and Karin M. Immunosuppression by glucocorticoids: inhibition of NF-kB activity through induction of IκB synthesis. Science 1995;270: 286-90.

116. Brostjan C, Anrather J, Csizmadia V, Stroka D, Soares M, Bach F, and Winkler H. Glucocorticoid-mediated repression of NFκB activity in endothelial cells does not involve induction of IκBα. J Biol Chem 1996;271: 19612-16.

117. Sheppard KA, Phelps KM, Williams AJ, Thanos D, Glass CK, Rosenfeld MG, Gerritsen ME, and Collins T. Nuclear integration of glucocorticoid receptor and nuclear factor-kappaB signaling by CREB-binding protein and steroid receptor coactivator-1. J Biol Chem 1998;273: 29291-4.

118. Newman P, To S, Robinson B, Hyland V, and Shrieber L. Effect of gold sodium thiomalate and its thiomalate component on the in vitro expression of endothelial adhesion molecules. J Clin Invest 1994;94: 1864-71.

119. Kapiotis S, Sengoelge G, Sperr WR, Baghestanian M, Quehenberger P, Bevec D, Li SR, Menzel EJ, Muhl A, Zapolska D, Virgolini I, Valent P, and Speiser W. Ibuprofen inhibits pyrogen-dependent expression of VCAM-1 and ICAM-1 on human endothelial cells. Life Sci 1996;58: 2167-81.

120. Bruynzeel I, van der Raaij LM, Willemze R, and Stoof TJ. Pentoxifylline inhibits human T-cell adhesion to dermal endothelial cells. Arch Dermatol Res 1997;289: 189-93.

121. Elferink JG, Huizinga TW, and de Koster BM. The effect of pentoxifylline on human neutrophil migration: a possible role for cyclic nucleotides. Biochem Pharmacol 1997;54: 475-80.

122. Sullivan GW, Carper HT, Novick WJ, Jr., and Mandell GL. Inhibition of the inflammatory action of interleukin-1 and tumor necrosis factor (alpha) on neutrophil function by pentoxifylline. Infect Immun 1988;56: 1722-9.

123. Krakauer T. Pentoxifylline inhibits ICAM-1 expression and chemokine production induced by proinflammatory cytokines in human pulmonary epithelial cells. Immunopharmacology 2000;46: 253-61.

124. Biswas D, Dezube B, Ahlers C, and Pardee A. Pentoxifylline inhibits HIV-1 LTR-driven gene expression by blocking NF-kappa B action. J Acquir Immune Defic Syndr 1993;6: 778-86.

125. Gille J, Paxton L, Lawley T, Caughman S, and Swerlick R. Retinoic acid inhibits the regulated expression of vascular cell adhesion molecule-1 by cultured dermal microvascular endothelial cells. J Clin Invest 1997;99: 492-500.

126. Argentieri D, Anderson D, Ritchi D, Rosenthal M, and Capetola R. Tepoxalin (RWJ 20485) inhibits prostaglandin (PG) and leukotriene (LT) production in adjuvant arthritis rats and in dog knee joints challenged with sodium urate and immune complexes. FASEB J 1990;4: A1142.

127. Kazmi S, Plante R, Visconti V, Taylor G, Zhou L, and Lau C. Suppression of NFkB activation and NFκB-dependent gene expression by tepoxalin, a dual inhibitor of cyclooxygenase and lipoxygenase. J Biol Chem 1995;57: 299-310.

128. Zhou L, Pope-BL, Chourmouzis E, Fung-Leung W, and Lau C. Tepoxalin blocks neutrophil migration into cutaneous inflammatory sites by inhibiting Mac-1 and E-selectin expression. Eur J Immunol 1996;26: 120-29.

129. Grunberger D, Banerjee R, Eisinger K, Oltz K, Efros E, Caldwell M, Estevez V, and Nakanishi K. Preferential cytotoxicity on tumor cells by caffeic acid phenylethyl ester isolated from propolis. Experientia 1988;44: 230-2332.

130. Natarajan K, Singh S, Burke Jr T, Grunberger D, and Aggarwal B. Caffeic acid phenylester is a potent and specific inhibitor of activation of nuclear transcription factor NF-κB. Prac Natl Acad Sci USA 1996;93: 9090-95.

131. Brouet I, and Ohshima H. Curcumin, an anti-tumor promoter and anti-inflammatory agent, inhibits induction of nitric oxide synthase in activated macrophages. Biochem Biophys Res Comm 1995;206: 533-40.

132. Chan M-Y. Inhibition of tumor necrosis factor by curcumin, a phytochemical. Biochemical Pharmacology 1995;49: 1551-56.

133. Razga Z, and Gabor M. Effects of curcumin and nordihydroguaiaretic acid on mouse ear oedema induced by croton oil or dithranol. Pharmazie 1995;50: 156-7.

134. Reddy S, and Aggarwal B. Curcumin is a non-competitive and selective inhibitor of phosphorylase kinase. FEBS Letters 1994;341: 19-22.

135. Reddy A, and Lokesh B. Studies on the inhibitory effects of curcumin and eugenol on the formation of reactive oxygen species and the oxidation of ferrous iron. Mol Cell Biochem 1994;137: 1-8.

136. Ruby A, Kuttan G, Babu K, Rajasekharan K, and Kuttan R. Anti-tumor and antioxidant activity of natural curcumoids. Cancer Letters 1995;94: 79-83.

137. Singh S, and Aggarwal B. Activation of transcription factor NF-κB is suppressed by curcumin. J Biol Chem 1995;42: 24995-5000.

138. Kakar S, and Roy D. Curcumin inhibits TPA induced expression of c-fos, c-jun and c-myc proto-oncogenes messenger RNAs in mouse skin. Cancer Letters 1994;87: 85-89.

139. Lu Y, Chang R, Lou Y, Huang M, Newmark H, Reuhl K, and Conney A. Effect of curcumin on 12-O-tetradecanoylphorbol-13-acetate and ultraviolet B light induced expression of c-Jun and c-Fos in JB6 cells and in mouse epidermis. Carcinogenesis 1994;15: 2363-70.

292

140.   Huang T-S, Lee S, and Lin J-K. Suppression of c-Jun/AP-1 activation by an inhibitor of tumor promotion in fibroblasts. Proc Natl Acad Sci USA 1991;88.

141.   Ferriola P, Cody V, and Middleton E. Protein kinase C inhibition by plant flavonoids. Kinetic mechanisms and structure-activity relationships. Biochem Pharmacol 1989;38: 1617-24.

142.   Baumann J, Bruchhausen F, and Wurm G. A structure activity study on the influence of phenolic compounds and bio-flavonoids on rat renal prostaglandin synthase. Naunyn-Schmiedeberg's Archives of Pharmacology 1980;307: 73-80.

143.   Loggia R, Ragazzi E, Tubaro A, Fassina G, and Vertua R. Anti-inflammatory activity of benzopyrones that are inhibitors of cyclo- and lipo-oxygenase. Pharmacol Res Comm 1988;20 (Suppl 5): 91-94.

144.   Mascolo N, Pinto A, and Capasso F. Flavonoids, leucocyte migration and eicosanoids. J Pharm Pharmacol 1987;40: 293-95.

145.   Gerdim B, and Svensjo E. Inhibitory effect of the flavonoid O-(B-hydroxyethyl)rutoside on increased microvascular permeability by various agents in rat skin. J Microcirc Clin Exper 1983;2: 298-302.

146.   Kim C-J, Su S-K, Joo J-H, and Cho S-K. Pharmacological activities of flavonoids. II Relationships of anti-inflammatory and antigranulomatous actions. Yakhak Hoeji 1990;34: 407-14.

147.   Kobuchi H, Roy S, Sen CK, Nguyen HG, and Packer L. Quercetin inhibits inducible ICAM-1 expression in human endothelial cells through the JNK pathway. Am J Physiol 1999;277: C403-11.

148.   Gerritsen M, Carley W, Ranges G, Shen C-P, Phan S, Ligon G, and Perry C. Flavonoids inhibit cytokine-induced endothelial cell adhesion protein gene expression. Am J Pathol 1995;147: 278-92.

149.   Kimura Y, Matsushita N, and Okuda H. Effects of baicalein isolated from Scutellaria baicalensis on interleukin 1 beta- and tumor necrosis factor alpha-induced adhesion molecule expression in cultured human umbilical vein endothelial cells. J Ethnopharmacol 1997;57: 63-7.

150.   Yuting C, Rongliang Z, Zhongjian J, and Yong Y. Flavonoids as superoxide scavengers and antioxidants. Free Radical Biology and Medicine 1990;9: 19-21.

151.   Middleton Jr E, and Kandaswami C. Effects of flavonoids on immune and inflammatory cell functions. Biochem Pharmacol 1992;43: 1167-79.

152.   Panes J, Gerritsen M, Anderson D, Miyasaka M, and Granger D. Apigenin inhibits TNF-induced ICAM-1 upregulation in vivo. Microcirculation 1996;3: 279-86.

153.   Tsai SH, Liang YC, Lin-Shiau SY, and Lin JK. Suppression of TNFalpha-mediated NFkappaB activity by myricetin and other flavonoids through downregulating the activity of IKK in ECV304 cells. J Cell Biochem 1999;74: 606-15.

154.   Wolle J, Hill RR, Ferguson E, Devall LJ, Trivedi BK, Newton RS, and Saxena U. Selective inhibition of tumor necrosis factor-induced vascular cell adhesion molecule-1 gene expression by a novel flavonoid. Lack of effect on transcription factor NF-kappa B. Arterioscler Thromb Vasc Biol 1996;16: 1501-8.

155. Kroes BH, van den Berg AJ, Quarles van Ufford HC, van Dijk H, and Labadie RP. Anti-inflammatory activity of gallic acid. Planta Med 1992;58: 499-504.

156. Bragt PC, Bansberg JI, and Bonta IL. Antiinflammatory effects of free radical scavengers and antioxidants: further support for proinflammatory roles of endogenous hydrogen peroxide and lipid peroxides. Inflammation 1980;4: 289-99.

157. Huang MT, Chang RL, Wood AW, Newmark HL, Sayer JM, Yagi H, Jerina DM, and Conney AH. Inhibition of the mutagenicity of bay-region diol-epoxides of polycyclic aromatic hydrocarbons by tannic acid, hydroxylated anthraquinones and hydroxylated cinnamic acid derivatives. Carcinogenesis 1985;6: 237-42.

158. Boyd L, and Beveridge EG. Antimicrobial activity of some alkyl esters of gallic acid (3,4,5,-trihydroxybenzoic acid) against Escherichia coli NCTC 5933 with particular reference to n-propyl gallate. Microbios 1981;30: 73-85.

159. Murase T, Kume N, Hase T, Shibuya Y, Nishizawa Y, Tokimitsu I, and Kita T. Gallates inhibit cytokine-induced nuclear translocation of NF-kappaB and expression of leukocyte adhesion molecules in vascular endothelial cells. Arterioscler Thromb Vasc Biol 1999;19: 1412-20.

160. Pan MH, Lin-Shiau SY, Ho CT, Lin JH, and Lin JK. Suppression of lipopolysaccharide-induced nuclear factor-kappaB activity by theaflavin-3,3'-digallate from black tea and other polyphenols through down-regulation of IkappaB kinase activity in macrophages. Biochem Pharmacol 2000;59: 357-67.

161. Sutton P, Newcombe N, Waring P, and Mullbacher A. In vivo immunosuppressive activity of gliotoxin a metabolite produced by human pathogenic fungi. Infect Immun 1994;62: 1192-98.

162. Mullbacher A, and Eichner R. Immunosuppression in vivo by a metabolite of a human pathogenic fungus. Proc Natl Acad Sci USA 1984;81: 3835-37.

163. Pahl H, Kraus B, Schulze-Osthoff K, Decker T, Traenckner E-M, Vogt M, Myers C, Parks T, Warring P, Muhlbacher A, Czerniolotsky A-P, and Baeuerle P. The immunosuppressive fungal metabolite gliotoxin specifically inhibits transcription factor NF-κB. J Exp Med 1996;183: 1829-40.

164. Kroll M, Arenzana-Seisdedos F, Bachelerie F, Thomas D, Friguet B, and Conconi M. The secondary fungal metabolite gliotoxin targets proteolytic activities of the proteasome. Chem Biol 1999;6: 689-98.

165. Erkel G, Anke T, and Sterner O. Inhibition of NF-kappa B activation by panepoxydone. Biochem Biophys Res Commun 1996;226: 214-21.

166. Hehner SP, Hofmann TG, Droge W, and Schmitz ML. The antiinflammatory sesquiterpene lactone parthenolide inhibits NF-kappa B by targeting the I kappa B kinase complex. J Immunol 1999;163: 5617-23.

167. Bork PM, Schmitz ML, Kuhnt M, Escher C, and Heinrich M. Sesquiterpene lactone containing Mexican Indian medicinal plants and pure sesquiterpene lactones as potent inhibitors of transcription factor NF-kappaB. FEBS Lett 1997;402: 85-90.

168. Jiang KY, Ruan CG, Gu ZL, Zhou WY, and Guo CY. Effects of tanshinone II-A sulfonate on adhesion molecule expression of endothelial cells and platelets in vitro. Chung Kuo Yao Li Hsueh Pao 1998;19: 47-50.

169. Wu TW, Zeng LH, Fung KP, Wu J, Pang H, Grey AA, Weisel RD, and Wang JY. Effect of sodium tanshinone IIA sulfonate in the rabbit myocardium and on human cardiomyocytes and vascular endothelial cells. Biochem Pharmacol 1993;46: 2327-32.

170. Chen F, Sun S, Kuhn DC, Lu Y, Gaydos LJ, Shi X, and Demers LM. Tetrandrine inhibits signal-induced NF-kappa B activation in rat alveolar macrophages. Biochem Biophys Res Commun 1997;231: 99-102.

171. Chang DM, Kuo SY, Lai JH, and Chang ML. Effects of anti-rheumatic herbal medicines on cellular adhesion molecules. Ann Rheum Dis 1999;58: 366-71.

172. Schreck R, Rieber P, and Baeuerle P. Dithiocarbamates as potent inhibitors of NF-kB activation in intact cells. J Exp Med 1992;175: 1181-94.

173. Roebuck K, Rahman A, Lakshminarayanan V, Janakidevi K, and Malik A. H_2O_2 and tumor necrosis factor a activate intercellular adhesion molecule 1 (ICAM-1) gene transcription through distinct cis-regulatory elements within the ICAM-1 promoter. J Biol Chem 1995;270: 18966-74.

174. Marui N, Offermann M, Swerlick R, Kunsch C, Rosen C, Ahmad M, Alexander R, and Medford R. VCAM-1 gene transcription and expression is regulated through an antioxidant sensitive mechanism in human vascular endothelial cells. J Clin Invest 1993;92: 1866-74.

175. Finco T, Beg A, and Baldwin A. Inducible phosphorylation of $I\kappa B\alpha$ is not sufficient for its dissociation from NF-κB and is inhibited by protease inhibitors. Prac Natl Acad Sci USA 1994;91: 11884-88.

176. Suzuki Y, and Packer L. Inhibition of NF-kappa B DNA binding activity by alpha tocopheryl succinate. Biochem Mol Biol Int 1993;31: 693-700.

177. Suzuki Y, and Packer L. Inhibition of NF-kappa B transcription factor by catechol derivatives. Biochem Mol Biol Int 1994;32: 299-305.

178. Sen C, Traber K, and Packer L. Inhibition of NF-kappa B activation in human T-cell lines by anetholdithiolthione. Biochem Biophys Res Comm 1996;218: 148-53.

179. Weber C, Erl W, Pietsch A, Strobel M, Ziegler-Heitbrock H, and Weber P. Antioxidants inhibit monocyte adhesion by suppressing nuclear factor-κB mobilization and induction of vascular cell adhesion molecule-1 in endothelial cells stimulated to generate radicals. Arterioscler Thromb 1994;14: 1665-73.

180. Matsubara T, and Ziff M. Increased superoxide anion release in response to cytokines. J Immunol 1986;137: 3295-98.

181. Suzuki Y, and Packer L. Inhibition of NF-kappa B DNA binding activity by alpha-tocopheryl succinate. Biochem Mol Biol Int 1993;31: 693-700.

182. Schiro J, Chan B, Roswit W, Kassner P, Pentland A, Hemler M, Eisen A, and Kupper T. Integrin a2b2 (VLA-2) mediates reorganization and contraction of collagen matrices by human cells. Cell 1991;67: 403-10.

183. Nobel CI, Kimland M, Lind B, Orrenius S, and Slater AF. Dithiocarbamates induce apoptosis in thymocytes by raising the intracellular level of redox-active copper. J Biol Chem 1995;270: 26202-8.

184. Galter D, Mihm S, and Droge W. Distinct effects of glutathione disulphide on the nuclear transcription factor kappa B and the activator protein-1. Eur J Biochem 1994;221: 639-48.

185. Das KC, Lewis-Molock Y, and White CW. Activation of NF-kappa B and elevation of MnSOD gene expression by thiol reducing agents in lung adenocarcinoma (A549) cells. Am J Physiol 1995;269: L588-602.

186. Friedrichs B, Muller C, and Brigelius-Flohe R. Inhibition of tumor necrosis factor-alpha- and interleukin-1-induced endothelial E-selectin expression by thiol-modifying agents. Arterioscler Thromb Vasc Biol 1998;18: 1829-37.

187. Pierce JW, Schoenleber R, Jesmok G, Best J, Moore SA, Collins T, and Gerritsen ME. Novel inhibitors of cytokine-induced IkappaBalpha phosphorylation and endothelial cell adhesion molecule expression show anti-inflammatory effects in vivo. J Biol Chem 1997;272: 21096-103.

188. Boschelli DH, Kramer JB, Connor DT, Lesch ME, Schrier DJ, Ferin MA, and Wright CD. 3-Alkoxybenzo[b]thiophene-2-carboxamides as inhibitors of neutrophil-endothelial cell adhesion. J Med Chem 1994;37: 717-8.

189. Boschelli DH, Kramer JB, Khatana SS, Sorenson RJ, Connor DT, Ferin MA, Wright CD, Lesch ME, Imre K, Okonkwo GC, and et al. Inhibition of E-selectin-, ICAM-1-, and VCAM-1-mediated cell adhesion by benzo[b]thiophene-, benzofuran-, indole-, and naphthalene-2-carboxamides: identification of PD 144795 as an antiinflammatory agent. J Med Chem 1995;38: 4597-614.

190. Cobb RR, Felts KA, McKenzie TC, Parry GC, and Mackman N. A benzothiophene-carboxamide is a potent inhibitor of IL-1beta induced VCAM-1 gene expression in human endothelial cells. FEBS Lett 1996;382: 323-6.

191. Daniel L, Civoli F, Rogers M, Smitherman P, Raju P, and Roederer M. ET-18-OCH3 inhibits nuclear factor-kappa B activation by 12-O-tetradecanoylphorbol-13-acetate but not by tumor necrosis factor-alpha or interleukin-1 alpha. Cancer Res 1995;55: 4844-49.

192. Goto M, Yamada K, Katayama K, and Tanaka I. Inhibitory effect of E3330, a novel quinone derivative able to suppress tumor necrosis factor-alpha generation, on activation of nuclear factor-kappa B. Molec Pharmacol 1996;49: 860-74.

193. Wasaki S, Sakaida I, Nagatomi A, Matsumaura Y, Yasunaga M, and Okita K. The effect of E3330 on active oxygen generation by isolated hepatic macrophages in rats. J Gastroenterol 1995;30: 273-74.

194. Bradley J, Johnson D, and Pober J. Four different classes of inhibitors of receptor-mediated endocytosis decrease tumor necrosis factor-induced gene expression in endothelial cells. J Immunol 1993;150: 5544-55.

195. de Waal Malefyt R, Abrams J, Bennett B, Figdor C, and De Vries J. Interleukin 10 (IL-10) inhibits cytokine synthesis by human monocytes: an autoregulatory role of IL-10 produced by monocytes. Journal of Experimental Medicine 1991;174: 1209-20.

196. Fiorentino D, Zlotnik A, Mosmann T, Howard M, and O'Garra A. IL-10 inhibits cytokine production by activated macrophages. J Immunol 1991;147: 3815-22.

197. Bogdan C, Paik J, Vodovotz Y, and Nathan C. Contrasting mechanisms for suppression of macrophage cytokine release by transforming growth factor-beta and interleukin-10. J Biol Chem 1992;267.

198. Wang P, Wu P, Siegel M, Egan, RW, and Billah M. IL-10 inhibits transcription of cytokine genes in human peripheral blood mononuclear cells. J Immunol 1994;153: 811-16.

199. Wang P, Wu P, Anthes J, Siegel M, Egan R, and Billah M. Interleukin-10 inhibits interleukin-8 production in human neutrophils. Blood 1994;83: 2678-83.

200. Krakauer T. IL-10 inhibits the adhesion of leukocytic cells to IL-1-activated human endothelial cells. Immunol Lett 1995;45: 61-65.

201. Shrikant P, Weber E, Jilling T, and Benveniste E. Intercellular adhesion molecule-1 gene expression by glial cells. Differential mechanisms of inhibition by IL-10 and IL-6. J Immunol 1995;155: 1489-501.

202. Willems F, Marchant A, Delville J, Gerard C, Delvaux A, Velu T, de-Boer M, and Goldman M. Interleukin-10 inhibits B7 and intercellular adhesion molecule-1 expression on human monocytes. Eur J Immunol 1994;24: 1007-09.

203. Wang P, Wu P, Siegel M, Egan R, and Billah M. Interleukin (IL)-10 inhibits nuclear factor kappa B (NF-κB) activation in human monocytes. IL-10 and IL-4 cyppress cytokine synthesis by different mechanisms. J Biol Chem 1995;270: 9558-63.

204. Garg U, and Hassid A. Nitric oxide generating vasodilators and 8-bromoguanosine monophosphate inhibit mitogenesis and proliferation of cultured rat vascular smooth muscle cells. J Clin Invest 1989;83: 1774-77.

205. Radomski M, Palmer R, and Moncada S. The anti-aggregating properties of vascular endothelium: interactions between prostacyclin and nitric oxide. Brit J Pharmacol 1987;92: 639-46.

206. Bath P, Hassall D, Gladwin A, Palmer R, and Martin J. Nitric oxide and prostacyclin. Divergence of inhibitory effects on monocyte chemotaxis and adhesion to endothelium in vitro. Arterioscler Thromb 1991;11: 254-60.

207. De Caterina R, Libby P, Peng H-B, Thannickal V, Rajavashisth T, Gimbrone M, Shin W, and Liao J. Nitric oxide decreases cytokine induced endothelial activation. J Clin Invest 1995;96: 60-68.

208. Peng H-B, Libby P, and Liao J. Induction and stabilization of IκBα by nitric oxide mediates inhibition of NF-κB. J Biol Chem 1995;270: 14214-19.

209. Thommesen L, Sjursen W, Gasvik K, Hanssen W, Brekke OL, Skattebol L, Holmeide AK, Espevik T, Johansen B, and Laegreid A. Selective inhibitors of cytosolic or secretory phospholipase A2 block TNF-induced activation of transcription factor nuclear factor-kappa B and expression of ICAM-1. J Immunol 1998;161: 3421-30.

210. Pietersma A, de Jong N, de Wit LE, Kraak-Slee RG, Koster JF, and Sluiter W. Evidence against the involvement of multiple radical generating sites in the expression of the vascular cell adhesion molecule-1. Free Radic Res 1998;28: 137-50.

211. Xin X, Yang S, Kowalski J, and Gerritsen ME. Peroxisome proliferator-activated receptor gamma ligands are potent inhibitors of angiogenesis in vitro and in vivo. J Biol Chem 1999;274: 9116-21.

212. Pasceri V, Wu HD, Willerson JT, and Yeh ET. Modulation of vascular inflammation in vitro and in vivo by peroxisome proliferator-activated receptor-gamma activators. Circulation 2000;101: 235-8.

213. Jackson SM, Parhami F, Xi XP, Berliner JA, Hsueh WA, Law RE, and Demer LL. Peroxisome proliferator-activated receptor activators target human endothelial cells to inhibit leukocyte-endothelial cell interaction. Arterioscler Thromb Vasc Biol 1999;19: 2094-104.

214. Stuhlmeier KM, Tarn C, Csizmadia V, and Bach FH. Selective suppression of endothelial cell activation by arachidonic acid. Eur J Immunol 1996;26: 1417-23.

215. Weber C, Erl W, Pietsch A, Danesch U, and Weber PC. Docosahexaenoic acid selectively attenuates induction of vascular cell adhesion molecule-1 and subsequent monocytic cell adhesion to human endothelial cells stimulated by tumor necrosis factor-alpha. Arterioscler Thromb Vasc Biol 1995;15: 622-8.

216. Stuhlmeier KM, Tarn C, and Bach FH. The effect of 5,8,11,14-eicosatetraynoic acid on endothelial cell gene expression. Eur J Pharmacol 1997;325: 209-19.

217. De Caterina R, Liao JK, and Libby P. Fatty acid modulation of endothelial activation. Am J Clin Nutr 2000;71: 213S-23S.

218. De Caterina R, Spiecker M, Solaini G, Basta G, Bosetti F, Libby P, and Liao J. The inhibition of endothelial activation by unsaturated fatty acids. Lipids 1999;34: S191-4.

219. De Caterina R, Bernini W, Carluccio MA, Liao JK, and Libby P. Structural requirements for inhibition of cytokine-induced endothelial activation by unsaturated fatty acids. J Lipid Res 1998;39: 1062-70.

220. De Caterina R, and Libby P. Control of endothelial leukocyte adhesion molecules by fatty acids. Lipids 1996;31: S57-63.

221. De Caterina R, Cybulsky MA, Clinton SK, Gimbrone MA, Jr., and Libby P. Omega-3 fatty acids and endothelial leukocyte adhesion molecules. Prostaglandins Leukot Essent Fatty Acids 1995;52: 191-5.

222. De Caterina R, Cybulsky MI, Clinton SK, Gimbrone MA, Jr., and Libby P. The omega-3 fatty acid docosahexaenoate reduces cytokine-induced expression of proatherogenic and proinflammatory proteins in human endothelial cells. Arterioscler Thromb 1994;14: 1829-36.

223. Hiromatsu Y, Sato M, Tanaka K, Ishisaka N, Kamachi J, and Nonaka K. Inhibitory effects of nicotinamide on intercellular adhesion molecule-1 expression on cultured human thyroid cells. Immunology 1993;80: 330-32.

224. Hiromatsu Y, Sato M, Yamada K, and Nonaka K. Inhibitory effects of nicotinamide on recombinant human interferon g-induced intercellular adhesion molecule-1 (ICAM-1) and HLA-DR antigen expression on cultured human endothelial cells. Immunol Lett 1991;31: 35-39.

225. Pober J, Lapierre L, Stolpen A, Brock T, Springer T, Fiers W, Bevilacqua M, Mendrick D, and Gimbrone MJ. Activation of cultured human endothelial cells by recombinant lymphotoxin: comparison with tumor necrosis factor and interleukin 1 species. J Immunol 1987;138: 3319-24.

226. Daneker G, Lund S, Caughman S, Staley C, and Wood W. Anti-metastatic prostacyclins inhibit the adhesion of colon carcinoma to endothelial cells by blocking E-selectin expression. Clin Exp Metastasis 1996;14: 230-38.

227. Ghersa P, van Huijsduijnen R, Whelan J, Camber Y, Pescani R, and DeLamarter J. Inhibition of E-selectin gene transcription through a cAMP-dependent protein kinase pathway. J Biol Chem 1994;269: 29129-37.

228. Panettieri R, Lazaar A, Pure E, and Albelda S. Activation of cAMP-dependent pathways in human airway smooth muscle cells inhibits TNF-alpha-induced ICAM-1 and VCAM-1 expression and T-lymphocyte adhesion. J Immunol 1995;154: 2358-65.

229. Oliver CJ, and Shenolikar S. Physiologic importance of protein phosphatase inhibitors. Front Biosci 1998;3: D961-72.

230. Goldberg A, and Rock K. Proteolysis, proteasomes and antigen presentation. Nature 1992;357: 375-79.

231. Peters J-M. Proteasomes: protein degradation machines of the cell. Trends Biochem Sci 1994;19: 377-82.

232. Ciechanover A. The ubiquitin-proteasome proteolytic pathway. Cell 1994;79: 13-21.

233. Palombella V, Rando O, Goldberg A, and Maniatis T. The ubiquitin-proteasome pathway is required for processing of the NF-κB precursor protein and the activation of NF-κB. Cell 1994;78: 773-85.

234. Read M, Neish A, Luscinskas F, Palombella V, Maniatis T, and Collins T. The proteasome pathway is required for cytokine-induced endothelial-leukocyte adhesion molecule expression. Immunity 1995;2: 1-20.

235. Cobb R, Felts K, Parry G, and Mackman N. Proteasome inhibitors block VCAM-1 and ICAM-1 gene expression in endothelial cells without affecting nuclear translocation of nuclear factor-κB. Eur J Immunol 1996;26: 839-45.

236. Lum RT, Kerwar SS, Meyer SM, Nelson MG, Schow SR, Shiffman D, Wick MM, and Joly A. A new structural class of proteasome inhibitors that prevent NF-kappa B activation. Biochem Pharmacol 1998;55: 1391-7.

237. Omura S, Fujimoto T, Otoguro K, Matsuzaki K, Moriguchi R, Tanaka H, and Sasaki Y. Lactacystin, a novel microbial metabolite, induces neuritogenesis of neuroblastoma cells [letter]. J Antibiot (Tokyo) 1991;44: 113-6.

238. Fenteany G, Standaert RF, Reichard GA, Corey EJ, and Schreiber SL. A beta-lactone related to lactacystin induces neurite outgrowth in a neuroblastoma cell line and inhibits cell cycle progression in an osteosarcoma cell line. Proc Natl Acad Sci U S A 1994;91: 3358-62.

239. Dick LR, Cruikshank AA, Destree AT, Grenier L, McCormack TA, Melandri FD, Nunes SL, Palombella VJ, Parent LA, Plamondon L, and Stein RL. Mechanistic studies on the inactivation of the proteasome by lactacystin in cultured cells. J Biol Chem 1997;272: 182-8.

240. Dick LR, Cruikshank AA, Grenier L, Melandri FD, Nunes SL, and Stein RL. Mechanistic studies on the inactivation of the proteasome by lactacystin: a central role for clasto-lactacystin beta-lactone. J Biol Chem 1996;271: 7273-6.

241. Craiu A, Gaczynska M, Akopian T, Gramm CF, Fenteany G, Goldberg AL, and Rock KL. Lactacystin and clasto-lactacystin beta-lactone modify multiple proteasome beta-subunits and inhibit intracellular protein degradation and major histocompatibility complex class I antigen presentation. J Biol Chem 1997;272: 13437-45.

242. Ditzel L, Stock D, and Lowe J. Structural investigation of proteasome inhibition. Biol Chem 1997;378: 239-47.

243. Ostrowska H, Wojcik C, Omura S, and Worowski K. Lactacystin, a specific inhibitor of the proteasome, inhibits human platelet lysosomal cathepsin A-like enzyme. Biochem Biophys Res Commun 1997;234: 729-32.

244. Imajoh-Ohmi S, Kawaguchi T, Sugiyama S, Tanaka K, Omura S, and Kikuchi H. Lactacystin, a specific inhibitor of the proteasome, induces apoptosis in human monoblast U937 cells. Biochem Biophys Res Commun 1995;217: 1070-7.

245. Li J, Post M, Volk R, Gao Y, Li M, Metais C, Sato K, Tsai J, Aird W, Rosenberg RD, Hampton TG, Sellke F, Carmeliet P, and Simons M. PR39, a peptide regulator of angiogenesis [published erratum appears in Nat Med 2000 Mar;6(3):356]. Nat Med 2000;6: 49-55.

246. Henkel T, Machleidt T, Alkalay I, Kronke M, Ben-Neriah Y, and Baeuerle P. Rapid proteolysis of IκBα is necessary in the activation of transcription factor NF-kB. Nature 1993;365: 182.

247. Chen CC, Rosenbloom CL, Anderson DC, and Manning AM. Selective inhibition of E-selectin, vascular cell adhesion molecule-1, and intercellular adhesion molecule-1 expression by inhibitors of IκBα phosphorylation. J Immunol 1995;155: 3538-45.

248. Lee C-H, Chen H, Hoke G, Jong J, White L, and Kang Y-H. Antisense gene suppression against human ICAM-1. ELAM-1 and VCAM-1 in cultured human umbilical vein endothelial cells. Shock 1995;4: 1-10.

249. Bennett C, Condon T, Grimm S, Chan H, and Chian M. Inhibition of endothelial cell adhesion molecule expression with antisense oligonucleotides. J Immunol 1994;152: 3530-40.

250. Kumasaka T, Quinlan W, Doyle N, Condon T, Sligh J, Takei F, Beaudet A, Bennet C, and Doerschuk C. Role of the intercellular adhesion molecule-1 (ICAM-1) in endotoxin-induced pneumonia evaluated using ICAM-1 antisense oligonucleotides, anti-ICAM-1 monoclonal antibodies and ICAM-1 mutant mice. J Clin Invest 1996;97: 2362-69.

251. Toda K, Kayano K, Karimova A, Naka Y, Fujita T, Minamoto K, Wang CY, and Pinsky DJ. Antisense intercellular adhesion molecule-1 (ICAM-1) oligodeoxyribonucleotide delivered during organ preservation inhibits posttransplant ICAM-1 expression and reduces primary lung isograft failure. Circ Res 2000;86: 166-74.

252. Cheng QL, Chen XM, Li F, Lin HL, Ye YZ, and Fu B. Effects of ICAM-1 antisense oligonucleotide on the tubulointerstitium in mice with unilateral ureteral obstruction. Kidney Int 2000;57: 183-90.

253. Feeley BT, Park AK, Alexopoulos S, Hoyt EG, Ennen MP, Poston RS, Jr., and Robbins RC. Pressure delivery of AS-ICAM-1 ODN with LFA-1 mAb reduces reperfusion injury in cardiac allografts. Ann Thorac Surg 1999;68: 119-24.

254. Dragun D, Lukitsch I, Tullius SG, Qun Y, Park JK, Schneider W, Luft FC, and Haller H. Inhibition of intercellular adhesion molecule-1 with antisense deoxynucleotides prolongs renal isograft survival in the rat. Kidney Int 1998;54: 2113-22.

255. Stepkowski SM, Wang ME, Condon TP, Cheng-Flournoy S, Stecker K, Graham M, Qu X, Tian L, Chen W, Kahan BD, and Bennett CF. Protection against allograft rejection with intercellular adhesion molecule-1 antisense oligodeoxynucleotides. Transplantation 1998;66: 699-707.

256. Weber MC, Groger RK, and Tykocinski ML. Antisense modulation of the ICAM-1 phenotype of a model human bone marrow stromal cell line. Exp Cell Res 1998;244: 239-48.

257. Dragun D, Tullius SG, Park JK, Maasch C, Lukitsch I, Lippoldt A, Gross V, Luft FC, and Haller H. ICAM-1 antisense oligodesoxynucleotides prevent reperfusion injury and enhance immediate graft function in renal transplantation. Kidney Int 1998;54: 590-602.

258. Yacyshyn BR, Bowen-Yacyshyn MB, Jewell L, Tami JA, Bennett CF, Kisner DL, and Shanahan WR, Jr. A placebo-controlled trial of ICAM-1 antisense oligonucleotide in the treatment of Crohn's disease. Gastroenterology 1998;114: 1133-42.

259. Stepkowski SM, Wang ME, Amante A, Kalinin D, Qu X, Blasdel T, Condon T, Kahan BD, and Bennett FC. Antisense ICAM-1 oligonucleotides block allograft rejection in rats. Transplant Proc 1997;29: 1285.

260. Bennett CF, Kornbrust D, Henry S, Stecker K, Howard R, Cooper S, Dutson S, Hall W, and Jacoby HI. An ICAM-1 antisense oligonucleotide prevents and reverses dextran sulfate sodium-induced colitis in mice. J Pharmacol Exp Ther 1997;280: 988-1000.

261. Haller H, Dragun D, Miethke A, Park JK, Weis A, Lippoldt A, Gross V, and Luft FC. Antisense oligonucleotides for ICAM-1 attenuate reperfusion injury and renal failure in the rat. Kidney Int 1996;50: 473-80.

262. Bielinska A, Shivdasani R, Zhang L, and Nabel G. Regulation of gene expression with double-stranded phosphorothionate oligonucleotides. Science 1990;250: 997-1000.

263. Wrighton C, Hofer-Warbinek R, Moll T, Eytner R, and Bach; F. Inhibition of endothelial cell activation by adenovirus-mediated expression of IκB, an inhibitor of the transcription factor, NF-κB. Journal of Experimental Medicine 1996;183: 11013-22.

264. Deisher T, Kaushansky K, and Harlan J. Inhibitors of toposiomerase II prevent cytokine-induced expression of vascular cell adhesion molecule-1, while augmenting the expression of endothelial leukocyte adhesion molecule-1 on human umbilical vein endothelial cells. Cell Adhesion and Communication 1993;1: 133-42.

265. Weber C, Negrescu E, Erl W, Pietsch A, Frankenberger M, Ziegler-Heitbrock H, Siess W, and Weber P. Inhibitors of protein tyrosine kinase suppress TNF-stimulated induction of endothelial cell adhesion molecules. J Immunol 1995;155: 445-51.

266. Mahon T, and O'Neill L. Studies into the effect of the tyrosine kinase inhibitor herbimycin A on NF-kappa B activation in T-lymphocytes. Evidence for covalent modification of the p50 subunit. J Biol Chem 1995;270: 28557-64.

267. Nishiya T, Uehara T, and Nomura Y. Herbimycin A suppresses NF-kappa B activation and tyrosine phosphorylation of JAK2 and the subsequent induction of nitric oxide synthase in C6 glioma cells. FEBS Letters 1995;371: 333-36.

268. Dhawan S, Singh S, and Aggarwal BB. Induction of endothelial cell surface adhesion molecules by tumor necrosis factor is blocked by protein tyrosine

phosphatase inhibitors: role of the nuclear transcription factor NF-kappa B. Eur J Immunol 1997;27: 2172-9.

269. Duff J, Quinlan K, Paxton L, Naik S, and Caughman S. Pervanadate mimics IFNγ-mediated induction of ICAM-1 expression via activation of STAT proteins. J Invest Dermatol 1997;108: 295-301.

270. Cobb R, Felts K, Parry G, and Mackman N. D609, a phosphatidylcholine-specific phospholipase C inhibitor, blocks interleukin-1 beta-induced vascular cell adhesion molceule 1 gene expression in human endothelial cells. Molecular Pharmacology 1996;49: 998-1004.

271. Kent KC, Collins LJ, Schwerin FT, Raychowdhury MK, and Ware JA. Identification of functional PGH2/TxA2 receptors on human endothelial cells. Circ Res 1993;72: 958-65.

272. Raychowdhury MK, Yukawa M, Collins LJ, McGrail SH, Kent KC, and Ware JA. Alternative splicing produces a divergent cytoplasmic tail in the human endothelial thromboxane A2 receptor. J Biol Chem 1995;270: 7011.

273. Raychowdhury MK, Yukawa M, Collins LJ, McGrail SH, Kent KC, and Ware JA. Alternative splicing produces a divergent cytoplasmic tail in the human endothelial thromboxane A2 receptor [published erratum appears in J Biol Chem 1995 Mar 24;270(12):7011]. J Biol Chem 1994;269: 19256-61.

274. Ishizuka T, Suzuki K, Kawakami M, Kawaguchi Y, Hidaka T, Matsuki Y, and Nakamura H. DP-1904, a specific inhibitor of thromboxane A2 synthesizing enzyme, suppresses ICAM-1 expression by stimulated vascular endothelial cells. Eur J Pharmacol 1994;262: 113-23.

275. Ishizuka T, Suzuki K, Kawakami M, Hidaka T, Matsuki Y, and Nakamura H. Thromboxane A2 receptor blockade suppresses intercellular adhesion molecule-1 expression by stimulated vascular endothelial cells. Eur J Pharmacol 1996;312: 367-77.

276. Ishizuka T, Kawakami M, Hidaka T, Matsuki Y, Takamizawa M, Suzuki K, Kurita A, and Nakamura H. Stimulation with thromboxane A2 (TXA2) receptor agonist enhances ICAM-1, VCAM-1 or ELAM-1 expression by human vascular endothelial cells. Clin Exp Immunol 1998;112: 464-70.

277. Yamaguchi M, Suwa H, Miyasaka M, and Kumada K. Selective inhibition of vascular cell adhesion molecule-1 expression by verapamil in human vascular endothelial cells. Transplantation 1997;63: 759-64.

278. Gerritsen ME, Nganele DM, and Rodrigues AM. Calcium ionophore (A23187)- and arachidonic acid-stimulated prostaglandin release from microvascular endothelial cells: effects of calcium antagonists and calmodulin inhibitors. J Pharmacol Exp Ther 1987;240: 837-46.

Chapter 9

THE NF-κB SYSTEM AND DRUG DISCOVERY

Anthony M. Manning

Arthritis and Inflammatory Diseases
Pharmacia Corporation
700 Chesterfield Parkway North
St. Louis, MO 63198

INTRODUCTION

The NF-κB signal transduction pathway plays an important role in leukocyte recruitment through the regulation of endothelial cell adhesion molecule and chemokine expression. NF-κB was originally identified as a transcription factor required for B cell-specific gene expression, and subsequent studies demonstrated that it is ubiquitously expressed and serves as a critical regulator of the inducible expression of many genes, particularly those associated with the pathogenesis of autoimmune diseases. For this reason, the pharmaceutical industry has focused significant attention on this pathway for the identification of novel therapeutic agents. However, much remains to be understood as to how NF-κB is regulated and what specific role this transcription factor plays in normal and diseased tissues. This chapter will briefly summarize recent findings on the mechanisms of NF-κB regulation in cells and on the potential role of NF-κB in human disease. This information suggests that NF-κB inhibitors will represent exciting new disease-modifying agents with broad potential in autoimmune, cardiovascular and neurological diseases, and in cancer. Challenges for drug discovery will be highlighted.

NF-κB AND IκB PROTEINS

NF-κB exists in the cytoplasm of the majority of cell types as homo- or heterodimers of a family of structurally related proteins.[1,2] Each member of this family contains a conserved N-terminal region called the Rel-homology domain (RHD), within which lies the DNA-binding and dimerization domains and the nuclear localization signal (NLS). To date, five proteins belonging to the NF-κB family have been identified in mammalian cells: RelA (also known as p65); c-Rel; RelB; NF-κB1 (p50/105); and NF-κB2 (p52/100). The first three are produced as transcriptionally active proteins. The latter are synthesized as longer precursor molecules of 105 and 100 kDa respectively, which are further processed to smaller, transcriptionally active forms. The classical NF-κB dimer contains RelA and NF-κB1, but a variety of other Rel-containing dimers are also known to exist.[3-5]

NF-κB exists in the cytoplasm in an inactive form associated with inhibitory proteins termed IκBs, of which the most important may be IκBα, IκBβ, and IκBε.[3,6] The IκB family members, which share common ankyrin-like repeat domains, regulate the DNA binding and subcellular localization of Rel/NF-κB proteins by masking an NLS located near the C-terminus of the Rel homology domain.[7]

NF-κB ACTIVATION PATHWAY

NF-κB can be activated in cells by a wide variety of stimuli associated with stress, injury, and inflammation. Potent inducers of NF-κB include: cytokines such as interleukin 1β (IL-1β) and tumor necrosis factor-α (TNF-α); bacterial and viral products such as lipopolysaccharide (LPS), sphingomyelinase, double-stranded RNA and the Tax protein from human T-cell leukemia virus 1 (HTLV-1); and pro-apoptotic and necrotic stimuli such as oxygen free radicals, UV light, and γ-irradiation.[8] This diversity of inducers highlights an intriguing aspect of NF-κB regulation, namely the ability of many different signal transduction pathways emanating from a wide variety of induction mechanisms to converge on a single target: the cytosolic NF-κB:IκB complex. The recent discovery of several key enzyme components of the NF-κB activation pathway suggests a molecular basis for signal integration in this pathway.

NF-κB activation is achieved through the signal-induced proteolytic degradation of IκB in the cytoplasm (Fig. 9-1). Extracellular stimuli initiate a signaling cascade leading to activation of two IκB kinases, IKK-1 (IKKα) and IKK-2 (IKKβ), which phosphorylate IκB at specific N-terminal serine residues (S32 and S36 for IκBα, S19 and S23 for IκBβ) (9-13, reviewed in 14-16). Phosphorylated IκB is then selectively ubiquitinated, presumably by an E3 ubiquitin ligase, the terminal member of a cascade of ubiquitin-conjugating enzymes.[17, 18] In the last step of this signaling cascade, phosphorylated and ubiquitinated IκB, which is still associated with NF-κB in the cytoplasm, is selectively degraded by the 26S proteasome.[19] This process exposes the NLS, thereby freeing NF-κB to interact with the nuclear import machinery and translocate to the nucleus,[7] where it binds its target genes to initiate transcription.

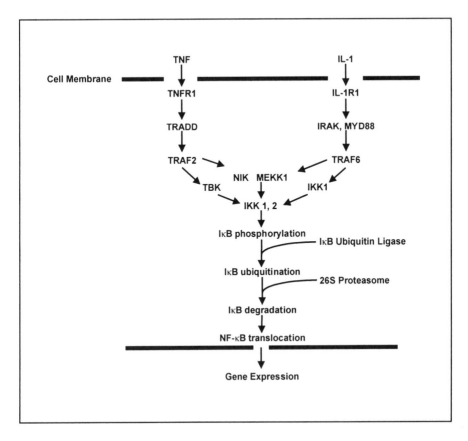

Figure 9-1: Schematic representation of components of the NF-κB signal transduction pathway leading from the TNF and IL-1 receptors. TNF-receptor associated factors 2 and 6 [TRAF2 and 6], death domain-containing proteins [TRADD and FADD], kinases associated with the IL-1 recpetor [IRAK1 and 2, and MYD88], ring finger interacting protein [RIP] mitogen activated protein kinase kinase kinase [MEKK], NF-κB inducing kinase [NIK], IκB kinase 1 and 2 [IKK1 and IKK2].

The IκB kinases IKK-1 (IKKα) and IKK-2 (IKKβ) are related members of a new family of intracellular signal transduction enzymes containing an amino-terminal kinase domain and a C-terminal region with two protein interaction motifs, a leucine zipper and a helix-loop-helix motif. These motifs mediate the heterodimerization of IKK-1 and IKK-2, although it is not yet known whether heterodimerization is essential for function. There is strong evidence that IKK-1 and IKK-2 are themselves phosphorylated and activated by one or more upstream activating kinases, which are likely to be members of the MAP kinase kinase kinase (MAPKKK) family of enzymes.[9-13, 20] One such upstream kinase, NF-κB Inducing Kinase (NIK), was identified by its ability to bind directly to TRAF2, an adapter protein thought to couple both TNF-α and IL-1 receptors to NF-κB activation.[21] A second MAPKKK,

MEKK-1, was shown to be present in the IKK signalsome complex.[12] Coexpression of either NIK or MEKK enhances the ability of the IKKs to phosphorylate IκB and activate NF-κB.[9, 10, 12, 13] The likely sites of activating phosphorylation on the IKKs have been identified as two serine residues within the kinase activation loop, which lie within a short region of homology to the MEK (MAP kinase kinase) family of proteins.[12] Phosphorylation of these two serine residues in the MEKs is required for their activation. In IKK-2, mutation of the two corresponding serine residues to alanine yields an inactive, "dominant-negative" protein capable of blocking the activation of endogenous NF-κB;[12] in both IKK-1 and IKK-2, mutation of these residues to glutamate yields a constitutively active kinase, presumably because the glutamate residues mimic to some degree the phosphoserines obtained after phosphorylation by the upstream activating kinase.[12]

The exact role of NIK and MEKK-1 in IKK activation has been the subject of several reports. NIK preferentially phosphorylates IKK1 on Ser 176 in the activation loop between subdomain VII and VIII, leading to the activation of IKK1 kinase activity.[22] Overexpression of NIK in cells leads to activation of both IKK1 and IKK2, suggesting that IKK1 activation somehow leads to activation of IKK2. Of note, MEKK-1 specifically phosphorylates the corresponding serines in the activation loop of IKK2, leading to activation of IKK2 kinase activity, but not IKK1 activity.[22,23] Because NIK and MEKK-1 are activated by discrete stimuli, they appear to provide a mechanism for the differential activation of members of the IKK complex. Extracellular stimuli which selectively activate IKK1 or IKK2 have not been identified, and the physiologic consequences of selective IKK activation are unknown. It is intriguing to speculate, however, that this mechanism may somehow play a role in the selective activation of different complexes of Rel family members within the cytoplasm. Clearly, additional characterization of the biochemical components of the IKK signalsome complex and the mechanism of recruitment and phosphorylation of different IκB-containing complexes is essential to further understanding the role of selective IKK activation. Recently, a non-kinase component of the IKK signalsome was identified through genetic complementation of a flat cellular variant of HTLV-1 Tax-transformed rat fibroblasts, known as 5R, which is unresponsive to all tested NF-κB activating stimuli.[24] This protein, named NEMO (NF-κB-Essential MOdulator), represents a 48 kDa protein, which contains a putative leucine zipper. NEMO is present in the IKK signalsome complex and is essential for its formation. In vitro, NEMO can homodimerize and directly interact with IKK2. The protein is absent from 5R cells, and another NF-κB-unresponsive cell line, 1.3E2. The NEMO protein was able to complement both cell lines, and to restore NF-κB activation induced by LPS, PMA, and IL-1. Using biochemical techniques, we recently identified the presence of two distinct IKK-containing complexes in cells; one a heterodimer containing IKK1 and IKK2, the other a homodimer containing IKK2.[25] Both complexes contained NEMO, presumably through its interaction with IKK2. The assembly of distinct IKK-containing complexes which can be differentially activated by upstream activators such as NIK or MEKK provides an additional mechanism for selective activation of NF-κB complexes in cells.

IκB phosphorylation serves as a molecular tag leading to rapid ubiquitination and degradation of IκB by components of the ubiquitin-proteasome system. The sites of ubiquitin conjugation are two adjacent lysine residues (Lys-20 and Lys-21 in IκBα) located just N-terminal to the two serines that are targets for phosphorylation by the IKKs.[26] Using peptides as substrate mimics, it was established that IκBα phosphorylation generates a binding site for the specific ubiquitin ligase(s) responsible for conjugation of ubiquitin to IκBα, and that microinjection of these phosphopeptides into cells could interfere with NF-κB activation.[27] These results emphasize the importance of the ubiquitin ligase system in IκB degradation. However, the IκB-specific E3 ligase remains to be molecularly identified.

The IκB kinases IKK1 and IKK2, their upstream activating kinases MEKK and NIK, and their downstream effector, the putative E3 ligase, all represent attractive targets for the discovery of drugs which selectively regulate NF-κB function. These proteins may all be part of a large 600-800 kDa protein complex, the IKK signalsome.[12] Particularly interesting is the crosstalk between the NF-κB and AP-1 activation pathways, exemplified by the homology between IKKs and MEKs, the presence of MEKK in the IKK signalsome complex, and the fact that the upstream kinases of the IKKs are members of the MAPKKK family. Additional insights into the mechanisms of IKK activation will expand our understanding of how specific signals generated in response to different stimuli can converge at the level of the IKK signalsome. One intriguing possibility is that there exist in cells distinct IKK signalsome complexes, composed of different IKK isoforms and associated proteins, which can be differentially activated by extracellular stimuli. In support of this hypothesis, we recently demonstrated that at least two distinct IKK signalsome complexes exist in Hela cells, and that the complex containing IKK2 homodimers but not IKK1 is differentially activated by TNF.[25] These findings were extended into primary human CD4+ T lymphocytes, where the IKK2 homodimeric complex was preferentially activated by stimulation through the T cell receptor.[28] The discovery of additional upstream activators of the IKK signalsome should facilitate the dissection of tissue- and stimulus-specific activation of NF-κB.

Post-translational modification of the NF-κB complex occurring subsequent to IκB degradation has been reported to enhance transcriptional activation of NF-κB-dependent genes. The transcriptional activity of NF-κB is stimulated upon phosphorylation of its p65 subunit on serine 276 by protein kinase A (PKA).[29] The transcriptional coactivator CBP/p300 was found to associate with NF-κB RelA through two sites, an N-terminal domain that interacts with the C-terminal region of unphosphorylated RelA, and a second domain that interacts with RelA phosphorylated on serine 276.[30,31] Accessibility to both sites is blocked in unphosphorylated RelA through an intramolecular masking of the N-terminus by the C-terminal region of RelA. Phosphorylation by PKA both weakens the interaction between the N- and C-terminal regions of RelA and creates an additional site for interaction with CBP/p300.[31] Because PKA phosphorylates a range of substrates *in vivo*, it is unlikely that PKA inhibitors could achieve selective modulation of NF-κB-dependent gene transcription. Of note, RelA is phosphorylated by IKK1 and IKK2 *in vitro* on several other residues in addition to Ser 276.[12,25] The exact residues phosphorylated by these enzymes and their effects on NF-κB-dependent transcription *in vivo*, however, have yet to be reported.

ROLE OF NF-κB IN DISEASE

Our understanding of the potential role of NF-κB in human disease has evolved from several sources. Initially, a pathogenic role for NF-κB in autoimmune disease was proposed based solely upon the finding of NF-κB binding sites in the promoters of genes involved in B and T cell differentiation and function. The definition of the role of NF-κB in the regulation of endothelial cell adhesion molecule expression clearly implicated this factor in the regulation of leukocyte-mediated inflammation. The development of specific reagents and techniques to detect NF-κB proteins and DNA-binding activity in tissue biopsies facilitated the direct demonstration of the presence of activated NF-κB in diseased tissues. Such studies documented a correlation between the presence of activated NF-κB, the expression of NF-κB-regulated genes, and the presence of disease. Recently, several studies in animal models of human disease have demonstrated that functional inhibition of NF-κB activation can result in either the prevention of disease or the resolution of established disease. These studies provide a critical "proof-of-concept" for NF-κB inhibition as a disease-modifying strategy. These and several other studies suggest that NF-κB plays a key role in a range of diseases in man not limited to autoimmune or inflammatory diseases, but including cardiovascular and neurological diseases, as well as cancer. A selection of such studies is discussed below.

Autoimmune and Inflammatory Diseases

A significant number of genes involved in the immune response are inducibly regulated via NF-κB, and include pro-inflammatory cytokines, hematopoietic growth factors, cell adhesion molecules, and a variety of enzymes capable of producing inflammatory mediators or inducing tissue destruction.

A role for NF-κB in rheumatoid arthritis (RA) is supported by the detection of activated NF-κB in human inflamed synovial tissue.

Nuclear extracts prepared from synovial membranes isolated from RA patients contain high levels of NF-κB DNA-binding activity.[32] Immunoreactive NF-κB is abundant in rheumatoid synovium.[33] The p50 and p65 NF-kB proteins are primarily localized to the nuclei of cells in the synovial intimal lining.[34] The same proteins are rife in osteoarthritis (OA) synovium, which is not surprising in light of the fact that NF-κB components are constitutively expressed in most cells. However, NF-κB activation as determined by EMSA is much greater in RA than OA,[32] probably due to phosphorylation and degradation of IκB in RA intimal lining cells. The time course of transcription factor activation and matrix metalloproteinase (MMP) gene expression in inflammatory arthritis has been evaluated in murine CIA using EMSA on joint extracts.[35] Notably, synovial NF-κB activation occurs before the onset of arthritis symptoms in this model. TNF-α has been shown to be a major therapeutic target in RA with the success of anti-TNF-α antibody clinical trials. In normal human macrophages and in human rheumatoid joint cell cultures, inhibition of NF-κB activation via adenoviral-mediated transfer of IκBα resulted in approximately 80% inhibtion in TNF-α expression.[36] Indeed, NF-κB was demonstrated to control the expression of multiple inflammatory molecules in synoviocytes, including TNF-α, IL-1β, IL-6 and VCAM-1.[37] Intraarticular administration of double-stranded oligonucleotides containing specific NF-κB binding sites decreased the severity of streptococcal wall-induced arthritis in rats, and prevented recurrence of disease. Of note, the severity of arthritis that occurs in contralateral, untreated joints was also diminished, demonstrating beneficial systemic effects of local suppression of NF-κB. As discussed later in this review, NF-κB protects certain cell types from TNF-α-induced cell death through induced transcription of a range of anti-apoptotic genes. Inhibition of NF-κB therefore represents a strategy to enhance the clearance of destructive cells that persist in disease tissues. In this regard, inhibition of NF-κB in rheumatoid synoviocytes *in vitro* enhanced TNF-α and Fas ligand cytotoxicity, while inhibition *in vivo* profoundly enhanced apoptosis in the synovium of rats with streptococcal cell wall- or pristane-induced arthritis.[37] These data suggest that NF-κB is a particularly attractive target for the development of disease-modifying anti-arthritic drugs.

Asthma is an inflammatory disorder of unknown cause. It is exceedingly common, with 5-7% of the population in the United States and Europe suffering from this disorder. Asthma in humans is considered a chronic inflammatory disease of the respiratory tract in which bronchial hyper-responsiveness results from the T lymphocyte-mediated recruitment of large numbers of activated, tissue-damaging eosinophils into the airways. Because of the key role NF-κB plays in the transcriptional activation of multiple genes central to allergic airway inflammation, it represents an attractive target for the development of anti-asthma agents. There have been no reports to date of studies analyzing NF-κB activation in bronchial biopsies from asthmatic patients. In an animal model of antigen-induced airway inflammation, we observed that NF-κB was rapidly activated in vascular endothelium and bronchial epithelium upon antigen challenge, an event associated with induction of E-selectin, VCAM-1 and ICAM-1 gene transcription, and the development of an eosinophil and T lymphocyte-rich leukocytic infiltrate.[38] Infections with viruses such as rhinovirus and influenza virus can trigger severe acute exacerbations of asthma by initiating a prolonged inflammatory response. Experimental infection with rhinovirus activates NF-κB and stimulates the secretion of interleukin-6 in nasal epithelial cells.[39] Oxidative stress may also exacerbate airway inflammation. In animal models, the inhalation of ozone induces inflammation in the lower respiratory tract and activates NF-κB and NF-κB-dependent genes.[40]

Anti-inflammatory glucocorticoids (GCs), used for decades as clinical tools to suppress both the immune response and the process of inflammation, represent the most efficacious agents currently available for the treatment of asthma.[41] The molecular mechanisms that underlie their therapeutic effects are poorly understood. GCs bind to a cytoplasmic glucocorticoid receptor (GR), a member of the steroid hormone receptor superfamily, promoting nuclear translocation and DNA binding. Cell adhesion molecule and cytokine gene expression is repressed by GCs even though these genes lack glucocorticoid responsive elements (GREs). Further support for NF-κB as a molecular target in asthma emerged from studies of the mechanism of action of these agents. GCs may inhibit activation of NF-κB through transcriptional activation of the IκBα gene,[42,43] although there is mounting evidence that GCs block NF-κB-dependent gene expression (see Chapters 4 and 8). GC inhibition of NF-κB activation could account for some of the anti-inflammatory properties of these molecules. Inhibition of NF-κB, therefore, represents a novel approach to anti-asthma drug development that might provide agents with the efficacy of GCs, but without their hormonal side effects.

Chronic intestinal inflammation is characteristic of both Crohn's disease and ulcerative colitis, although both diseases display distinct etiologies and disease cycles. Chronic intestinal inflammation can be induced in animals by 2,4,6-trinitrobenzene sulfonic acid (TNBS), and is characterized by a transmural granulomatous colitis that mimics some features of Crohn's disease. NF-κB was activated in TNBS-colitis and in colitis of interleukin-10-deficient mice.[44] Local adminstration of p65 antisense phosphothiorate oligonucleotides abrogated clinical and histological signs of colitis and was more effective at treating TNBS-induced colitis than single or daily administration of GCs. These data provide direct evidence for the central importance of NF-κB in chronic intestinal inflammation.

Injection of nephrotoxic serum (NTS) in rats induces a syndrome with features similar to glomerulonephritis in man. NF-κB DNA-binding activity in glomeruli increased on day 1 after NTS injection, and persisted to at least day 14, in this model.[45] Pyrrolidine dithiocarbamate (PDTC), an inhibitor of NF-κB activation in cultured cells, inhibited the NTS-induced increase of glomerular NF-κB DNA-binding activity, and induction of mRNA expression of IL-1β, MCP-1, ICAM-1, and iNOS, which are known to be regulated by NF-κB. PDTC also prevented urinary protein excretion which is a pathophysiological parameter for glomerulonephritis. In this same animal model, treatment with prednisolone, a glucocorticorid hormone, also prevented activation of both NF-κB and AP-1 in glomeruli and subsequent mRNA expression of NF-κB- and AP-1-regulated genes.[46] Prednisolone was also effective therapeutically and reduced DNA-binding activities of NF-κB and AP-1 that are already activated in nephritic glomeruli. These results suggest that activated NF-κB and AP-1 may play an important pathogenic role in glomerulonephritis and the anti-nephritic action of glucocorticoids may be mediated through the suppression of these transcription factors.

Bacterial cell wall products, including lipopolysaccharide, are potent inducers of NF-κB activation *in vitro* and *in vivo* and cause disseminated acute inflammation and septic shock in animals and man. Administration of LPS to animals results in rapid activation of NF-κB in multiple tissues and in peripheral blood mononuclear cells, and is correlated with transcriptional activation of endothelial adhesion molecule genes and leukocyte infiltration and activation.[47] NF-κB DNA-binding activity was studied in nuclear extracts from peripheral blood mononuclear cells of 15 septic patients (10 surviving and 5 not surviving).[48] Non-survivors could be distinguished from survivors by an increase in NF-κB binding activity during the observation period, with increases in NF-κB activity being comparable to changes in the APACHE-II score as a predictor of outcome. In a mouse model of endotoxemia, somatic gene therapy with IκBα attenuated renal NF-κB activity and enhanced survival.

Cardiovascular Diseases

NF-κB is probably involved in the pathogenesis of atherosclerosis. Activated NF-κB is present in the fibrotic thickened intima-media and theromatous areas of the atherosclerotic lesion, within smooth muscle cells, macrophages and endothelial cells, whereas little or no activated NF-κB can be detected in vessels lacking atherosclerosis.[49] A variety of molecules have been identified in the atherosclerotic environment that are able to activate NF-κB *in vitro*. Furthermore, an increased expression of numerous genes known to be regulated by NF-κB has been found in the atherosclerotic lesion. Oxidized LDL induced NF-κB activation in fibroblasts, endothelial and smooth muscle cells.[50] The extent of NF-κB activation was proportional to the degree of LDL oxidation, as assessed by the lipid peroxidation product and the conjugated diene level, consistent with evidence linking NF-κB activation to intracellular oxidative stress. NF-κB activation is a hallmark of endothelial dysfunction. It plays a key role in the regulation of endothelial genes associated with vascular pathologies, including atherosclerosis.[51,52]

Myocardial infarction is a leading cause of sudden death in the Western world. Ischemia and reperfusion activates NF-κB in the myocardium and other organs.[53, 54] Treatment before and after transient coronary artery occlusion by transfection of NF-κB double-stranded oligodeoxynucleotide decoys, but not scrambled decoy oligodeoxynucleotides, significantly reduced the area of myocardial infarction.[55] NF-κB inhibitors may have additional utility in organ transplantation, where transient ischemia associated with organ harvesting contribute to delay of graft function.

NF-κB likely plays a role in stroke through its dual role in atherosclerosis and in ischemia and reperfusion. The distribution of NF-κB was investigated immunohistochemically in recently infarcted areas of postmortem human brain.[56] Immunoreactive NF–κB was observed in glial cells of infarcted areas, but not in the unaffected surrounding areas. Prominent staining for NF-κB was seen in some astrocytes, particularly in the penumbra or border zone between ischemic and non-ischemic areas. In some cases, positively stained macrophages were also observed in affected areas. These results suggest that enhanced expression of astrocytic NF-κB occurs in cerebral infarcted areas. In rats subjected to 30 minutes of four-vessel occlusion levels, NF-κB immunoreactivity localized to nuclei of CA1 neurons at 72 hours after reperfusion, and correlated with induction of neuronal cell death.[57] Induction of the activated p50 and p65 subunits was confirmed by Western blot and electromobility shift analysis. The exact stimuli that activate NF-κB in response to ischemia and reperfusion *in vivo* are unknown, although oxidative stress and local cytokine release from tissue mast cells are likely contributors.

Central Nervous System Disease

Evidence has begun to accumulate that NF-κB is critically involved in brain function, particularly following injury and in neurodegenerative conditions such as Alzheimer's disease and amyotrophic lateral sclerosis. A wide range of stimuli activate NF-κB in brain-derived cells, and a number of NF-κB genes are induced. NF-κB complexes are present in neurons, glial cells, and astrocytes. Activation of NF-κB in these cells was found to be regulated in a similar fashion to that of cells in the periphery. Brain-specific activators of NF-κB include glutamate (via both AMPA/KA and NMDA receptors) and neurotrophins, suggesting an involvement in synaptic plasticity (reviewed in-depth in 58).

NF-κB may play a role in a range of neurodegenerative diseases, including Alzheimer's disease and multiple sclerosis. The neurotoxic peptide Aβ, which is deposited in plaques of Alzheimer's patients, can activate NF-κB in neuroblastoma cells and in cerebellar granule cells.[59] This activation appears dependent on the formation of reactive oxygen intermediates. In the same study, NF-κB was activated in regions around early plaque stages in Alzheimer's patients. Target genes with potential relevance for Alzheimer's disease include the Aβ precursor APP, which can be induced via glutamate and IL-1,[60] possibly suggesting a positive feedback loop. Activated NF-κB has been detected in neuronal tissues in experimental autoimmune encephalomyelitis (a mouse model of multiple sclerosis),[61] and in HIV-induced encephalitis.[62]

Cancer

Malignant tumor cells have usually accumulated mutations that affect a variety of processes, including those that sustain cell growth, that block growth inhibition and apoptosis, that affect DNA repair, and that allow the tumor to escape immune surveillance. Several lines of evidence suggest that NF-κB family members are involved in tumor growth and metastasis. The genes for c-rel, NFKB1 (p105), NFKB2 (p52), RelA and Bcl-3 are located at sites of recurrent tranlsocation and genomic rearrangements in cancer.[63-65] Several Hodgkin's lymphoma cell lines contain constitutively nuclear NF-κB complexes, and inhibition of NF-κB activation by overexpression of a non-degradable IκBα molecule inhibits proliferation and tumorigenicity of these cells.[66] Many breast cancer cells also contain high levels of nuclear NF-κB DNA-binding activity, which is essential for survival of these cells in culture.[67] Of note, chemically induced breast cancers in rats and many primary human breast cancer samples have high levels of nuclear NF-κB. High levels of activated NF-κB are also associated with progression of breast cancer cells to estrogen-independent growth.[68] High levels of NF-κB activity have been reported in diverse solid tumor-derived cell lines.[69] Human thyroid carcinoma cells require nuclear NF-κB activity for proliferation and malignant transformation *in vitro*.[70] The Tax protein from the leukemogenic virus HTLV-1 is a potent activator of NF-κB.[71] Oncogenic forms of ras, the most common defect in human tumors, can activate NF-κB, and NF-κB inhibition can block ras-mediated cellular transformation.[72] Insights into the molecular mechanisms by which NF-κB promotes these effects are being elucidated. Many cancer therapies function to kill transformed cells through apoptotic mechanisms. Resistance to apoptosis provides protection against cell killing by these therapies and may represent a major mechanism of tumor drug resistance.[73] The activation of NF-κB by tumor necrosis factor-α (TNF-α), ionizing radiation, and chaemotherapeutic agents was found to protect tumor cells from cell killing.[74-76] Inhibition of NF-κB activation in this setting enhanced apoptotic killing by these reagents but not by apoptotic stimuli that do not activate NF-κB. NF-κB can induce the expression of anti-apoptotic genes such as IAP proteins (cellular inhibitor of apoptosis-1),[77] superoxide dismutase and A20, or genes that contribute to proliferation such as c-myc and Bcl-2. NF-κB activation was shown to result in transcriptional activation of the genes encoding a novel factor IEX-1L[78] and TRAF1 and 2, and to result in suppression of caspase 8 activation.[79] In this way, persistent activation of NF-κB represents a unique mechanism by which these cells express genes that protect against apoptotic stimuli. NF-κB also regulates the expression cell adhesion molecules that facilitate cell adhesion and migration (51, and Chapters 2, 4, and 5). Inhibition of NF-κB activation, using either antisense oligonucleotides or transcription factor decoys, results in loss of adhesion in tumor cells in culture,[80] and reduction in tumor formation in nude mice.[81] Inhibition of NF-κB represents a novel approach to cancer drug development, and when used as combination therapy with established therapeutics, these agents could provide a more effective treatment for drug-resistant forms of cancer.

SUMMARY AND FUTURE DIRECTIONS

Translating new scientific insights into novel pharmaceutical products represents a challenging and often frustrating endeavor. In recent years, pharmaceutical companies have focused on identifying key steps in disease processes that can be targeted by highly specific pharmacologic or biologic agents. Historically tractable drug targets include enzymes and receptors, as these proteins contain high-affinity binding sites for substrates and ligands that can be effectively mimicked by small organic compounds. Our understanding of the NF-κB signal transduction pathway reveals several enzymes which appear as attractive drug targets, including receptor-associated kinases such as IRAK or MEKKs, the IKKs themselves, and downstream components including the ubiquitin ligases. Modern drug discovery technologies, including biochemical and cell-based high-throughput screening systems and structure-based drug design based on the atomic structure of each target, can be applied to identify chemical series of compounds with high specificity for inhibition of each target. These compounds can then be examined in cellular and animal systems to determine efficacy and side effect profiles, to predict the potential therapeutic index (efficacy vs. side effects) in man. If, following extensive safety testing in multiple animal species, an acceptable therapeutic index can be demonstrated, human clinical trials can be initiated. Several challenges must be overcome to successfully translate our understanding of the role of NF-κB in disease into novel therapeutics.

Clearly the first issue to resolve is whether inhibition of NF-κB-mediated gene transcription will provide efficacy with acceptable side effect profiles. This therapeutic index will vary somewhat in different disease states, due to the perceived risk-benefit ratio to patients. For example, cancer treatment has traditionally tolerated additional risk of side effects in an attempt to combat an aggressive proliferative disease. In contrast, rheumatologic diseases such as osteoarthritis require a greater therapeutic index, since these patients are at lesser risk of mortality and are predominantly managing a chronic degenerative disease. As discussed earlier, there is a substantial and growing body of knowledge that inhibition of NF-κB function can result in significant disease-modifying activity. Unfortunately, we know less about the potential side effects of such interventions. One approach to determine the impact of NF-κB inhibition has been to generate strains of mice in which the genes comprising the NF-κB system have been deleted via homologous recombination or where tissue-specific mutations have been introduced via transgenic technology. Several observations can be made from such studies. The first is that it is clear that NF-κB has many physiologic functions beyond just the regulation of inflammation and leukocyte trafficking, and that it will be important to monitor these functions during the drug development process. Of note, mice with homozygous deletions of either the gene for the p65 subunit of NF-κB, of IKK2, or of NEMO, have specific defects in liver development in utero resulting in fetal death at approximately day 14 of embryogenesis.[82-84] In each case, the liver abnormality results from TNF-α-induced apoptosis of developing hepatocytes. Additional studies defined a key function for the NF-κB system in protecting hepatocytes from stress-induced apoptosis (see earlier discussion). These observations highlight the potential side effects of NF-κB inhibitors in man. Additional experimentation is clearly required to determine if NF-κB plays a similar role in human adults, and whether pharmacologic agents would recapitulate the effects of complete gene deletion seen in mice.

Additional issues for drug discovery include identifying the preferred site for targeted intervention on the NF-κB activation cascade. Is it better, for example, to develop a MAPKKK inhibitor or an IKK inhibitor, for example? There are reasons to suggest that inhibitors of each step of the cascade may provide a different profile of activity based on the cellular environment, activation and amplification steps. Likewise, protein kinases are challenging drug targets but despite the conventional dogma that the catalytic site inhibitors are non-specific the identification of very selective kinase inhibitors has created considerable optimism for the future. There is also the promise that protein substrate binding sites will provide additional opportunities for kinase inhibitor design. Finally, the rapid progress made in the production of crystal structures of a number of serine/threonine and tyrosine-specific protein kinases has identified the catalytic core of these important enzymes. As greater knowledge of active site inhibitors for p38 and other kinases emerges, it is becoming clearer how to modify compounds to create greater potency and specificity. A debate also continues as to the benefits of developing reversible inhibitors that block the enzyme only for a few hours, in contrast to irreversible inhibitors that provide greater duration of inhibition. Since protein kinases are intracellular enzymes, issues related to cell penetration, selectivity, and *in vivo* efficacy and safety remain the challenge for the medicinal chemist. In this age of increased chemical diversity it is anticipated that new kinase inhibitor templates will emerge from chemical libraries, from natural product screening, and from the availability of massive combinatorial libraries to address these issues. Biologically, enhanced screening capacities resulting from high-throughput screening as well as the availability of multiple recombinant human kinases is expediting selective kinase inhibitor identification. There are numerous kinases within the cell and, consequently, rapid and broad profiling remains an important goal. Understanding of the secondary and tertiary events that are modified by selective kinase inhibition will be greatly facilitated by gene chip methodologies that will allow transcript profiles to be obtained. Such profiles will be extremely useful in evaluating the selectivity of drug candidates.

In the ten years since the identification of NF-κB, this transcription factor has been revealed as an important regulator of a broad range of genes implicated in human disease. Recent studies using human tissues and animal models of human disease have substantiated a key role for NF-κB in autoimmune disease and cancer. The elucidation of enzyme components of the NF-κB signal transduction pathway is providing molecular targets for the development of potent and selective small molecular inhibitors of NF-κB activation. The availability of these inhibitors will permit a complete evaluation of the potential of NF-κB as a novel target for disease-modifying therapeutics.

REFERENCES

1.   Baldwin AS. The NF-κB and IκB proteins: New discoveries and Insights. Annu Rev Immunol 1996:14:649-681.

2.   Kopp E, Ghosh S. NF-κB and rel proteins in innate immunity. Adv. Immunol. 1995 58, 1-27

3.   Baeuerle PA, Baltimore D. NF-κB: Ten years after. Cell 1996;87:13-20.

4.   Baeuerle PA, Henkel T. Function and activation of NF-κB in the immune system. Annu Rev Immunol 1994;12:141-179.

5.   May MJ, Ghosh S. Signal transduction through NF-κB. Immunology Today (1998) 19, 80-88

6.   Whiteside ST, Epinat JC, Rice NR, Israel A. IκB epsilon, a novel member of the IκB family, controls RelA and cRel NF-κB activity. EMBO J 1997;16:1413-1426.

7.   Henkel T, Zabel U, Van Zee K, Muller JM, Fanning E, Baeuerle P. Intramolecular masking of the nuclear location signal and dimerization domain in the precursor for the p50 NF-κB subunit. Cell 1992;68:1121-1133.

8.   Bauerle PA, Baichwal VR. NF-κB as a frequent target for immunosuppressive and anti-inflammatory molecules. Adv. Immunol. 1997 65: 111-136.

9.   Regnier CH, Song H, Gao H, Goeddel DV, Cao Z, Rothe M. Identification and characterization of an IκB Kinase. Cell 1997;90:373-383.

10.   DiDonato JA, Hayakawa M, Rothwarf DM, Zandi E, Karin M. A cytokine-responsive IκB kinase that activates the transcription factor NF-κB. Nature 1997;388:853-862.

11. Zandi E, Rothwarf DM, Delhasse M, Hayakawa M, Karin M. The IκB kinase complex (IKK) contains two kinase subunits, IKKα and IKKβ, necessary for IκB phosphorylation and NF-κB activation. Cell 1997;91:243-252.

12. Mercurio F, Zhu H, Murray BW, Shevchenko A, Bennett BL, Li J, Young D, Barbosa M, Mann M, Manning AM, Rao A. IKK-1 and IKK-2: Cytokine-activated IκB kinases essential for NF-κB activation. Science 1997;278:860-866.

13. Woronicz JD, Gao X, Cao Z, Rothe M, Goeddel DV. IκB kinase-β: NF-κB activation and complex formation with IκB kinase-α and NIK. Science 1997;278:866-869.

14. Stancovski I, Baltimore D. NF-κB activation: the IκB kinase revealed? Cell 1997; 91:299-302.

15. Maniatis T. Catalysis by a multiprotein IκB kinase complex. Science 1997; 278:818-819.

16. Verma IM, Stevenson J. IκB kinase: beginning, not end. Proc Nat Acad Sci USA 1997;94: 11758-11760.

17. Alkalay I, Yaron A, Hatzubai A, Orian A, Ciechanover A, Ben-Neriah Y. Stimulation-dependent IκBα phosphorylation marks the NF-κB inhibitor for degradation via the ubiquitin-proteasome pathway. Proc Natl Acad Sci USA 1995; 92:10599-10603.

18. Yaron A, Alkalay I, Hatzubai A, et al. Inhibition of NF-κB cellular function via specific targeting of the IκB ubiquitin ligase. EMBO J 1997;16:101-107.

19. Chen Z, Hagler J, Palombella VJ, Melandri F, Scherer D, Ballard D, Maniatis T. Signal-induced site-specific phosphorylation targets IκBα to the ubiquitin-proteasome pathway. Genes Dev 1995;9:1586-159.

20. Lee FS, Hagler J, Chen ZJ, Maniatis T. Acitvation of the IκBα complex by MEKK1, a kinase of the JNK pathway. Cell 1997;88:213-222.

21. Malinin NL, Boldin MP, Kovalenko AV, Wallach D. MAP3K-related kinase involved in NF-κB induction by TNF, CD95 and IL-1. Nature 1997;385:540-544.

22. Hiroyasu Nakano, Masahisa Shindo, Sachiko Sakon, Shigeyuki Nishinaka, Motoyuki Mihara, Hideo Yagita, and Ko Okumura. Differential regulation of IκB kinase α and β by two upstream kinases, NF-κB-inducing kinase and mitogen-activated protein kinase/ERK kinase kinase-1. Proc Natl Acad Sci USA 1998 95: 3537-3542

23. Min-Jean Yin, Lori B. Christerson, Yumi Yamamoto, Youn-Tae Kwak, Shuichan Xu, Frank Mercurio, Miguel Barbosa, Melanie H. Cobb, and Richard B. Gaynor. HTLV-I Tax Protein Binds to MEKK1 to Stimulate IκB Kinase Activity and NF-κB activation. Cell 1998 93: 875-884

24. Shoji Yamaoka, Gilles Courtois, Christine Bessia, Simon T. Whiteside, Robert Weil, Fabrice Agou, Heather E. Kirk, Robert J. Kay, and Alain Israël. Complementation Cloning of NEMO, a Component of the IκB Kinase Complex Essential for NF-κB Activation. Cell 1998 93: 1231-1240

25. Mercurio F, Murray B, Shevchenko A, Bennett BL, Young DB, Li JW, Pascaul G, Motiwala A, Zhu H, Mann M, Manning AM. IkappaB kinase (IKK)-associated protein 1, a common component of the heterogenous IKK complex. Mol Cell Biol 1999; 19:1526-1538.

318

26. Orian A, Whiteside S, Israel A, Stancovski I, Schwartz AL, Ciechanover A. Ubiquitin-mediated processing of NF-κB transcriptional activator precursor p105. Reconstitution of a cell-free system and identification of the ubiquitin-carrier protein, E2, and a novel ubiquitin-protein ligase, E3, involved in conjugation. J Biol Chem 1995;270:21707-21714.

27. Yaron A, Alkalay I, Hatzubai A, Jung S, Beyth S, Mercurio F, Manning AM, Gonen H, Ciechanover A, Ben-Neriah Y. Inhibition of NF-κB cellular function via specific targeting of the IκB ubiquitin ligase. EMBO J 1997 16: 101-107

28. Khoshnan A, Kempiak SJ, Bennett BL, Bae D, Xu W, Manning AM, June CH, Nel AE. Primary human CD4+ T cells contain heterogeneous IκB kinase complexes: role in activation of the IL-2 promoter. J Immunol 1999; 163: 5444-5452

29. Zhong H, Yang HS, Erdjument-Bromage H, Tempst P, Ghosh S. The transcriptional activity of NF-κB is regulated by the IκB-associated PKAc subunit through a cyclic AMP-independent mechanism. Cell 1997 89: 413-424

30. Perkins, ND, Felzien, LK, Betts JC, Leung K, Beach DH, Nabel GJ. Regulation of NF-κB by cyclin-dependent kinases associated with the p300 coactivator. Science 1997; 275: 523-527

31. Zhong H, Voll RE, Ghosh S. Phosphorylation of NF-κB p65 by PKA stimulates transcriptional activity by promoting a novel bivalent interaction with the coactivator CBP/p300. Mol Cell 1998 1: 661-671

32. Handel ML, McMorrow LB, Gravallese EM. Nuclear factor-κB in rheumatoid synovium. Localization of p50 and p65. Arth & Rheum. 1995 38: 1762-1770

33. Marok R, Winyard PG, Coumbe A, Kus ML, Gaffney K, Blades S, Mapp PI, Morris CJ, Blake DR, Kaltschmidt C, Baeuerle PA. Activation of the transcription factor nuclear factor-κB in human inflamed synovial tissue. Arth & Rheum. 1996 39: 583-591

34. Fujisawa K, Aono H, Hasunuma T, Yamamoto K, Mita S, Nishioka K. Activation of transcription factor NF-κB in human synovial cells in response to tumor necrosis factor α. Arth & Rheum. 1996 39: 197-203

35. Han Z, Boyle DL, Manning AM, Firestein GS. AP-1 and NF-κB regulation in rheumatoid arthritis and murine collagen-induced arthritis. Autoimmunity 1998; 28:197-208.

36. Foxwell B, Browne K, Bondenson J, Clarke C, De Martin R, Brennan F, Feldman M. Efficient adenoviral infection with IκBα reveals that macrophage tumor necrosis a production in rheumatoid arthritis is NF-κB dependent. Proc Nat Acad Sci USA 1998 95: 8211-8215.

37. Migakov AV, Kovalenko DV, Brown CE, Didsbury JR, Cogswell JP, Stimpson SA, Baldwin AS, Makarov SS. NF-κB activation provides the link between inflammation and hyperplasia in the arthritic joint. Proc Nat Acad Sci USA, in press.

38. Bell FP, Bennett BL, Rosenbloom CL, Kolbasa KP, Fidler SD, Brashler JR, Chosay JG, Chin JE, Dunn CJ, Richards IM, Manning AM. Activation of the nuclear transcription factor NF-κB during antigen-induced airway inflammation. Am J Physiol, in press.

39. Zhu Z, Tang W, Ray A, et al. Rhinovirus stimulation of interleukin-6 in vivo and in vitro: evidence for nuclear factor-κB dependent transcriptional activation. J Clin Invest 1996 97: 421-430.

40. Haddad E-B, Salmon M, Koto H, Barnes PJ, Adcock I, Chung KF. Ozone induction of cytokine-induced neutrophil chemoattractant (CINC) and nuclear factor-κB in rat lung: inhibition by glucocorticoids. FEBS Lett 1996 379: 265-268.

41. Barnes PJ, Adcock I. Anti-inflammatory actions of steroids: molecular mechanisms. Trends Pharmacol Sci 1993 14: 436-441.

42. Auphan N, DiDonato JA, Rosette C, Helmberg A, Karin M. Immunosuppression by glucocorticoids: Inhibition of NF-κB activity through induction of IκBα synthesis. Science 1995 270, 286-290.

43. Scheinman RI, Crosswell PC, Loftquist AK, Baldwin AS. Role of transcriptional activation of IκBα in mediation of the immunosuppression by glucocorticoids. Science 1995 270: 283-286.

44. Neurath MF, Petterson S, Buschenfelde K-H, Strober W. Local administration of antisense phosphothiorate oligonucleotides to the p65 subunit of NF-κB abrogates established experimental colitis in mice. Nature Med 1996 2:

45. Sakurai H, Hisada Y, Ueno M, Sugiura M, Kawashima K, Sugita T. Activation of transcription factor NF-κB in experimental glomerulonephritis in rats. Biochim Biophys Acta 1996 1316: 132-8

46. Sakurai H, Shigemori N, Hisada Y, Ishizuka T, Kawashima K, Sugita T. Suppression of NF-κB and AP-1 activation by glucocorticoids in experimental glomerulonephritis in rats: molecular mechanisms of anti-nephritic action. Biochim Biophys Acta 1997 1362: 252-62

47. Manning AM, Bell FP, Rosenbloom CL, Chosay JG, Simmons CA, Hoover JL, Shebuski J, Dunn CJ, Anderson DC. NF-κB is activated during acute inflammation in vivo in association with endothelial cell adhesion molecule gene expression and leukocyte recruitment. J. Inflammation 1995 45: 283-296.

48. Bohrer H, Qiu F, Zimmerman T, Zhang Y, Jilmer T, Mannel D, Bottiger BW, Stern DM, Waldherr R, Saeger H-D, Ziegler R, Bierhaus A, Martin E, Naworth PP. Role of NF-κB in the mortality of sepsis. J. Clin. Invest. 1997 100: 972-985

49. K Brand, S Page, G Rogler, A Bartsch, R Brandl, R Knuechel, M Page, C Kaltschmidt, P A Baeuerle, D Neumeier. Activated Transcription Factor Nuclear Factor-κB Is Present in the Atherosclerotic Lesion. J. Clin. Invest. 1996 97: 1715-1722.

50. Maziere C, Auclair M, Djavaheri-Mergny M, Packer L, Maziere JC. Oxidized low denstiy lipoprotein induces activation of the transcription factor NF-κB in fibroblasts, endothelial and smooth muscle cells. Biochem Mol Biol Int 1996 39:6 1201-7

51. Chen CC, Manning AM. Transcriptional regulation of endothelial cell adhesion molecules: A dominant role for NF-κB. Agents & Actions 1995 S47: 135-141.

52. Collins T, Cybulsky MA. NF-κB: pivotal mediator or innocent bystander in atherosclerosis. J. Clin. Invest., in press.

53. Kukielka GL, Lacson RG, Manning AM, Anderson DC, Becker LC. NF-κB is activated and is associated with upregulation of intercellular adhesion

molecule-1 [ICAM-1] and interleukin-8 [IL-8] genes in reperfused canine myocardium. American Heart Association, 71st Annual Meeting.

54. Bell FP, Essani NA, Manning AM, Jaeschke H. Ischemia-reperfusion activates the nuclear transcription factor NF-κB and upregulates messenger RNA synthesis of adhesion molecules in the liver in vivo. Hepatology Res 1997 8: 178-188

55. Morishita R, Sugimoto T, Aoki M, Kida I, Tomita N, Moriguchi A, Maeda K, Sawa Y, Kaneda Y, Higaki J, Ogihara. In vivo transfection of cis element "decoy" against nuclear factor-κB binding site prevents myocardial infarction. Nature Med 1997 3: 894-899

56. Terai K, Matsuo A, McGeer EG, McGeer PL. Enhancement of immunoreactivity for NF-κB in human cerebral infarctions. Brain Res 1996 739:1-2 343-9

57. Clemens JA, Stephenson DT, Smalstig EB, Dixon EP, Little SP. Global ischemia activates nuclear factor-κB in forebrain neurons of rats. Stroke 1997 28:5 1073-80

58. O'Neill LAJ, Kaltschmidt C. NF-κB: A crucial transcription factor for glial and neuronal cell function. TINS 1997 20: 252-258.

59. Kaltschmidt B, Uherek M, Volk B, Baeuerle PA, Kaltschmidt C. Transcription factor NF-kappaB is activated in primary neurons by amyloid beta peptides and in neurons surrounding early plaques from patients with Alzheimer disease. Proc Nat Acad Sci USA 1997 94: 2642-2647.

60. Grilli M, Goffi F, Memo M, Spano P. Interleukin-1beta and glutamate activate the NF-kappaB/Rel binding site from the regulatory region of the amyloid precursor protein gene in primary neuronal cultures. J. Biol. Chem. 1996 271: 15002-15007.

61. Kaltschmidt C, Kaltschmidt B, Lannes-Vieira J, Kreutzberg GW, Wekerle H, Baeuerle PA, Gehrmann J. Transcription factor NF-kappa B is activated in microglia during experimental autoimmune encephalomyelitis. J. Neuroimmunol.1994 55: 99-106.

62. Dollard SC, James HJ, Sharer LR, Epstein LG, Gelbard HA, Dewhurst S. Activation of NF-κB in the brains of children with HIV-1 encephalitis. Neuropathol Appl Neurobiol 1995 21: 518-528

63. Matthew S, Murty, VV, Dalla-Favera R, Chaganti, R.S.Chromosomal localization of genes encoding the transcription factors c-Rel, NF-κB p50, NF-κB p65 and lyt10 by fluoresence in situ hybridization. Oncogene 1993 8: 191-193

64. Gilmore TD, Morin PJ. The IκB proteins: members of a multifunctional family. Trends Genet. 1993 9: 427433.

65. Bargou RC, Emmerich F, Krappmann D, Bommert K, Mapara MY, Arnold W, Royer HD, Grinstein E, Greiner A, Scheidereit C, Dorken B. Constitutive NF-κB RelA activation is required for proliferation and survival of Hodgkin's disease tumor cells. J. Clin. Invest. 1997 100: 2961-2969.

66. Sovak MA, Bellas RE, Kim DW, Zanieski GJ, Rogers AE, Traish AM, Sonenshein GE. Aberrant NF-κB expression and the pathogenesis of breast cancer. J. Clin. Invest. 1997 100: 2952-2960.

67. Nakshatri H, Bhat-Nakshatri P, Martin D, Goulet R, Sledge G. Constitutive activation of NF-κB during progression of breast cancer to hormone-independent growth. Mol Cell Biol 1997 17: 3629-3639.

68. Bours V, Dejardin E, Goujon-Letawe F, Merville MP, Castronovo V. The NF-κB transcription factor and cancer: high expression of NF-κB and IκB-related proteins in tumor cell lines. Biochem. Pharmacol. 1994 47: 145-149.

69. Visconti R, Cerutti J, Battista S, Fedele M, Trapasso F, Zeki K, Miano MP, de Nigris F, Casalino L, Curcio F, Santoro M, Fusco A. Expression of the neoplastic phenotype by human thyroid carcinoma cell lines requires NF-κB p65 protein expression. Oncogene 1997 15: 1987-1994.

70. Hammarskjold ML, Simurda MC. Epstein Barr virus latent membrane protein transactivates the human immunodeficiency virus type 1 long terminal repeat through induction of NF-κB activity. J. Virol. 1992 66: 6496-6501.

71. Finco TS, Westwick JK, Norris JL, Beg AA, Der CJ, Baldwin AS. Oncogenic Ha-Ras –induced signalling activates NF-κB transcriptional activity, which is required for cellular transformation. J. Biol. Chem. 1997 272: 24113-24116.

72. Fisher DE. Apoptosis in cancer therapy: crossing the threshold. Cell 1994 78: 539 - 542.

73. Wang C, Mayo M, Baldwin AS. TNF and cancer therapy-induced apoptosis: potentiation by inhibition of NF-κB. Science 1996 274: 784-787.

74. Van Antwerp DJ, Martin SJ, Kafri T, Green DR, Verma IM. Suppression of TNFα-induced apoptosis by NF-κB. Science 1996 274: 787-789.

75. Beg AA, Baltimore D. An essential role for NF-κB in preventing TNFα-induced cell death. Science 1996 274: 782-784.

76. Chu ZL, McKinsey TA, Liu L, Gentry JJ, Malim MH, Ballard DW. Suppression of TNF-induced cell death by inhibitor of apoptosis c-IAP2 is under NF-κB control. Proc Natl Acad Sci USA 1997 94: 10057-10062.

77. You M, Ku PT, Hrdlickova R, Bose HR. ch-IAP1, a member of the inhibitor of apoptosis family, is a mediator of the anti-apoptotic activity of the v-Rel oncoprotein. Mol Cell Biol 1997 17: 7328-7341.

78. Wu MX, Ao Z, Prasad KV, Wu R, Schlossman SF. IEX-1L, an apoptosis inhibitor involved in NF-κB-mediated cell survival. Science 281: 998-1001.

79. Wang C-Y, Mayo MW, Korneluk RG, Goeddell DV, Baldwin AS. NF-κB antiapoptosis: induction of TRAF1 and TRAF2 and c-IAP1 and c-IAP2 to suppress caspase 8 activation. Science 1998; 281:1680-1683.

80. Sokoloski JA, Sartorelli AC, Rosen CA, Narayanan R. Antisense oligonucleotides to the p65 subunit of NF-κB block CD11b expression and adhesion properties of differentiated HL-60 granulocytes. Blood 1993 82: 625-632.

81. Higgins KA, Perez JR, Coleman TA, Dorshkind K, McComas WA, Sarmiento UM, Rosen CA, Narayanan R. Antisense inhibition of the p65 subunit of NF-κB blocks tumorigenicity and causes tumor regression. Proc. Natl. Acad. Sci. USA 1993 90: 9901-9905.

82. Beg AA, Sha WC, Bronson RT, Ghosh S, Baltimore D. Embryonic lethality and liver degeneration in mice lacking the RelA component of NF-κB. Nature 1995; 376: 167-169.

83. Li Q, Van Antwerp D, Mercurio F, Lee K, Verma, IM. Severe Liver Degeneration in Mice Lacking IB Kinase 2 Gene. Science 1999;284: 321-5.

84. Makris C, Godfrey VL, Krähn-Senftleben G, Takahashi T, Roberts JL, Schwarz T, Feng L, Johnson RS, Karin M. Female Mice Heterozygous for IKK /NEMO Deficiencies Develop a Dermatopathy Similar to the Human X-Linked Disorder Incontinentia Pigmenti. Molecular Cell 2000 5: 969-979.

INDEX